TELEPEN

7719407032

7719407032

**Books are to be returned on or before
the last date below**

ON DISPLAY
18 JUL 1978
SCIENCE AND
ENGINEERING FLOOR

27 MAR 2001

24 MAR 2003

26. NOV

13 JAN 86

17. NOV

NOV 89

25 NOV 1997

IEX —

D1357271

MODERN RESERVOIR ENGINEERING—
A Simulation Approach

MODERN RESERVOIR ENGINEERING–
A Simulation Approach

HENRY B. CRICHLOW

School of Petroleum and Geological Engineering
University of Oklahoma

Prentice-Hall, Inc.
Englewood Cliffs, New Jersey 07632

Library of Congress Cataloging in Publication Data

CRICHLOW, HENRY B.
    Modern reservoir engineering.

    Includes bibliographies.
    1. Oil reservoir engineering—Mathematical
models. 2. Oil reservoir engineering—Data
processing. I. Title.
TN871. C694   622'. 33'80184   76–19004
ISBN 0–13–597468–2

10   9   8   7   6   5   4   3   2   1

Printed in the United States of America

PRENTICE-HALL INTERNATIONAL, INC., *London*
PRENTICE-HALL OF AUSTRALIA PTY. LIMITED, *Sydney*
PRENTICE-HALL OF CANADA, LTD., *Toronto*
PRENTICE-HALL OF INDIA PRIVATE LIMITED, *New Delhi*
PRENTICE-HALL OF JAPAN, INC., *Tokyo*
PRENTICE-HALL OF SOUTHEAST ASIA PTE. LTD., *Singapore*
WHITEHALL BOOKS LIMITED, *Wellington, New Zealand*

To my son,

Richard

and the women in my life,

my wife, *Teddie*
my daughter, *Reneé*

Contents

11 OPTIMIZATION AND SIMULATION, *307*

Preface

This book is written primarily for the engineer in practice in the industry and the engineering student who is about to embark on a career in petroleum reservoir engineering. It is also aimed at managers, educators, and other professional men who interface with engineers doing modern reservoir simulation and who would like to understand more fully some of the concepts involved in this particular discipline. The scope of topics covered is designed to refresh those who have forgotten that there is a basis for almost all the analytical techniques finding their way into simulation methods in reservoir engineering.

The main theme of this text is to focus on the essentials without too much detail to cloud the issue. In those areas where detail is required, the author has attempted to use rigor to crystallize an idea or concept. The more adventurous practitioner is amply provided with several hundred references to quench his eager thirst. The author presupposes that the engineer has had some exposure to petroleum reservoir work or at the very least that he is familiar with the terminology used by that dedicated group of men and women who try to get all the oil out!

SUGGESTIONS TO READERS

In a work of this kind, there will be areas of emphasis that different groups will find too shallow or too deep. You can't win them all! However, the reader should realize his needs, and then select those topics that interest him most. The mathematically inclined, who would not be concerned with flow concepts and data preparation, could well omit several chapters; the staff engineer in the field office would not be overly excited by the mathematical analysis of stability, so he can skip this area. The following diagram gives the prospective reader some clue as to where his interests may be in the chapters that lie ahead. In the final analysis, only the student really knows his needs, and he can omit many sections within several chapters without any loss of continuity.

	Practicing Engineer	Developer of Models	Simulation Practitioner	Neophyte Practitioner	Student Engineer
Chapters	2	2	2	2	2
	3	4	7	7	3
	4	5	9	8	4
	7	6	11	9	5
	8	11		10	6
	9				7
					8
					9
					10
					11

HENRY B. CRICHLOW

1 Introduction: The Age of Simulation

1.1 INTRODUCTION

The dictionary defines *simulate* as simply "to give an appearance of." To the engineer or analyst, *simulation* involves the utilization of a model to obtain some insight into the behavior of a physical process. It is a process or mechanism by which a particular problem can be studied in varying depths of detail to obtain answers or to confirm hypotheses. Simulation has long been recognized in many applied science disciplines as the final resort; as Wagner[1] aptly says: "When all else fails, . . . simulate." In operations research, extensive use has been made of simulation studies; some examples are:

1. Transportation model networks
2. Stock market performance
3. Telephone system design
4. Supermarket checkout counters

Because of the widespread need in some of these areas, special-purpose languages have been developed to meet the particular demands for simulation.

Simulation, however, involves a lot more than just the design and use of a good model to analyze a process, be it an oil reservoir system or a network switching problem. The word *simulation* conjures up different things to different people. Some people's concept of simulation borders on the incredible: the simulator is a black box of unknowns which miraculously produces results that are in some way sacred, numbers that are infallible to all their significant

1

digits. This is the blue-sky approach to simulation. More realistically, simulation is a process wherein the engineer integrates several factors to produce information on the basis of which managers can make intelligent decisions. He begins by selecting the best vehicle for this project, that being the best model. Added to this is his expertise, his knowledge of the quality of data, and in particular the data sources; he then produces a finished product in the form of recommendations and conclusions which are usable within the realm of managerial activity. At all points along the way the engineer is on top of the situation. Nothing the simulation process does can improve the quality of his work, but it can certainly give him a great insight into the interrelationships of the processes which are occurring in his project.

The growth in computer systems has been a necessary precursor to the development of simulation. Engineers have long recognized the guiding principles for most of the physical phenomena they study, but the tools to solve these problems were lacking. As the computer evolved, simulation has expanded with it, almost like the inner portion of two expanding concentric waves. At times it seems that the simulation needs would outrun the computational resources; however, in these two dynamic areas there always seems to be an operating medium. The engineer could always produce effectively with what he had available. No doubt as computer technology continues to create, the engineer will be right in step to push its usefulness to the limit in applying his expertise to the solution of ever-increasing problems.

The Necessity for Simulating

The classical approach to solving a problem has been to formulate the problem and then try to make as many simplifying assumptions as possible to produce a new problem which is manageable. What happens if even after all these simplifying assumptions the problem still remains rather intractable? The individual could solve it in two ways at this point. First, he can define the problem as having no solution, somewhat like the alchemists of old did when they developed the phlogiston theory of burning. They knew the theory was wrong, but they still gave it a name. Giving the difficulty a name does no more to solve it than leaving it alone in the first place. As a second alternative, he can attempt a solution with the best available technical help and at some point come to an answer which is satisfactory to him. The knowledge that this is not the full answer would in no way detract from him utilizing the results. There are very few cases in nature where *no answer* is better than an approximate one. The point is clear that analytical tools become less effective as problems begin to increase in complexity. In the petroleum engineering discipline, complexity in physical processes is more the rule than the exception. The engineer today is required not only to determine the best performance based on physical behavior of the system, but to become increasingly aware of the interaction of the economic, regulatory, legal, and environmental impacts of his decisions. All these forces acting together have produced such a complex pattern that any useful analysis must necessarily incorporate them all. Such built-in complexity naturally lends itself to some simulation process whereby the effect of various parameters on the solution can be examined rather critically. It is as though the whole process has to be relived several times in a simulation mode before it is put into practice.

The Modeling Approach

Someone once said, "The human mind has difficulty in considering more than 10 to 20 factors at the same time in making a decision." The decision-making problem faced by the engineer in developing and producing a petroleum resource of any size involves several hundred variables. These variables may not be quantified or cataloged in an easily definable form, but they do exist. The engineer thus has to decide on the producing characteristics, completion techniques, pump sizes, we'l locations, and operating characteristics for each well, and all this information has to be determined over a time horizon which involves continual change. Good engineers and managers have used the intuitive approach in the past, and in many cases they have prospered; those whose intuition was not as brilliant or whose "logical" deductions lacked that subtle quality, "insight," are not being heard from today. In order to add some structure to the manager's or engineer's methodology, we need a disciplined technique which allows us to determine the relevant factors and their interrelationships to a given solution. Furthermore, this technique should allow us to implement decisions effectively and provide some means for updating, modifying, and redefining our systems and objectives as we go along. The modeling approach most nearly combines all these attributes.

Models are basically of two types; in a very pedestrian way they are simply the ones you can touch and the ones you cannot. The former are *physical* models, the latter *mathematical* models.

1. Physical models are essentially scaled-down reproductions of the original, as evidenced in pilot plants, prototypes, and the like, or models constructed to duplicate a process which is physically similar to the original although it may operate under a different set of physical laws. The best example of this is the potentiometric model used to predict reservoir flow by capitalizing on the one-to-one correspondence between flow in porous media and the flow of ions in an electric potential field.

2. Mathematical models are systems of mathematical equations describing the physical behavior of the process under investigation. In petroleum reservoir work, these equations are generally very complicated partial differential equations, but they could be rather easy equations system in other

fields. Because of the size and the complexity of these mathematical models, a computer is requred to solve the system.

Throughout this book the word *model* refers to mathematical model and is used interchangeably with *simulator* or *simulation*.

The technique of mathematical modeling and the role played by the engineer can be visualized by the block diagram shown in Fig. 1.1. The central

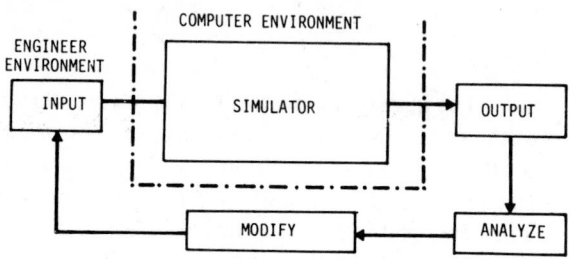

Figure 1.1: Mathematical modeling.

box is the simulator; its formulation and development require substantial background in mathematics and the applied sciences. The use of it, however, requires only good engineering skills and common sense. As indicated by the figure, there is a feedback loop in mathematical modeling. The simulator operates in a computer environment (to use the term rather loosely), and everything else operates in an engineering setting. The process begins with the input provided by the engineer; this is processed by the simulator and the output is obtained. At this point the information is analyzed for the effects of previous changes on the operating characteristics, and if modifications are needed, they are made and the process repeated. As the engineer cycles through this loop, his input, by virtue of his expertise, continuously upgrades the results, and as more and better information becomes available as time passes, he can produce an efficient and reasonably accurate predictive tool for his process.

The engineer is using proven technology in trying to make decisions quantitatively and optimize his projects. Modeling techniques today have expanded, improved, and become more applications-oriented to such a degree that the engineer or scientist who has not begun to utilize these methods may find himself trying to communicate with his peers across an ever-widening chasm. The stress today for economic justification and the need to "back up" all decisions with technical support inexorably pushes him to the use of modeling strategy.

Reservoir Simulation

The area of reservoir simulation applies the concepts and techniques of mathematical modeling to the analysis of the behavior of petroleum reservoir systems. In a narrower sense the term *reservoir simulation* refers only to the hydrodynamics of flow within the reservoir, but in a larger sense it can and more often does refer to the total petroleum system which includes the reservoir, the surface facilities, and any interrelated significant activity. The basic flow model consists of the partial differential equations which govern the unsteady-state flow of all fluid phases in the reservoir medium. Incorporated into the model are all the algorithms needed to solve these equations. The simulator is then a collection of computer programs which implement the mathematical model on a particular digital machine. The origin of the simulator and the synthesis into a coherent whole are shown in Fig. 1.2.

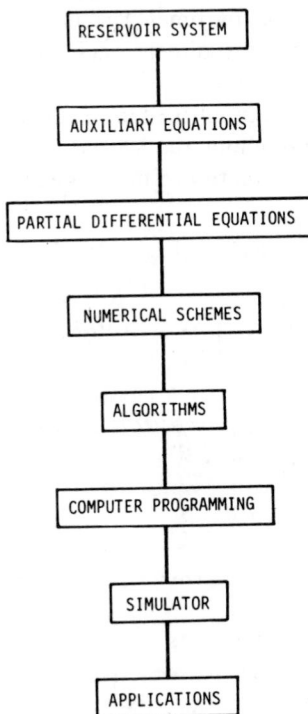

Figure 1.2: The origin of the simulator.

1.2 GROWTH OF SIMULATION

The growth in reservoir simulation has proceeded parallel to the upsurge in technology over the last 30 years. The engineer has strived at all times to use the best tools available to him to understand the mechanics of petroleum reservoirs and petroleum production and to apply these to the efficient operation of reservoirs. Today the use of simulation has made the computer as much an everyday tool as the slide rule and desk calculator were 20 years ago. In the following pages we shall explore some of the techniques used to simulate reservoir performance during its growth and indicate how the weaknesses in each of these methods were resolved by the new methods. Some of these methods are still being used because they still are economically justifiable and technically correct. A notable example is the material balance equation.

The Material Balance Equation

In 1936 Schilthuis[2] developed a conservation equation for a hydrocarbon reservoir. This equation is derived by considering the whole reservoir to be a homogenous tank of uniform rock and fluid properties. The mass balance was made by accounting for all quantities which may enter or leave the reservoir over a period of time. The material balance equation is sometimes referred to as the zero-dimensional simulator since there are no changes in any direction within the system. The saturations and pressures are distributed continuously throughout the tank, and any changes in pressures are instantaneously felt throughout the system. Figure 1.3 illustrates the basis of the material balance equation. The complete equation is shown below:

$$N_p[B_t + B_g(R_p - R_{s_i})] + W_p = N\left[(B_t - B_{t_i}) + \frac{B_{t_i}}{1 - S_{w_i}}(C_f + S_w C_w)\Delta P\right.$$

$$\left. + \frac{mB_{t_i}}{B_{g_i}}(B_g - B_{g_i})\right] + W_e + W_i + G_i B_g \tag{1.1}$$

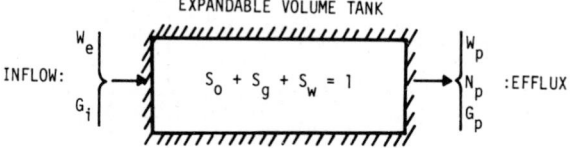

EXPANDABLE VOLUME TANK

INFLOW: W_e, G_i $S_o + S_g + S_w = 1$ W_p, N_p, G_p : EFFLUX

Figure 1.3: Material balance equation.

The left-hand side represents the production of oil, gas, and water, while the right-hand side refers to the expansion of the rock and fluids in place and influx and injection terms. By various algebraic manipulations the basic equation can be reordered to solve for any of the following parameters:

1. Oil in place
2. Water influx
3. Gas cap size and gas in place
4. Oil production

The material balance approach has been solved either graphically or computationally—more recently the material balance equation has been analyzed as a straight line by Odeh and Havlena[8]—but the basic premise behind the material balance approach involves the following rather serious drawbacks:

1. There is no allowance for variation of fluid and rock properties with location within the reservoir.
2. The dynamic effects of fluid movement within the system are overlooked.

Further developments in reservoir analysis evolved as these drawbacks in the material balance equation were resolved. The next approach involved the use of resistance-capacitance networks.

Analog Resistance-Capacitance Networks[3]

Analog resistance-capacitance networks, usually called electrical analyzers, employ the similarity between electrical flow and fluid flow in a reservoir to develop an electrical analog of the petroleum reservoir. By analyzing the variation of the electrical parameters with time under different operating conditions, the behavior of the reservoir can be computed using suitable conversion factors. The analogy between the two systems can be seen from the following two equations:

Fluid Flow:

$$q = \frac{kA}{\mu L}(P_1 - P_2)$$

$$q_1 - q_2 = V_c \frac{\partial P}{\partial t}$$

(1.2)

Current Flow:

$$i = \frac{1}{R}(E_1 - E_2)$$

$$i_1 - i_2 = C_E \frac{\partial E}{\partial t}$$

(1.3)

The correspondence in parameters is shown in Table 1.1.

The R-C network is usually a two-dimensional grid of the reservoir as shown in the example in Figs. 1.4, 1.5, and 1.6, where the Woodbine Basin[4] reservoir is modeled by the use of an R-C network. Equations (1.2), and (1.3) indicate the one-to-one correspondence between the following:

$$q \simeq i$$

$$\frac{1}{R} \simeq \frac{kA}{\mu L}$$

(1.4)

$$P \simeq E$$

TABLE 1.1
Correspondence Between Fluid and Electrical Systems

Fluid System		Electrical System	
Item	Units	Item	Units
Pressure	psi (P)	Voltage	Volts (E)
Production/Injection	B/D (q)	Current	μamps (i)
Fluid Capacitance (storage)	B/psi	Capacitance	μfarads
Transmissibility (kh/μ)	Darcy-ft. cp	Conductivity	Mhos
Real Time	months	Model Time	Seconds

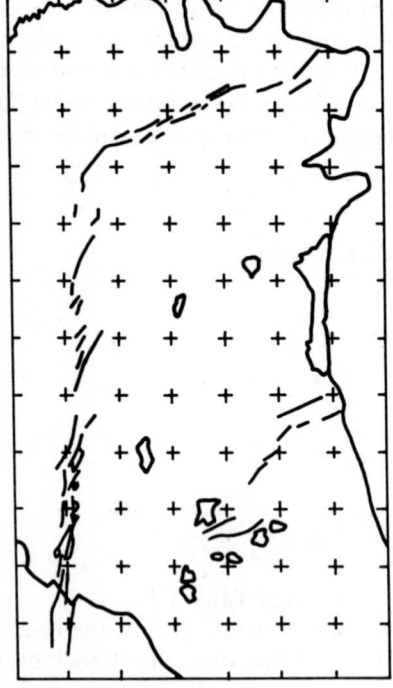

Figure 1.4: Subdivision of Woodbine Basin.

The resistances in the given network are calculated from the existing rock permeability in the given sector of the field. The electrical parameters measured are voltage and current as the capacitances are varied in the circuits.

The Electrolytic Model

Steady-state electrolytic models have been developed by several investigators—e.g., Botset[5], Wyckoff, Muskat—in an effort to analyze the movement

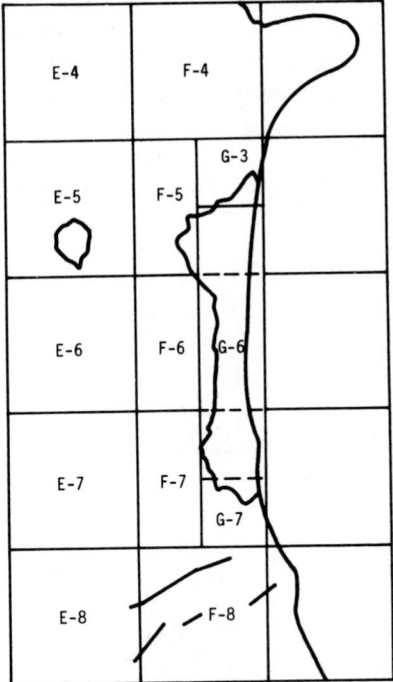

Figure 1.5: Subdivision of area around East Texas field.

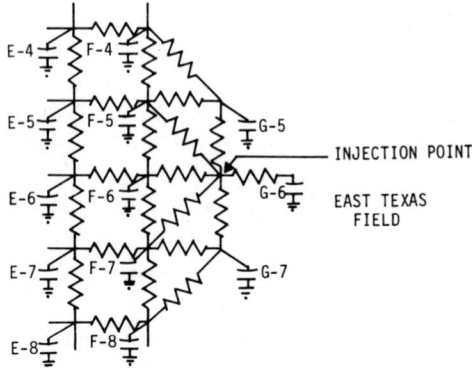

Figure 1.6: Resistor network.

of fluid fronts with the reservoir. These models are based on the analog between Ohm's law for flow in a conductor and Darcy's law for flow in a porous medium. If the sources and sinks in a fluid flow process and the boundaries of the medium are adequately defined, then a steady-state model can be made, usually on blotting paper or agar gelatin, to analyze the potential distribution. The model is scaled geometrically except that the vertical scale is exaggerated. A voltage is applied at well locations (in this case, copper electrodes) and the movement of the front traced by the progress of colored copper ammonium ions which move away from the negative electrode to the positive electrode. The medium is impregnated with zinc ammonium ions which are colorless. The copper ions move at right angles to the isopotential lines set up by the potential field. Figure 1.7 illustrates the growth of a displacement pattern in a given mode.

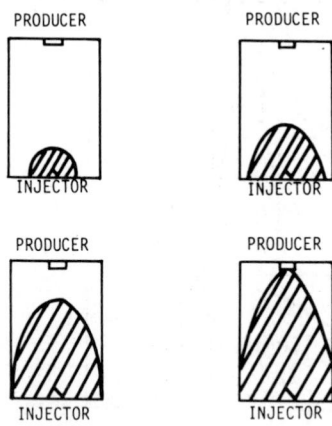

Figure 1.7: Electrolytic model.

The Potentiometric Model[6]

The potentiometric model is a steady-state model that uses a container sculptured to conform to the boundaries and permeability-thickness products of the reservoir under study. The wells are represented by copper electrodes placed within the medium, which consists of an electrolyte like potassium chloride. The production and injection rates are modeled by using alternating currents (to prevent electrolysis) of given magnitudes. The objective of the potentiometric model is to determine the steady-state potential distribution in the model. Since this distribution is analogous to the pressure distribution in a reservoir, the streamlines can be determined by plotting a set of points at right angles to these isopotential lines.

In practice the isopotential lines were determined using a movable probe controlled by a servomechanism. When a position was found along a given isopotential, the location and direction of the streamline vector was immediately plotted by a pair of points which were permanently fixed at right angles to the potential points on the probe head. Thus, at the end of a run the potential distribution and the streamline distribution were found simultaneously.

Once the streamlines were obtained, the engineer had to calculate the location of the flood front by calculating the distance traveled along every streamline emanating from a point source. The locus of all points from a given source at a specified time gives the location of that front, as illustrated in Fig. 1.8.

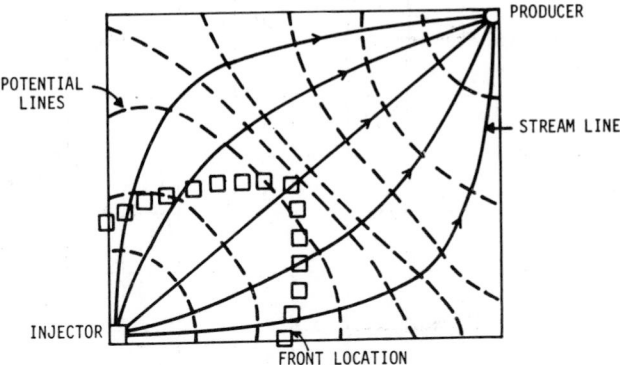

PRODUCER

POTENTIAL LINES

STREAM LINE

INJECTOR

FRONT LOCATION

Figure 1.8: Potentiometric model.

All the models so far described suffer from several restrictions or weaknesses. The biggest problem was the fact that each reservoir had its own unique model which had to be built literally "from the ground up" every time. The custom-building was expensive for large models and lacked adaptability. Modifications of the completed model were difficult to make and entailed physically reworking the circuits or the systems. There was also an inherent problem in the components of the system circuits caused by faulty equipment. These include leakages in condensers, meter malfunctions, and other similar problems. Finally, the R-C networks could be so large that they occupied rooms at a time, and the engineer literally had to walk through the model to adjust resistors and capacitors. Working with a model this size does provide certain insurmountable problems.

Numerical Models[7]

Numerical models utilize digital machines to solve the mathematical equations which govern the behavior of the fluids in porous media. They provide a generalized approach using a gridded format which can accomodate any reservoir description just by a reordering of the indices of the grids. The numerical models originated in the middle 1950s with Peaceman and Rachford[7] and have evolved extremely rapidly to the point where almost every conceivable reservoir behavior pattern can be simulated. The procedure involved consists of discretizing the reservoir into blocks and performing mass and energy balances on all these blocks simultaneously. This gridding of cells allows a more realistic representation of rock and fluid properties which can vary in any manner. A typical grid is shown in Fig. 1.9.

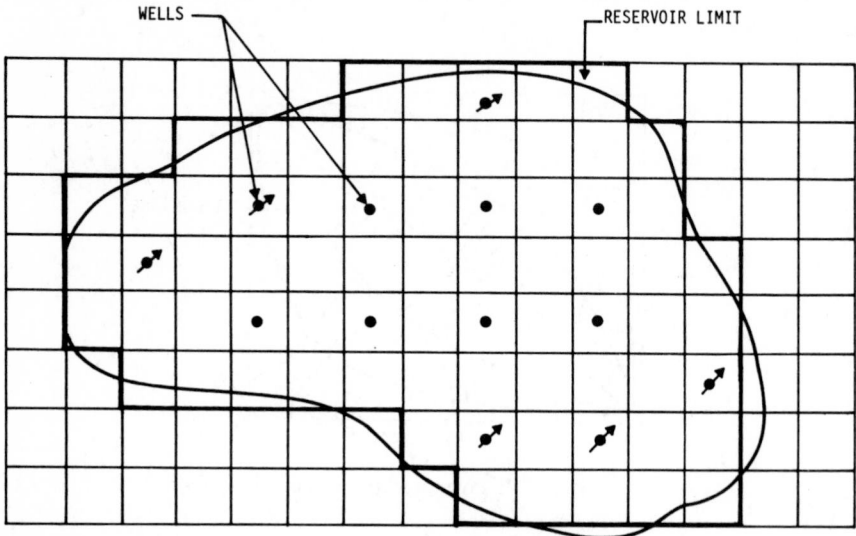

Figure 1.9: Grid of reservoir.

1.3 PURPOSE OF RESERVOIR SIMULATION

The simulator produces a lot of output data, and the engineer should be able to analyze this data to obtain the results he needs. The simulation program can be used to study a reservoir containing a single well, a group of wells, or several wells interacting as a complex. Simulators have also become quite popular as educational tools where the mechanics of fluid flow in porous media can be examined. As an engineering tool the broad objectives of simulation are shown in Fig. 1.10.

The original oil in place is an important and necessary objective in any study; this quantity is usually required as a reservoir total—e.g., Fig. 1.11.

In the case of multiple-zoned reservoirs the productivity and oil in place of a given horizon or zone may be needed. Modeling the reservoir as a series of zones be as shown in Fig. 1.12 can produce this information and allow the engineer to schedule production and completion operations for these zones more effectively.

At other times there may be several leases or units contributing to the total reservoir. The appropriate oil-in-place figure is essential in planning an equitable unitization program, and the simulation study allows a breakdown of oil in place by leases or units with very little additional work.

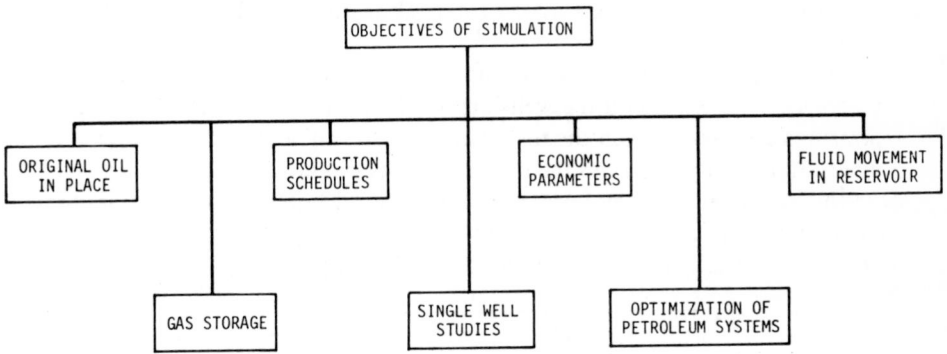

Figure 1.10: Objectives of simulation.

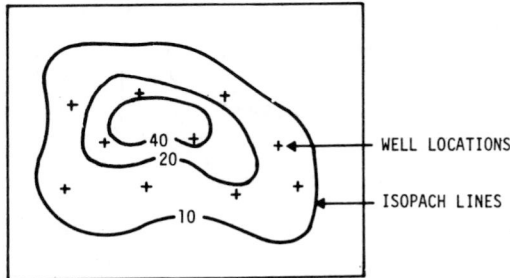

Figure 1.11: Oil in place in total reservoir.

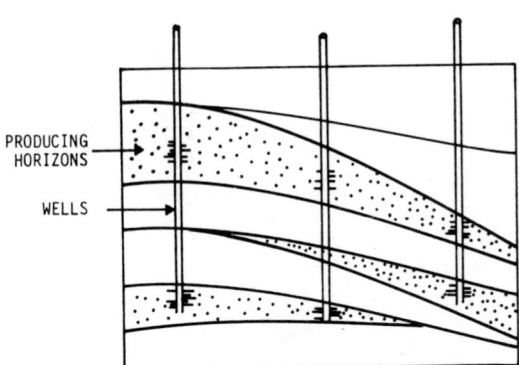

Figure 1.12: Oil in place in horizon or zone.

In developing a cash flow for a project, the basic data required include income generated, expenses, and capital investments over the planning horizon. The income-generating parameters are the oil and gas production that are essential outputs of the simulator. These are available on a per-well basis, by lease (Fig. 1.13), or by reservoir total. The typical result is shown in Fig. 1.14.

At the same time that the production rates are determined, the flowing bottom hole pressures on the wells are available. These flowing bottom hole pressures are used to plan the installation of downhole or surface lift equipment. Figure 1.15 illustrates the behavior of several production wells.

In a secondary recovery project, be it water injection or gas injection, the engineer needs to know the volumes of material injected and the injection pressures. These parameters are needed to design the size of the injection

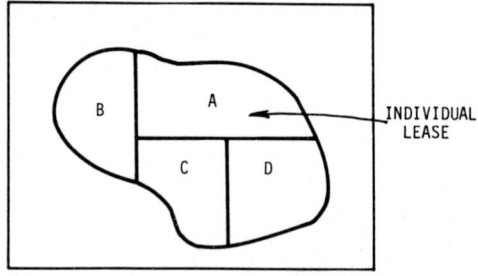

Figure 1.13: Oil in place by lease.

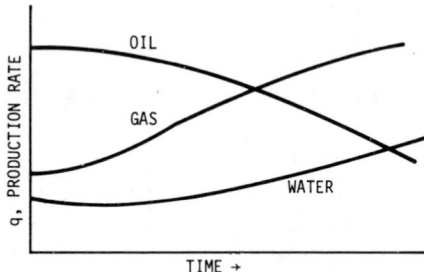

Figure 1.14: Production schedules.

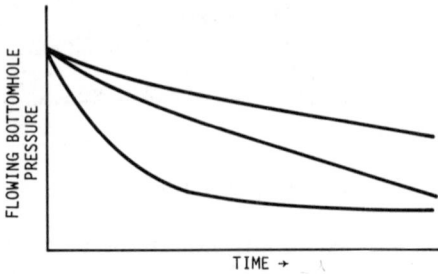

Figure 1.15: Flowing bottom hole pressures.

units, water supply, water-treatment or gas-processing plant size. The typical results are shown in Fig. 1.16.

As the engineer obtains the production and injection data he can develop the necessary parameters required to formulate his economic analysis. From the cash flow stream, he can determine any of his economic indicators such as payout time, profitability ratio, and present worth value of the project under study. This type of analysis, indicated graphically in Figs. 1.17 and 1.18, is

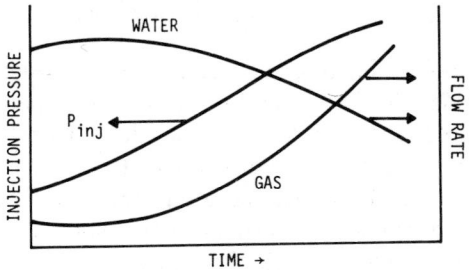

Figure 1.16: Injection rates and pressures.

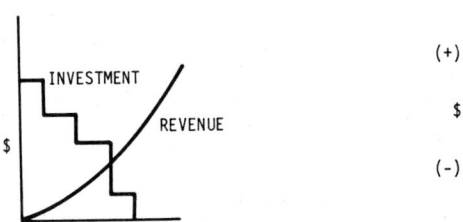

Figure 1.17: Cash flow.　　　　　**Figure 1.18:** Payout time.

the basis for comparing the merits of various operational schemes. The sensitivity of the various parameters to modifications in the way the reservoir is operated allows a certain amount of slack in the decision process, since the actual implementation of the project may differ somewhat from that recommended by the study.

In engineering larger reservoirs or reservoirs which are common to several operators, it is possible that during the life of the project significant quantities of fluid will move large distances from one lease to another. Since the reservoir sand is continuous, it is obvious that there will be movement based on pressure gradients regardless of what subjective boundaries are placed on the surface. The migration of fluids as illustrated in Fig. 1.19 can be monitored and the location of wells and the required production rates selected to control migration.

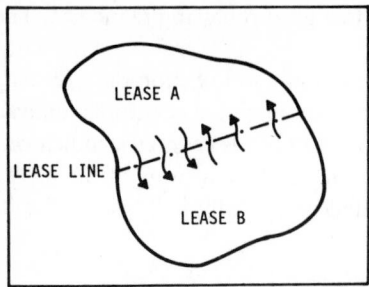

Figure 1.19: Migration across lease lines.

In addition to monitoring migration patterns of the fluids in the reservoir, the simulator enables the determination of sweep-out patterns around the injectors as indicated in Fig. 1.20. Once the flood fronts are located the movable oil in the unswept areas can be calculated and the location of new producers determined to maximize the total recovery. This process, in addition to locating new producers, also indicates the optimal drilling sequences, (i.e., number of wells drilled in each period, as shown in Fig. 1.21.) the conversion sequence from production to injection, and the optimal water-oil ratio at which wells are shut-in or converted to injectors.

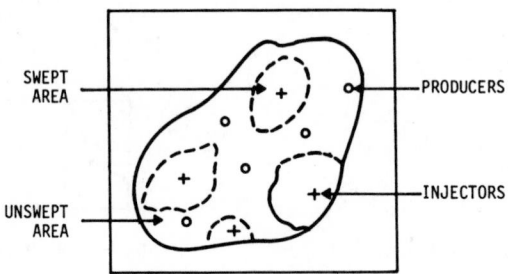

Figure 1.20: Sweep-out patterns.

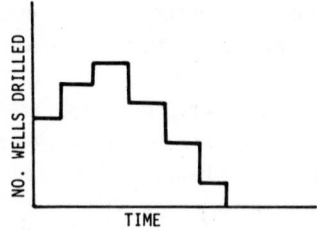

Figure 1.21: Drilling sequences.

Planning of Storage Requirements

In gas storage systems the engineer delivers gas to an underground storage reservoir from the remote producing areas during the off-season. This gas is withdrawn during the heating season (Fig. 1.22). In designing the overall storage facilities the engineer must be able to determine withdrawal rates, replenishment rates, makeup gas, and the effect of seasonal fluctuations and scheduling on the performance of the facilities. The typical configuration is shown in Fig. 1.23. When reservoir behavior is simulated the effects of inter-

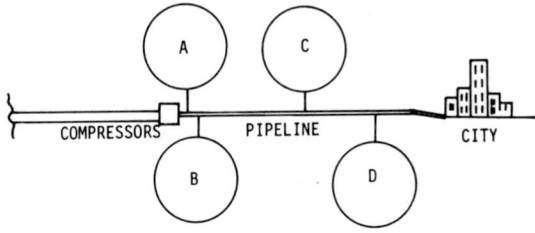

Figure 1.22: Storage system.

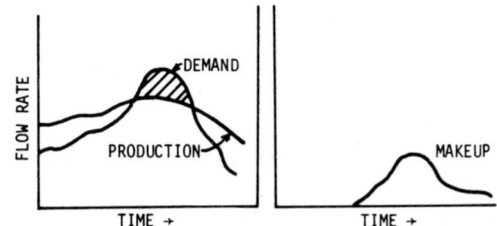

Figure 1.23: Deliverability and makeup requirements.

ference of well behavior can be included, and a more realistic analysis can be made of the process variables, thereby leading to better predictions. In addition, sensitivity analyses of the effects of variations in the predicted climatic factors can be studied.

Single-Well Studies

The ability to design an optimum completion program is essential for the proper exploitation of a reservoir. In some operations it is not feasible to carry out a full-blown reservoir simulation study, and a single-well study (Fig. 1.24) can be ideally employed to obtain parameters that allow the engineer to determine the following:

1. Critical flow rates required to prevent coning of gas or water
2. Maximum efficient rates to ensure optimum well response
3. Effects of perforation intervals and fracture penetration on well productivity (Fig. 1.25)

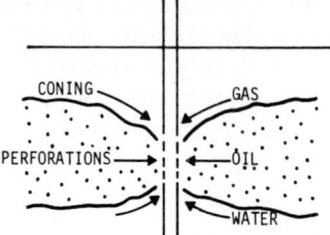

Figure 1.24: Single-well study.

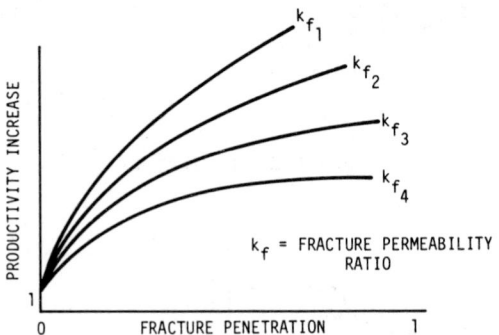

Figure 1.25: Fracture effect on productivity.

These single-well studies, sometimes referred to as coning models because of the implied presence of gas/water coning, are reasonably economical to use as a design tool.

Overall decision making in reservoir operations can be done more accurately and with firmer conviction if the manager has a feel for the operation parameters. The decision-making process, as indicated earlier, is at best difficult under the most simplified conditions; the present-day complexities make the need for a more viable data base, from which plans can be drawn, all the more necessary.

1.4   BENEFITS OF SIMULATION

The engineer knows he has a *single* opportunity to produce the reservoir; any mistakes made in this process will be around forever. However, the simulation study can be made *several* times and the alternatives examined. When the simulation study is used as a management tool, the efficient utilization of available energy within the reservoir can lead to greater ultimate production and certainly a more economical operation. In the more complex systems— for example, layered heterogeneous reservoirs with commingled production— it has been previously impossible to handle all these variables; today the engineer can examine such systems without undue difficulty to predict their behavior. One benefit of simulation which in reality was not designed into the process at the start but has evolved as a fruitful by-product is the presence now of a common ground between companies and regulatory bodies and other agencies which deal with petroleum resources. This commonality is the knowledge that all these groups are now using simulators to determine reservoir performance, and the differences between two opposing groups can be narrowed down to the data used rather than to the calculation procedure itself. The calculation procedures do not differ by very much, and if need be a standardized approach can be used in which the data can be run by a third-party system for comparison purposes. Finally, it can be said in a rather laconic way that even if the results of the simulator study were inconclusive, the mechanics of simulation have compiled all the data pertinent to that reservoir into one compact data base which is probably now in better shape than it ever was before.

REFERENCES

1. H. M. WAGNER, *Principles of Operations Research* (Englewood Cliffs, N.J.: Prentice Hall, 1969).

2. R. J. SCHILTHUIS, "Active Oil and Reservoir Energy," *Trans. AIME* (1936), **118**, 33.

3. W. A. BRUCE, "An Electrical Device for Analyzing Oil Reservoir Behavior," *Trans. AIME* (1943), **146**, 112.

4. R. C. RUMBLE, H. H. SPAIN, and H. E. STAMER III, "A Reservoir Analyzer Study of the Woodbine Basin," *Pet. Trans.* Reprint Series, No. 4, 123.

5. H. G. BOTSET, "The Electrolytic Model and Its Application to the Study of Recovery Problems," *Trans. AIME* (1946), **165**, 15.

6. B. D. LEE, "Potentiometric Model Studies of Fluid Flow in Petroleum Reservoirs," *Trans. AIME* (1948), **174**, 41.

7. G. H. BRUCE, D. W. PEACEMAN, H. H. RACHFORD, JR., and J. D. RICE, "Calculations of Unsteady-state Gas Flow through Porous Media," *Trans. AIME* (1953), **198**, 79.

8. A. S. ODEH and D. HAVLENA, "The Material Balance as an Equation of a Straight Line," *Trans. AIME* (1963), **228**, 896.

BIBLIOGRAPHY

COATS, K. H., "Use and Misuse of Reservoir Simulation Models," *J. Pet. Tech.* (Nov. 1969), 1391–98.

FERGUSON D. S., and H. D. ATTRA, "The Uses and Limitations of Computers in Petroleum Engineering Work," *J. Pet. Tech.* (July 1961), 625–28.

STAGGS H. M., and E. F. HERBECK, "Reservoir Simulation Models—An Engineering Overview," *J. Pet. Tech.* (Dec. 1971), 1428–36.

——————, "Reservoir Simulation Models—Mythology or Methodology?" SPE 3304, American Institute of Mining, Metallurgical, and Petroleum Engineers, Dallas, May 1971.

THACHUK A. R., and R. A. WATTENBARGER, "The What, Why, When, and How of Reservoir Simulation," *Canad. Pet.* (April 1970), 86–92.

2 Reservoir Engineering Concepts in Simulation

2.1 INTRODUCTION

Flow in porous media is a very complex phenomenon and as such cannot be described as explicitly as flow through pipes or conduits. It is rather easy to measure the length and diameter of a pipe and compute its flow capacity as a function of pressure; however, in porous media flow is different in that there are no clear-cut flow paths which lend themselves to measurement.

The analysis of fluid flow in porous media has evolved throughout the years along two fronts—the experimental and the analytical. Physicists, engineers, hydrologists, and the like have examined experimentally the behavior of various fluids as they flow through porous media ranging from sand packs to fused Pyrex glass. On the basis of their analyses they have attempted to formulate laws and correlations which can then be utilized to make analytical predictions for similar systems.

Flow in porous media is described by a dictionary of new concepts which must first be elucidated and understood before we can adequately formulate the equations to be used in a simulator. These concepts include permeability, flow potential, single-phase, multiphase, relative permeability, and fluid compressibility. The objective of this chapter is to describe qualitatively and sometimes quantitatively these ideas.

Darcy's Law[1]—The Concept of Permeability

The ability to predict the behavior of petroleum reservoirs hinges around the ability of the engineer to predict the flow characteristics of the fluids in the reservoir. After all the measurements of porosity and fluid saturations have

21

been made, we still have to determine at what rate the reservoir fluids can be produced.

In order to quantitatively define the ability of a rock to transmit fluid, we must introduce a new concept. This is the concept of permeability of a rock, which is a petrophysical constant defined by Darcy's law:

> The rate of flow of a homogenous fluid through a porous medium is proportional to the pressure or hydraulic gradient and to the cross-sectional area normal to the direction of flow and inversely proportional to the viscosity of the fluid.

Mathematically:

$$V_s = -\frac{k}{\mu v}\left(\bar{v}\frac{\partial P}{\partial s} + \frac{\partial z}{\partial s}\right) \tag{2.1}$$

where

V_s = macroscopic velocity in positive s

μ = absolute viscosity

k = homogenous fluid permeability

z = elevation

$\bar{v}$ = specific volume = $\dfrac{1}{\rho g}$

ρ = density

g = gravitational acceleration

This is the definitive equation for the permeability of a porous medium.

The quantity in parentheses is the potential of the fluid, and Eq. (2.1) can be written as:

$$V_s = -\frac{k}{\mu \bar{v}}\frac{\partial \Phi}{\partial s} \tag{2.2}$$

where Φ is the total fluid potential.[2] This will be discussed later. Darcy's law is an empirical law, and as written in Eq. (2.1) or Eq. (2.2) it is a differential equation relating to a point. It is possible that every term in the equation k, Φ, μ, $\bar{v}$ can vary with location and these variations must be accounted for in the use of the equation.

In Darcy's experiments there were certain limitations and assumptions to his work; these limit the area of applicability. The assumptions are:

1. Fluid—homogenous and single-phase
2. No chemical reaction between media and fluid

3. Permeability independent of fluid, temperature, pressure, and location
4. Laminar flow—i.e., no turbulence
5. No electrokinetic effect*
6. No Klinkenberg effect†

Darcy's work was essentially for linear systems; however, this work has been extended to multidimensional systems, not because it has been proven applicable but because no one has yet been able to prove it inapplicable.

The nature of the units of the permeability can be determined by dimensional analysis. The unit of permeability is called the darcy.

By dimensional analysis:

$$V_s = -\frac{k}{\mu \bar{v}} \frac{\partial \Phi}{\partial s} \tag{2.3}$$

The units on the left must be the same as those on the right for "dimensional homogeniety." In the MLT system:

$$V_s = \frac{L}{T}, \qquad \rho = \frac{M}{L^3}, \qquad \frac{\partial P}{\partial s} = \frac{M}{L^2 T^2}, \qquad \mu = \frac{M}{LT}, \qquad g = \frac{L}{T^2} \tag{2.4}$$

Making the above substitutions in Eq. (2.1), we have:

$$\frac{L}{T} = \frac{k}{M/LT}\left(\frac{M}{L^2 T^2} - \frac{M}{L^3}\frac{L}{T^2}\right)$$

$$= \frac{KLT}{M}\left(\frac{M}{L^2 T^2} - \frac{M}{L^2 T^2}\right)$$

$$= \frac{k}{LT} \tag{2.5}$$

If k/LT is to be identical to L/T, then $k = L^2$;

$$\therefore \quad \frac{k}{LT} = \frac{L^2}{LT} = \frac{L}{T} \tag{2.6}$$

The unit of permeability is thus (length)^2.

The Flow Potential

A fundamental tenet of fluid mechanics of porous media is that the macroscopic fluid velocity vectors are always normal to the equipotential surfaces

---

*Streaming potential—production of potential difference when a liquid is forced through a porous membrane or capillary. This can be measured and is commonly called zeta potential.

†When pore sizes approach the mean free path of molecules, slippage at wall begins to occur.

and that the magnitudes of these vectors are proportional to the gradients of these potentials (see Fig. 2.1). Since the distribution of potential within a fluid determines the macroscopic velocity of the fluid and also the overall flow, the investigation of the flow potential is warranted. Hubbert[2] defines the potential Φ as mechanical energy per unit mass of fluid at any location. To get the

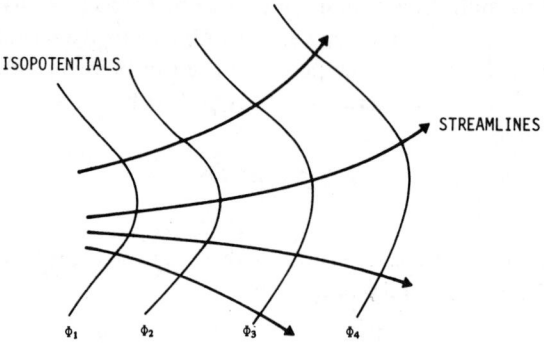

Figure 2.1: Isopotentials and streamlines.

fluid to this location, several kinds of work must be done on the fluid. The sum total of this work done on the fluid reflects the mechanical energy within the fluid. Consider a particle of fluid at some datum with zero potential ($\Phi = 0$). Then the potential associated with this fluid in moving to a new location (1) is Φ_1 (Fig. 2.2); Φ_1 is computed by detailing all the work done on the fluid:

$$
\begin{aligned}
\Phi_1 = & -P'\bar{V}_1 & \text{Collect} \\
& + \int_{V_1'}^{V_1} P\, d\bar{V} & \text{Compress} \\
& + z_1 & \text{Elevate} \\
& + P_1\bar{V}_1 & \text{Eject} \\
& + \frac{\mu_1^2}{2g} & \text{Accelerate}
\end{aligned}
\tag{2.7}
$$

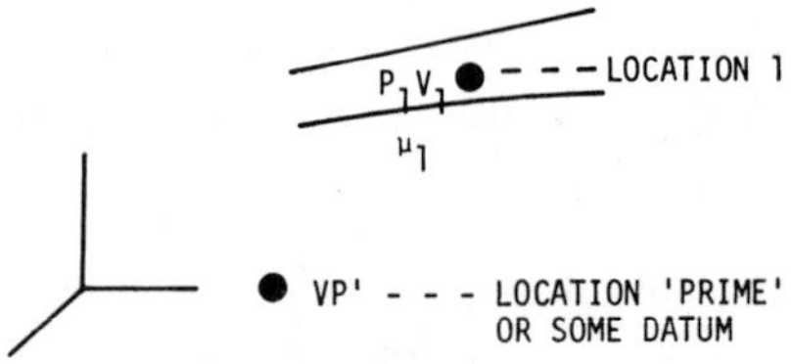

$P_1 V_1 \bullet\ -\ -\ -$ LOCATION 1

μ_1

$\bullet$ VP' $-\ -\ -$ LOCATION 'PRIME' OR SOME DATUM

Figure 2.2: Particle location.

This can be simplified by calculus to:

$$\Phi_1 = \int_{P'}^{P_1} \bar{V}\,dP + z_1 + \frac{\mu_1^2}{2g} \tag{2.8}$$

Since the velocity term is negligible in porous media, we have:

$$\Phi_1 = \int_{P'}^{P_1} \bar{V}\,dP + z_1 \tag{2.9}$$

Assuming incompressible flow, then $\bar{V}$ is not a function of pressure and:

$$\Phi_1 = \bar{V} \int_{P'}^{P_1} dP + z_1 \tag{2.10}$$

Thus,

$$\Phi_1 = \bar{V}(P_1 - P') + z_1 \tag{2.11}$$

for incompressible flow.

Let us consider some examples of the use of the flow potential in some simple systems.

EXAMPLE [1]   FREE DOWNWARD FLOW: Note in Fig. 2.3 that the flow direction s and the coordinate z are decreasing in the same direction. Then, using Eq. (2.11):

$$\Phi = \bar{V}(P_1 - P') + z$$
$$= \frac{(P_1 - P')}{\rho g} + z$$

Note that Φ *must* decrease in direction of flow. Therefore, from the geometry of the system, we can arrive at Fig. 2.4. If flow direction s is same as coordinate direction z, then

$$ds = dz$$

$$\therefore \quad V_{+z} = -\frac{k}{\mu\bar{V}} \frac{d\Phi}{dz}$$

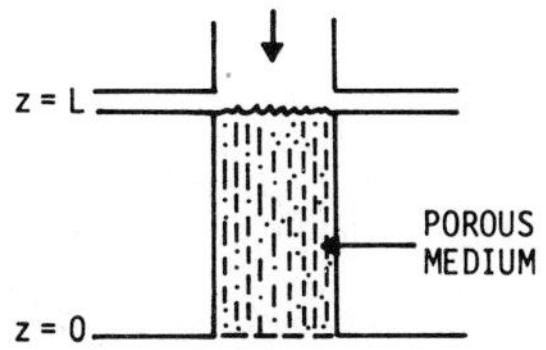

Figure 2.3: Free downward flow.

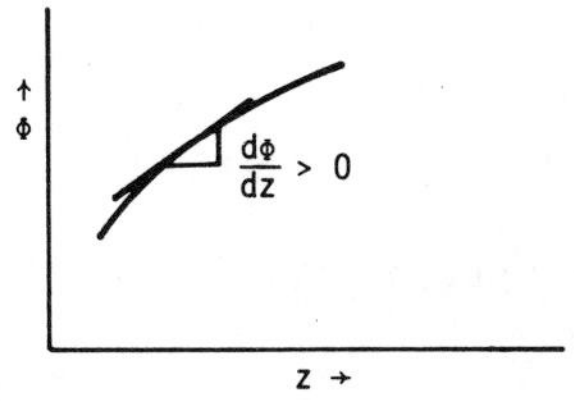

Figure 2.4: Potential change with distance.

If flow direction s is opposite to coordinate direction z, then

$$ds = -dz$$

$$\therefore \quad V_{-z} = \frac{k}{\mu \bar{V}} \frac{d\Phi}{dz}$$

In Example [1],

$$V_{-z} = \frac{k}{\mu \bar{V}} \frac{d\Phi}{dz} \quad \text{Potential flow}$$

$$= \frac{q}{A} \qquad \qquad \text{Pipe flow}$$

Therefore, setting up limits and integrating:

$$\frac{q}{A} \int_0^L dz = \frac{k}{\mu \bar{V}} \int_{\Phi_0}^{\Phi_L} d\Phi$$

Thus:

$$\frac{qL}{A} = \frac{k}{\mu \bar{V}} (\Phi_L - \Phi_0)$$

From Eq. (2.11):

$$\Phi_L = \bar{V}(P' - P') + L = L$$
$$\Phi_0 = \bar{V}(P' - P') + 0 = 0$$

Then:

$$\frac{qL}{A} = \frac{k}{\mu \bar{V}} L$$

$$q = \frac{kA}{\mu \bar{V}}$$

Flow rate is

$$q = \frac{kA}{\mu} \rho g \qquad \qquad (2.12)$$

This equation can be rearranged to solve for permeability:

$$k = \frac{q \mu \bar{V}}{A} \qquad \qquad (2.13)$$

EXAMPLE [2]   FLOW DOWNWARD WITH HEAD (FIG. 2.5):

$$\Phi_L = \bar{V}(P_L - P_{atm}) + L$$

$$= \bar{V}\left[P_{atm} + \frac{(H-L)}{\bar{V}} - P_{atm}\right] + L$$

$$= H$$

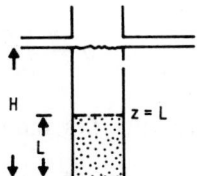

Figure 2.5: Downward flow with head.

The potential at points $z = L$ and $z = 0$ are first determined, using Eq. (2.11):

$$\Phi_0 = \bar{V}[P_{atm} - P_{atm}] + 0$$

$$= 0$$

At any instant:

$$V = \frac{q}{A} = \frac{k}{\mu\bar{V}}\frac{d\Phi}{dz}$$

Then;

$$q = \frac{Ak}{\mu\bar{V}}\frac{d\Phi}{dz}$$

which gives:

$$q\int_0^L dz = \frac{Ak}{\mu\bar{V}}\int_{\Phi_0}^{\Phi_L} d\Phi$$

Finally:

$$q = \frac{Ak}{\mu\bar{V}}\frac{H}{L}$$

The procedure for solving flow potential problems is straightforward and is summarized below:

1. Select two points, usually one on either side of the porous medium, for which to write the potential equations.

2. Write the potential terms using Eq. (2.11) and the equation for the

pressure due to hydrostatic head:

$$\Phi = \bar{V}(P - P') + z$$

and

$$P = P' + \rho g(h - z)$$

3. Invoke the pipe flow equation to get another equation if needed:

$$q = A\frac{dh}{dt} \tag{2.14}$$

4. Equate flow rates or velocities and solve.

Real Gas Flow—Real Gas Potential

Under ideal conditions the properties of most gases are assumed independent of pressure. This tacit assumption allows the use of straightforward ideal gas laws to analyze the behavior of gases. However, under reservoir conditions no gases behave ideally, and the engineer must account for the variations in gas properties with pressure. The major variations usually included in a study are:

1. Viscosity variation with pressure
2. Variation of gas deviation factor z with pressure

Until now, analysis of gas flow was based upon linearizations which required physical properties evaluated at some average flowing pressure. This assumption implied that flowing gradients were small, a situation not usually met with in real reservoir situations. In order to simplify gas flow analysis and incorporate some of the above variations, Al-Hussainy et al.[3] developed a function called the real gas potential. The real gas potential includes pressure, viscosity, and z-factor as one variable. It is mathematically defined as

$$m(P) = 2\int_{P_m}^{P} \frac{P'}{\mu(P')z(P')}\,dP' \tag{2.15}$$

where

P_m = an arbitrary datum pressure

P = pressure of gas

μ = viscosity

z = gas deviation factor

P' = a dummy variable of integration

This function is used primarily in gas well–testing analysis and in single-phase simulators for dry gas. It is not used in the typical reservoir models where gas, oil, and water are flowing. The efficacy of this equation is seen by comparing the following two equations:

$$q = \lambda \frac{\partial m}{\partial r} \tag{2.16}$$

$$q = \frac{cP}{z\mu} \frac{\partial P}{\partial r} \tag{2.17}$$

Equation (2.16) states that the flow rate is a function of λ, a constant dependent on rock and spatial dimensions only and the potential gradient, while Eq. (2.17) states that the flow rate is a function of some pressure P, the viscosity, and the deviation factor in addition to the pressure gradient. The real gas potential, in addition to being more realistic, simplifies the equations required.

Concepts in Steady and Unsteady Flow

One of the more puzzling concepts to understand both by the experienced engineer and the neophyte student is that of steady and unsteady flow. The engineer sometimes ponders why he cannot start producing 1000 BPD of fluid from a reservoir *the day after* he starts injecting 1000 BPD of water. This and many more problems are caused by the behavior of fluids within the pore space of the rock and are indicated by the way the pressure responds. Since pressure is an easily measurable and readily recognizable parameter, we shall restrict our treatment of these concepts to the way in which the pressure is affected. This discussion could just as well be made using the density of the fluid as a parameter.

To begin, let us trace the path of a fluid particle meandering through the pore spaces of the rock as shown in Fig. 2.6. The velocity of the particle is shown in the diagram by V_s. The acceleration of the particle can be obtained

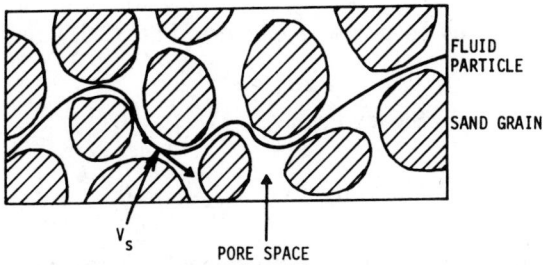

Figure 2.6: Particle moving through porous medium.

by determining the rate of change of velocity. For example, since $V = f(s, t)$—i.e., a function of two variables—then

$$dV = \left(\frac{\partial V}{\partial t}\right)_s dt + \left(\frac{\partial V}{\partial s}\right)_t ds \qquad (2.18)$$

The equation for the total acceleration can be determined:

$$\frac{dV}{dt} = \left(\frac{\partial V}{\partial t}\right)_s + \left(\frac{\partial V}{\partial s}\right) \frac{ds}{dt} \qquad (2.19)$$

Since ds/dt = velocity, Eq. (2.19) can be written:

$$\frac{dV}{dt} = \left(\frac{\partial V}{\partial t}\right)_s + \left(\frac{\partial V}{\partial s}\right)_t v \qquad (2.20)$$

The first term on the right-hand side is the acceleration at a point, while the second term is the convectional acceleration. In words, Eq. (2.20) is as follows:

$$
\begin{array}{ccc}
\text{Total} & \text{Local} & \text{Convectional} \\
\text{Acceleration} = & \text{Acceleration} + & \text{Acceleration} \\
| & | & | \\
\left(\begin{array}{l}\text{substantial} \\ \text{derivative or} \\ \text{derivative} \\ \text{following the} \\ \text{fluid motion}\end{array}\right) & \text{(at a point)} & \left(\begin{array}{l}\text{acceleration} \\ \text{experience if} \\ \text{you moved} \\ \text{with the fluid}\end{array}\right)
\end{array}
$$

By inspection of the two terms that correspond to the total acceleration in Eq. (2.20), we can predict whether a flow regime is steady or unsteady. If

$$\left(\frac{\partial V}{\partial t}\right)_s = 0 \qquad (2.21)$$

then flow is steady. If

$$\left(\frac{\partial V}{\partial t}\right)_s \neq 0 \qquad (2.22)$$

then flow is unsteady. In terms of pressure, Eqs. (2.21) and (2.22) can be written:

Steady flow: $\quad \left(\frac{\partial P}{\partial t}\right)_s = 0 \qquad (2.23)$

Unsteady flow: $\quad \left(\frac{\partial P}{\partial t}\right)_s \neq 0 \qquad (2.24)$

Let us consider a reservoir to be represented by a well of radial symmetry with a finite well bore radius and some finite outer radius as shown in Fig. 2.7.

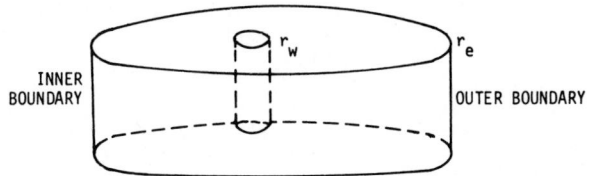

Figure 2.7: Radial reservoir system.

The reservoir remains at equilibrium unless some disturbance occurs at one of the boundaries. Depending on the nature of the disturbance, the system may or may not reach a steady state. The possible conditions are:

At the inner boundary:
Constant well bore pressure:

$$\text{1. } P(r_w, t) = \text{Constant} \tag{2.25}$$

Constant flow rate:

$$\text{2. } r\frac{\partial P(r_w, t)}{\partial r} = \text{Constant} \tag{2.26}$$

Variable well bore pressure:

$$\text{3. } P(r_w, t) = f_1(t) \tag{2.27}$$

Variable flow rate:

$$\text{4. } r\frac{\partial P(r_w, t)}{\partial t} = g_1(t) \tag{2.28}$$

Shut-in well:

$$\text{5. } r\frac{\partial P(r_w, t)}{\partial t} = 0 \tag{2.29}$$

At the outer boundary:
Constant pressure:

$$\text{6. } P(r_e, t) = \text{Constant} \tag{2.30}$$

Constant influx across the boundary:

$$\text{7. } \frac{\partial P(r_e, t)}{\partial r} = \text{Constant} \tag{2.31}$$

Variable influx rate:

$$8. \quad \frac{\partial P(r_e, t)}{\partial r} = f_2(t) \qquad\qquad (2.32)$$

Closed outer boundary:

$$9. \quad \frac{\partial P(r_e, t)}{\partial t} = 0 \qquad\qquad (2.33)$$

Infinite reservoir system:

$$10. \quad \lim_{r \to \infty} P(r, t) = P_i \qquad\qquad (2.34)$$

As the well is produced, the pressure around the inner radius begins to drop and the decreased pressure wave moves outward to the limits of the reservoir. The pressure profile as a function of time is shown in Fig. 2.8.

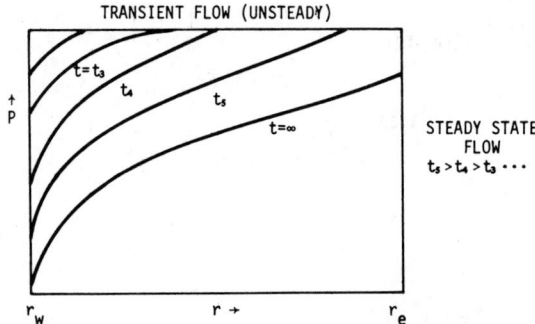

Figure 2.8: Radial pressure profile.

Several combinations of inner and outer boundary conditions could produce a steady-state flow pattern; in contrast, there are some conditions which preclude the existence of steady-state flow. Under the following conditions, steady-state flow cannot be obtained. If the outer boundary is closed, i.e.,

$$\frac{\partial P(r_e, t)}{\partial r} = 0 \qquad\qquad (2.35)$$

then no mass crosses the boundary and the reservoir will continue to deplete at all times.

To achieve a steady-state flow regime, there must be some support for the system in terms of influx or a constant pressure. This is achieved in practice by the presence of an aquifer adjacent to the oil reservoir.

2.2 FLUID TYPES[4]

Reservoir fluids are classified into three groups depending on their compressibility. In some cases these classifications are arbitrary and are only made for the purpose of simplifying the assumptions. The groups are:

1. Incompressible
2. Slightly compressible
3. Compressible

Incompressible fluids have a constant density. Slightly compressible fluids have a measurable change of density with pressure, and compressible fluids have a significant density change with pressure. See Fig. 2.9. In reservoir

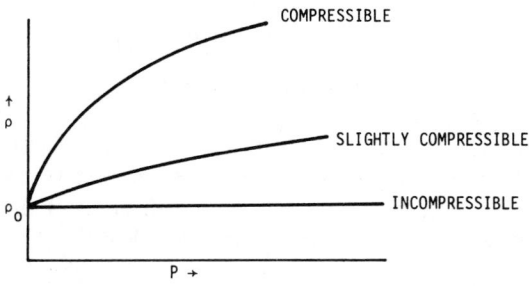

Figure 2.9

calculations the compressibility term is evidence by the formation volume factors.

The equation of state used in the development of the diffusion-type equation later involves the density/pressure relation

$$\rho = \rho_0 e^{c(P-P_0)} \tag{2.36}$$

where

$\quad c = $ compressibility

$\quad P_0 = $ datum pressure

$\quad P = $ any pressure

For incompressible fluids:

$$c = 0$$

Then:

$$\rho = \rho_0, \quad \text{for all } P$$

For slightly compressible fluids:

$$c \simeq 0$$

Then:

$$\rho = \rho_0 e^{c(P-P_0)} \qquad (2.37)$$

Note:

$$e^x = 1 + x + \frac{x^2}{2!} + \frac{x^3}{3!} + \ldots \qquad (2.38)$$

$$e^{c(P-P_0)} = 1 + c(P - P_0) + \left[\frac{c(P - P_0)}{2!}\right]^2 + \ldots \qquad (2.39)$$

Since $c \simeq 0$, neglect higher-order terms:

$$e^{c(P-P_0)} = 1 + c(P - P_0)$$
$$\rho = \rho_0[1 + c(P - P_0)]$$
$$= \rho_0 + \rho_0 c(\Delta P) \qquad (2.40)$$

Fluids which can be represented by Eq. (2.40) are classified as slightly compressible. These include most reservoir oils and reservoir waters.

For compressible fluids—e.g. gases—the truncation of the series expansion of the exponential is not valid, and the complete equation is used.

2.3　FLOW IN POROUS MEDIA

Multiphase Flow:　In fluid-saturated porous media, there can be as many as three fluid phases present. To understand more fully the behavior of these fluids within the porous medium, we derive the system of equations which govern the motion of these fluids. The multiphase flow equations are non-linear partial differential equations which are not capable of being integrated by analytical means. To fully develop the flow equations we should first define some new concepts.

Relative Permeability:　In rocks saturated with more than one fluid, the ability of each fluid to move under an applied pressure gradient is a function of the relative permeability of that phase. The relative permeability is defined as the ratio of the permeability of the rock to the fluid at a given saturation to the permeability when 100% saturated with the given fluid:

$$k_{ro_{(0.5)}} = \frac{k_{0_{(0.5)}}}{k_{0_{(1.00)}}}$$

Relative permeability is a function of fluid saturation, and relative permeability curves have a characteristic shape (see Fig. 2.10). Below a given value of saturation the relative permeability for either the wetting or nonwetting phase is zero. Therefore, up to and including this point of saturation, there will be no flow of that particular phase since its mobility will be zero:

$$V_0 = k \frac{k_{ro}}{u_0} \frac{\partial p}{\partial s} = 0, \quad \text{since} \quad k_{ro} \equiv 0$$

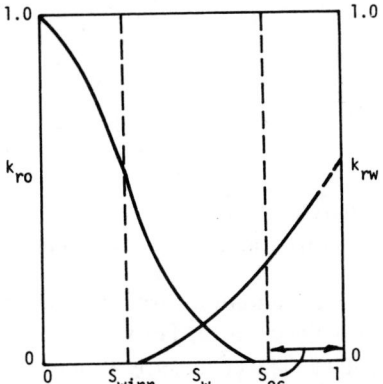

Figure 2.10: Relative permeability curve.

This saturation is called the critical point or critical saturation (S_{oc}). This feature should be remembered, and it will be discussed later in Section 9.5 on history matching.

Two-Phase Relative Permeability: Relative permeability data are usually obtained from laboratory investigations on suitable cores. However, this source may be lacking and some suitable approximations must be derived. These approximations are determined for and depend on the process which the reservoir is undergoing. Two approximations often used are the following:

1. Corey approximation[5]
Displaced-phase relative permeability:

$$k_0 = (1 - S)^4$$

Displacing-phase relative permeability:

$$k_D = S^3(2 - S)$$

where

$$S = \frac{S_D}{1 - S_{wc}} \quad \text{a normalized saturation function}$$

This approximation is good for drainage processes—e.g., a gas drive where the saturation of the wetting phase is being decreased.

2. Naar-Henderson approximation[6]

$$k_0 = \frac{(1 - 2S)^{3/2}}{2 - (1 - 2S)^{1/2}}$$

$$k_D = S^4$$

where

$$S = \frac{S_D - S_{wc}}{1 - S_{wc}}$$

This approximation is good for imbibition processes—e.g., water drives where the saturation of the wetting phase is being increased.

These approximations are all functions of a "normalized" saturation S which was defined above. The type of function depends on the type of system being modeled. The engineer is at liberty to modify the equations for relative permeability by making the required changes in the exponents which would make the relative permeability data more closely duplicate that of the reservoir. A general equation could be the following:

$$\left.\begin{array}{l} k_0 = (1 - S)^n \\ k_{0D} = S^k(2 - S) \end{array}\right\} \quad \text{Drainage processes}$$

and

$$\left.\begin{array}{l} k_0 = \dfrac{(1 - 2S)^m}{[2 - (1 - 2S)]^p} \\ k_{0D} = S^q \end{array}\right\} \quad \text{Imbibition process}$$

where n, k, m, p, and q are exponents which can be appropriately determined by a trial-and-error process.

This trial-and-error process will be explored further in history matching when the correct relative permeability curve is being sought to match the reservoir performance.

Three-phase Relative Permeability: Up to this time we have considered only two fluids flowing simultaneously, as evidenced by the typical relative permeability plots. In a simulation model we must be able to predict the behavior of all three phases flowing simultaneously within the porous medium.

Stone[7] has developed a very elegant model of three-phase flow in which he combines the theory of channel flow in porous media with probability concepts to obtain a simple result for determining the relative permeability to oil in the presence of water and gas flowing. This model has enjoyed relatively wide acceptance because of its ability to reproduce measured data and the simplicity of its form.

The three-phase model is developed from two-phase data. The required data consist of a set of oil/water relative permeability data and oil/gas relative permeability data. From the two sets of data the values of k_{rg}, k_{rw}, and k_{ro} are determined.

The values of k_{rw} obtained from Fig. 2.11 and k_{rg} obtained from Fig. 2.12 are used directly in the three-phase model (Fig. 2.13):

$$k_{rg} = f(S_g)$$
$$k_{rw} = f(S_w)$$

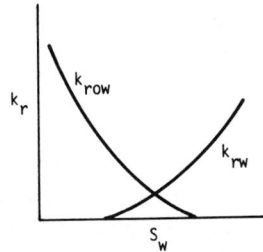

Figure 2.11: Oil/water relative permeability.

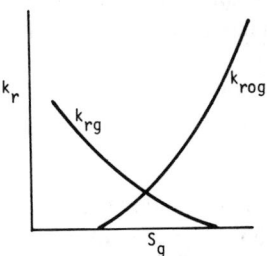

Figure 2.12: Oil/gas relative permeability.

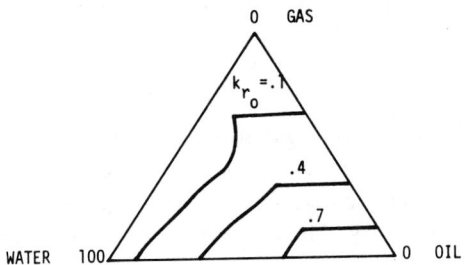

Figure 2.13: Composite three-phase curve.

The oil relative permeability is obtained with the following equation:

$$k_{ro} = (k_{row} + k_{rw})(k_{rog} + k_{rg}) - (k_{rw} + k_{rg})$$

such that $k_{ro} \geq 0$.

The inequality must be satisfied. It is possible that the calculated value for k_{ro} is less than zero, in which case $k_{ro} \equiv 0$ and no oil flows where:

k_{ro} = relative permeability to oil

k_{rg} = relative permeability to gas

k_{rw} = relative permeability to water

k_{row} = relative permeability to oil in oil/water system

k_{rog} = relative permeability to oil in gas/oil system

The defining equation for k_{ro} is developed as follows:

The channel flow theory postulates that one and only one mobile phase exists in each channel at a given time: The wetting phase in the smaller channels and the nonwetting phase in the larger channels, these two phases being separated by the intermediate phase. Extending this concept to the porous medium as a whole, the relative permeability of a phase is the composite of an infinite number of contributions of each little channel.

Let us introduce the probability concept by defining:

$$\sigma_w = k_{row} + k_{rw}$$

which is the summation of the relative permeability to oil and relative permeability to water in an oil/water system. The value σ_w is equal to 1.0 at $S_w = 1 - S_{wc}$. As the water saturation increases, σ_w will tend to change; thus, σ_w is saturation-dependent. The term $\sigma_w(S_w)$ is a fraction of the total relative permeability at S_w, and it can be considered the probability that these contributions are still open to flow at a given S_w. Similarly, σ_g and $\sigma_g(S_g)$ are defined:

$$\sigma_g = k_{rog} + k_{rg}$$

Since water displacing oil and gas displacing oil are occurring at different places at the same time, these two processes are assumed to be independent events, and the total probability of the events occurring is then the product of each individual probability. Thus, the fraction of the total relative permeability remaining is:

$$\begin{aligned} k_{ro} + k_{rw} + k_{rg} &= \sigma_w \sigma_g \\ &= (k_{row} + k_{rw})(k_{rog} + k_{rg}) \end{aligned}$$

This equation is solved for k_{ro}:

$$k_{ro} = (k_{row} + k_{rw})(k_{rog} + k_{rg}) - (k_{rw} + k_{rg})$$

This equation provides a good fit to experimental data except in the regions of high water saturation and low gas saturation.

REFERENCES

1. H. DARCY, *Les Fontaines publiques de la ville de Dijon* (Paris: Victor Dalmont, 1856).

2. M. KING HUBBERT, "Darcy's Law and the Field Equations of the Flow of Underground Fluids," *Trans. AIME* (1956), **207**, 222–39.

3. R. AL-HUSSAINY, H. J. RAMEY, JR., and P. B. CRAWFORD, "The Theory of the Real Gas Potential," SPE Paper 1243-A. Society of Petroleum Engineers of AIME, Denver (Oct. 1965); R. AL-HUSSAINY and H. J. RAMEY, JR., "Application of the Real Gas Potential," SPE Paper 1243-B, SPE of AIME, Denver (Oct. 1965).

4. M. B. STANDING, *Volumetric and Phase Behavior of Oil Field Hydrocarbon Systems* (New York: Reinhold, 1952).

5. A. T. COREY, C. H. RATHJENS, J. H. HENDERSON, and R. M. J. WYLLIE, "Three Phase Relative Permeability," *Trans. AIME* (1956), **207** 349–51.

6. J. NAAR, R. J. WYGAL, and J. H. HENDERSON, "Imbibition Relative Permeability in Unconsolidated Porous Media," Soc. of Pet. Eng. Journal, AIME (1962), 11–13.

7. H. L. STONE, "Probability Model for Estimating Three-phase Relative Permeability," *J. Pet. Tech.* (1970), I-214–18.

BIBLIOGRAPHY

AL-HUSSAINY, R., H. J. RAMEY, JR., and P. B. CRAWFORD, "The Flow of Real Gases Through Porous Media," SPE of AIME (1966), I-624.

ANDRE, H., and D. W. BENNION, "A Transform Approach to the Simulation of Transient Gas Flow in Porous Media," *Soc. Pet. Eng. J.* (June 1970), 135–39.

ARONOFSKY, J. S., and R. A. JENKINS, "A Simplified Analysis of Unsteady Radial Gas Flow," *J. Pet. Tech.* (July 1954), 23–35.

AUFRICHT, W. R., and E. H. KOEPT, "The Interpretation of Capillary Pressure Data from Carbonate Reservoirs," SPE of AIME (1957), 402.

BAPTIST, OREN C., and ELIOT J. WHITE, "Clay Content and Capillary Behavior of Wyoming Reservoir Sands," SPE of AIME (1957), 414.

BLAIR, P. M., "Calculation of Oil Displacement by Countercurrent Water Imbibition," *Soc. Pet. Eng. J.* (Sept. 1964), 195–202.

BOURGOYNE, A. T. JR., B. H. CAUDLE, and O. K. KIMBLER, "The Effect of Interfacial Films on the Displacement of Oil by Water in Porous Media," SPE of AIME (1972), II-60.

CARTER, R. D., "Solution of Unsteady-state Radial Gas Flow," *J. Pet. Tech.* (May 1962), 549–54; *Trans. AIME*, **225**.

COATS, K. H., M. R. TEK, and D. L. KATZ, "Unsteady-state Liquid Flow Through Porous Media Having Elliptic Boundaries," *Trans. AIME* (1959), **216**, 460–64.

COLONNA, J., F. BRISSAND, and J. L. MILLET, "Evolution of Capillarity and Relative Permeability Hysteresis," SPE of AIME (1972), II-28.

COMBARNOUS, M. A., and P. BIA, "Combined Free and Forced Convection in Porous Media," SPE of AIME (1971), II-399.

COREY, A. T., C. H. RATHJENS, J. H. HENDERSON, and M. R. J. WYLLIE, "Three-phase Relative Permeability," SPE of AIME (1956), 349.

CRICHLOW, HENRY B., and PAUL J. ROOT, "A Numerical Study of the Effect of Completion Technique on Gas Well Deliverability," SPE 2809, Second Symposium on Numerical Simulation of Reservoir Performance, Dallas, Texas, Feb. 5–6, 1970.

DOUGHERTY, E. L., "Mathematical Models of an Unstable Miscible Displacement," *Soc. Pet. Eng. J.* (June 1963), 155–63.

DOUGHERTY, E. W., and J. W. SHELDON, "The Use of Fluid-Fluid Interfaces to Predict the Behavior of Oil Recovery Processes," *Soc. Pet. Eng. J.* (June 1964), 171–82.

DOUGLAS, J., JR., P. M. BLAIR, and R. J. WAGNER, "Calculation of Linear Water-flood Behavior Including the Effects of Capillary Pressure," *Trans. AIME* (1958), **213**, 96–102.

DUMORE J. M., and R. S. SCHOLS, "Drainage Capillary-pressure Functions and Their Computation from One Another," SPE 4096, 47th Annual Meeting, San Antonio, Texas, Oct. 8–11, 1973.

FATT, I., "The Network Model of Porous Media, I: Capillary Pressure Characteristics," SPE of AIME (1956), 144.

FAYERS, F. J., and J. W. SHELDON, "The Effect of Capillary Pressure and Gravity on Two-phase Fluid Flow in a Porous Medium," *Trans. AIME* (1959), **216**, 147–55.

FOSTER, W. R., J. M. McMILLEN, and A. S. ODEH, "The Equations of Motion of Fluids in Porous Media, I: Propagation Velocity of Pressure Pulses," SPE of AIME (1967), II-333.

HAWTHORNE, R. C., "Two-phase Flow in Two-dimensional Systems—Effects of Rate, Viscosity, and Density of Fluid Displacement in Porous Media," *Trans. AIME* (1960), **219**, 81–87.

HOLM, L. W., "The Mechanism of Gas and Liquid Flow Through Porous Media in the Presence of Foam," SPE of AIME (1968), II-359.

HOVANESSIAN, S. A., and F. J. FAYERS, "Linear Water Flood with Gravity and Capillary Effects," *Soc. Pet. Eng. J.* (March 1961), 32–36; *Trans. AIME*, **249**.

JACKS, H. H., O. J. E. SMITH, and C. C. MATTAX, "Modeling of Three-dimensional Reservoirs with Two-dimensional Reservoir Simulators—The Use of Dynamic Pseudo Functions," SPE 4071, 47th Annual Meeting, San Antonio, Texas, Oct. 8–11, 1972.

JAVANDEL, I., and P. A. WITHERSPOON, "Application of the Finite Element Method to Transient Flow in Porous Media," SPE of AIME (1968), II-241.

JONES-PARRA, J., "Comments on Capillary Equilibrium," SPE of AIME (1953), 314.

KYLE, C. R., and R. L. PERRINE, "Turbulent Dispersion in Porous Materials as Modeled by a Mixing Cell With Stagnant Zone," *Soc. Pet. Eng. J.* (March 1971), 57–62; *Trans. AIME*, **251**.

LAND, C. S., "Calculation of Imbibition Relative Permeability for Two- and Three-phase Flow From Rock Properties," SPE of AIME (1968), II-149.

LAND, C. S., "Comparison of Calculated with Experimental Imbibition Relative Permeability," SPE of AIME (1971), II-419.

MARTIN, J. C., "Partial Integration of Equations of Multi-phase Flow," *Soc. Pet. Eng. J.* (Dec. 1968), 370–80; *Trans. AIME*, **243**.

MARTIN, J. C., and D. M. JAMES, "Analysis of Pressure Transients in Two-phase Radia Flow," *Soc. Pet. Eng. J.* (June 1963), 116–26.

McEWEN, C. R., "A Numerical Solution of the Linear Displacement Equation With Capillary Pressure," *Trans. AIME* (1959), **216**, 412–15.

MUNGAN, N., "Relative Permeability Measurements Using Reservoir Fluids," SPE of AIME (1972), II-398.

NAAR, J., and R. J. WYGAL, "Three-phase Imbibition Relative Permeability," SPE of AIME (1961), II-254.

NAAR, J., R. J. WYGAL, and J. H. HENDERSON, "Imbibition Relative Permeability in Unconsolidated Porous Media," SPE of AIME (1962), II-13.

OWENS, W. W., and D. L. ARCHER, "The Effect of Rock Wettability on Oil-Water Relative Permeability Relationships," SPE of AIME (1971), I-873.

PERKINS T. K., and O. C. JOHNSTON, "A Review of Diffusion and Dispersion in Porous Media," SPE of AIME (1963), II-70.

ROWAN, G., and M. W. CLEGG, "An Approximate Method for Non-Darcy Radial Gas Flow," *Soc. Pet. Eng. J.* (June 1964), 96.

SCHNEIDER F. N., and W. W. OWENS, "Sandstone and Carbonate Two- and Three-phase Relative Permeability Characteristics," SPE of AIME (1970), II-75.

SHEFFIELD, M., "Three-phase Fluid Flow Including Gravitational, Viscous, and Capillary Forces," SPE of AIME (1969), II-255.

STONE, H. L., "Probability Model for Estimating Three-phase Relative Permeability," SPE of AIME (1970), I-214.

SWIFT, G. W., and O. G. KIEL, "The Prediction of Gas-well Performance Including the Effect of Non-Darcy Flow," *J. Pet. Tech.* (July 1962), 791–98; *Trans. AIME*, **225**.

TEK, M. R., K. H. COATS, and D. L. KATZ, "The Effect of Turbulence on Flow of Natural Gas Through Porous Reservoirs," *J. Pet. Tech.* (July 1962), 799–806; *Trans. AIME*, **225**.

TEMPLETON, CHARLES C., "A Study of Displacements in Microscopic Capillaries," SPE of AIME (1954), 162.

TEMPLETON C. C., and S. S. RUSHING, JR., "Oil-Water Displacements in Microscopic Capillaries," SPE of AIME (1956), 211.

VAN POOLLEN, H. K., and J. R. JARGON, "Steady-state and Unsteady-state Flow of Non-Newtonian Fluids Through Porous Media," SPE of AIME (1969), II-80.

WALLICK, GEORGE C., "The Steady-state Flow of Gas Through Glass Capillary Tubes," SPE of AIME (1953), 331.

WARREN, J. E., and H. S. PRICE, "Flow in Heterogeneous Porous Media," *Soc. Pet. Eng. J.* (Sept. 1961), 153–69; *Trans. AIME*, 222.

WARREN J. E., and F. F. SKIBA, "Macroscopic Dispersion," *Soc. Pet. Eng. J.* (Sept. 1964), 215.

3 Formulation of Reservoir Simulation Equations

3.1 INTRODUCTION

Some engineers experience "conceptual shock" in going from the real-world reservoir to the simulation model within the computer. This experience is a by-product of a need perpetuated by engineers for some physical working base, a need nurtured by many years of "shop" operators who stressed, to the point of dogma, that a model is a small-scale reproduction. In practice we now know that a model can be "anything" that will allow us to infer the behavior of a system from its performance (as discussed in Chapter 1).

In order to allay this shock let us unfold the basis of reservoir simulation in common language and then extend this by the use of more definitive mathematical terminology. Consider a system represented by Fig. 3.1. This system consists of a portion of the universe which is separated from the rest by a definite boundary. The system exists in space (x-, y-, z-dimensions) and in time (t). This is a finite system. We can make several observations about this system.

1. Anything that enters or leaves the system must cross the boundary.
2. At some initial time the system could be described by some set of conditions.
3. The processes which occur within the system obey some known physical laws and consequently can be described by some set of conditions.

The above observations allow us to describe in abstract terms the behavior of the system. Observation 1 gives us the *boundary conditions*. These spell out

43

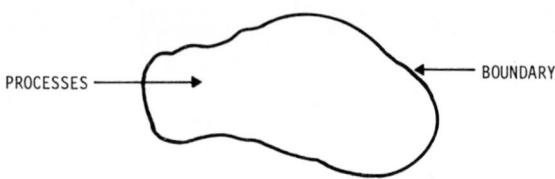

Figure 3.1: System.

the interaction between the problem domain and the rest of the world. It can be visualized from the following. Consider some independent parameter P of the system shown in Fig. 3.2.

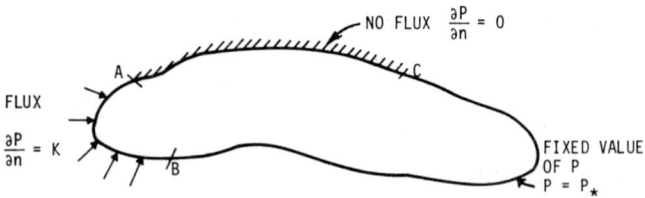

Figure 3.2: Boundary conditions.

Within A to C we have observed that no flux crosses the boundary, implying a zero gradient or physically some "insulated" surface. Between A and B we have a known flux entering, and this flux is represented by some value K. From C to B the boundary is defined by a fixed value of the independent parameter. Thus specifying conditions $A \rightarrow C \rightarrow B \rightarrow A$, we have completely described the contact the system makes with the rest of the world.

Observation 2 allows us to describe the state of the system at zero time. All systems in equilibrium at zero time will remain that way unless some disturbance occurs. A classically appropriate example is an oil reservoir; it remains undisturbed and at rest until the first well is drilled. This causes a disturbance (a local pressure sink) of the normal equilibrium, to which the reservoir begins to react by readjusting its pressure and flow patterns across the reservoir. This initial state is described by *initial conditions*, which take the following general form:

$$P(x, y, z) = \Phi$$

where Φ is some constant or a function of the space dimensions describing the parameter distribution at zero time.

Observation 3 is more than an observation in the strictest sense. It more closely resembles a hypothesis in that we have to describe the behavior of the processes within the system to the best of our ability and in that we do not always have complete knowledge of the minutest working of the system. We sometimes hypothesize. Consider the system again as shown in Fig. 3.3.

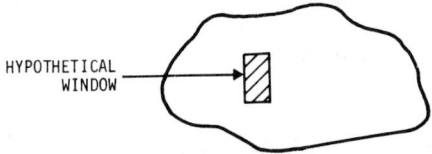

HYPOTHETICAL WINDOW

Figure 3.3: System processes.

If there were a hypothetical "window," we could look into the system at random locations and record exactly what we see and then try to associate these processes with the physical laws that apply. These laws may govern fluid flow, energy conservation, and the like. By defining the physical laws that apply, we can then formulate the mathematical equations which govern the processes within the system. These *governing equations* form part of the model of the system.

The complete mathematical model is then a combination of:

1. Governing equations
2. Boundary conditions
3. Initial conditions

3.2 DERIVATIONS OF EQUATIONS[1-5]

To understand the flow of fluids in porous media we must be able to postulate some system of equations which govern the behavior of these fluids. Having developed such a system of equations, we can then analyze the effect of varying conditions on the flow behavior.

The basic equations are obtained by combining several physical principles, namely:

1. Conservation of mass
2. Conservation of momentum
3. Conservation of energy (first law of thermodynamics)
4. Rate equations—Darcy's law
5. Equations of state

As indicated in the previous section, the governing equations together with the necessary boundary conditions and initial conditions form the mathematical model for our system. To solve this mathematical model we need to determine the values of the independent parameters which satisfy all the governing equations and boundary conditions simultaneously. In general, we have two choices: analytical or numerical methods. The former are not used because the governing equations are highly nonlinear and are impossible of solution by today's theoretical methods. Numerical methods are more adaptable to solving these equations.

Derivation Overview: The process involved in derivation of many of these equations consists of the following steps:

1. Select an elemental volume of the system (see Fig. 3.4).

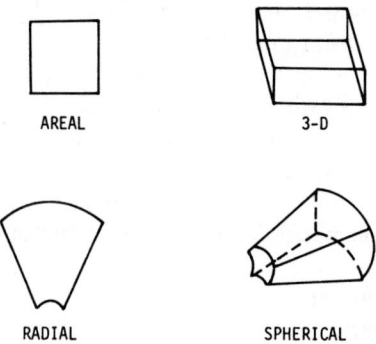

Figure 3.4: Elemental volumes.

2. Write all the fluxes into and out of the elemental volume over an interval of time, keeping a strict sign convention (Fig. 3.5).

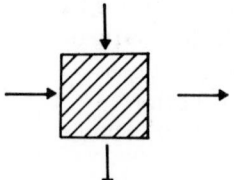

Figure 3.5: Flux directions.

3. Equate the fluxes to the changes within the system during this time—i.e., conserve mass within the system.

4. Take the limit over an instant as the elemental volume shrinks to an infinitesimal size. This involves:

$$\lim \Delta t \longrightarrow 0$$
$$\Delta x \longrightarrow 0$$

5. The resulting differential equation is the required governing equation.

The total process involved in setting up the simulation equations is summarized in Fig. 3.6.

PROCESS OVERVIEW

DARCY'S LAW

CONSERVATION OF MASS

THREE PHASE FLOW

PRESSURE EQUATION

SATURATION EQUATION

Figure 3.6: Process overview.

Single-Phase Flow

The equation governing the single-phase flow of a fluid through a porous medium is developed by combining the following:

1. Conservation of Mass
2. Rate equation
3. Equation of State

Conservation of Mass: Consider an element of a reservoir through which a single phase is flowing in the x-direction (Fig. 3.7). Than at any instant:

Mass rate in — Mass rate out = Mass rate of accumulation

$$(v_x \rho_x \, \Delta y \, \Delta z) - (v_{x+\Delta x} \rho_{x+\Delta x} \, \Delta y \, \Delta z) = (\Delta x \, \Delta y \, \Delta z) \phi \frac{(\rho_{t+\Delta t} - \rho_t)}{\Delta t} \qquad (3.1)$$

Dividing Eq. (3.1) by $\Delta x, \Delta y \, \Delta z$:

$$-\frac{(v_{x+\Delta x} \rho_{x+\Delta x}) - (v_x \rho_x)}{\Delta x} = \frac{\phi(\rho_{t+\Delta t} - \rho_t)}{\Delta t} \qquad (3.2)$$

Take the limit as $\left\{ \begin{matrix} \Delta x \\ \Delta t \end{matrix} \right\}$ go to zero simultaneously:

$$\frac{\partial(v\rho)}{\partial x} = -\phi \frac{\partial \rho}{\partial t} \qquad (3.3)$$

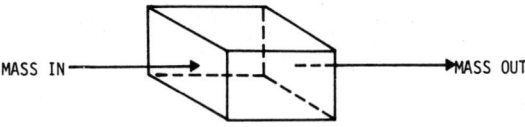

MASS IN — → MASS OUT

Figure 3.7: Mass balance on element.

This is the *continuity equation* in a linear system. Similarly:

$$\frac{\partial(v\rho)}{\partial y} = -\phi\frac{\partial\rho}{\partial t} \tag{3.4}$$

$$\frac{\partial(v\rho)}{\partial z} = -\phi\frac{\partial\rho}{\partial t} \tag{3.5}$$

Then for three-dimensional flow:

$$\frac{\partial(v\rho)}{\partial x} + \frac{\partial(v\rho)}{\partial y} + \frac{\partial(v\rho)}{\partial z} = -\phi\frac{\partial\rho}{\partial t} \tag{3.6}$$

Rate Equation: Darcy's law relates the velocity to the pressure gradient:

$$v = -\frac{k}{\mu}\frac{\partial P}{\partial x} \tag{3.7}$$

Then, substituting Eq. (3.7) into Eq. (3.3):

$$\frac{\partial\left(-\dfrac{k}{\mu}\dfrac{\partial P}{\partial x}\rho\right)}{\partial x} = -\phi\frac{\partial\rho}{\partial t} \tag{3.8}$$

Equation of State: The equation of state is needed to express the density in terms of pressure. Most oil field liquid systems are considered to be slightly compressible. In this case, the equation of state is:

$$\rho = \rho_0 e^{c(P-P_0)} \tag{3.9}$$

where

ρ = density at pressure P

ρ_0 = density at pressure P_0

c = isothermal compressibility factor

$$c \equiv -\frac{1}{V}\left(\frac{dV}{dP}\right)_T$$

Equation 3.8 can be written as follows by expanding the left-hand side:

$$-\left(\frac{k}{\mu}\frac{\partial^2 P}{\partial x^2}\rho + \frac{k}{\mu}\frac{\partial P}{\partial x}\frac{\partial\rho}{\partial x}\right) = -\phi\frac{\partial\rho}{\partial t} \tag{3.10}$$

Note that:

$$\frac{\partial\rho}{\partial x} = \frac{\partial\rho}{\partial P}\frac{\partial P}{\partial x}$$

and

$$\frac{\partial \rho}{\partial t} = \frac{\partial \rho}{\partial P} \frac{\partial P}{\partial t}$$

Therefore,

$$-\left(\frac{k}{\mu} \frac{\partial^2 P}{\partial x^2} \rho + \frac{k}{\mu} \frac{\partial P}{\partial x} \frac{\partial P}{\partial x} \frac{\partial \rho}{\partial P}\right) = -\phi \frac{\partial \rho}{\partial P} \frac{\partial P}{\partial t} \tag{3.11}$$

$$-\left[\frac{k}{\mu} \frac{\partial^2 P}{\partial x^2} \rho + \frac{k}{\mu} \frac{\partial \rho}{\partial P} \left(\frac{\partial P}{\partial x}\right)^2\right] = -\phi \frac{\partial \rho}{\partial P} \frac{\partial P}{\partial t} \tag{3.12}$$

Neglecting the $(\partial P/\partial x)^2$ term, since we are going to assume small pressure gradients, Eq. (3.12) becomes, by multiplication through by -1:

$$\frac{k}{\mu} \frac{\partial^2 P}{\partial x^2} \rho = \phi \frac{\partial \rho}{\partial P} \frac{\partial P}{\partial t} \tag{3.13}$$

Dividing both sides by density:

$$\frac{k}{\mu} \frac{\partial^2 P}{\partial x^2} = \phi \frac{1}{\rho} \frac{\partial \rho}{\partial P} \frac{\partial P}{\partial t} \tag{3.14}$$

By definition, the compressibility is as follows:

$$c = \frac{1}{\rho} \frac{\partial \rho}{\partial P}$$

This is indicated in the graph of ρ versus P in Fig. 3.8. Then:

$$\frac{k}{\mu} \frac{\partial^2 P}{\partial x^2} = \phi c \frac{\partial P}{\partial t} \tag{3.15}$$

Since k/μ was considered independent of spatial dimension:

$$\frac{\partial^2 P}{\partial x^2} = \frac{\phi \mu c}{k} \frac{\partial P}{\partial t} \tag{3.16}$$

If k/μ were a function of the spatial dimension, then:

$$\frac{\partial \left(\frac{k}{\mu} \frac{\partial P}{\partial x}\right)}{\partial x} = \phi c \frac{\partial P}{\partial t} \tag{3.17}$$

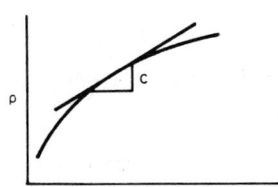

Figure 3.8: ρ versus P.

Equation (3.16) is generally called the *diffusivity equation* because of its resemblance to the diffusivity equation for heat transfer:

$$\frac{\partial^2 T}{\partial x^2} = \frac{1}{\alpha} \frac{\partial T}{\partial t} \tag{3.18}$$

Other Coordinate Systems:

$$\frac{\partial^2 P}{\partial r^2} + \frac{1}{r} \frac{\partial P}{\partial r} = \frac{\phi \mu c}{k} \frac{\partial P}{\partial t} \qquad \text{Radial flow}$$

$$\frac{\partial^2 P}{\partial x^2} + \frac{\partial^2 P}{\partial y^2} = \frac{\phi \mu c}{k} \frac{\partial P}{\partial t} \qquad \text{Two-dimensional} \tag{3.19}$$

$$\frac{\partial^2 P}{\partial x^2} + \frac{\partial^2 P}{\partial y^2} + \frac{\partial^2 P}{\partial z^2} = \frac{\phi \mu c}{k} \frac{\partial P}{\partial t} \qquad \text{Three-dimensional}$$

The typical reservoir configurations for the above equations are shown in Fig. 3.9.

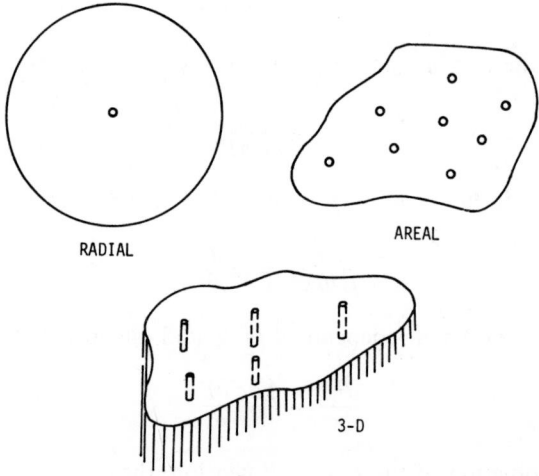

RADIAL   AREAL

3-D

Figure 3.9: Radial, areal, and three-dimensional systems.

3.3 DERIVATION OF MULTIPHASE FLOW EQUATIONS[2]

The flow for each phase is developed identically to that scheme outlined for a single-phase fluid.

Oil: Starting with an element of the reservoir, the basic equation for oil flow is derived by combining the continuity equation, the Darcy flow equation, and equation of state (see Fig. 3.10). Using a balance on the STB oil

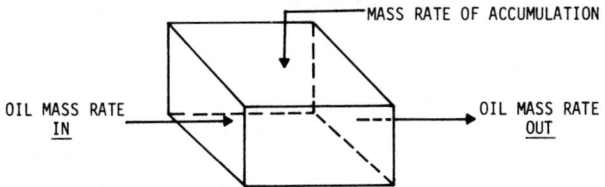

Figure 3.10: Oil mass balance on element.

flowing in a linear system:

Mass rate in − Mass rate out = Mass rate of accumulation

Thus:

$$\left(-A\frac{k_o}{\mu_o B_o}\frac{\partial P}{\partial x}\right)_x - \left(-A\frac{k_o}{\mu_o B_o}\frac{\partial P}{\partial x}\right)_{x+\Delta x} = V\left[\frac{\left(\frac{\phi S_o}{B_o}\right)^{n+1} - \left(\frac{\phi S_o}{B_o}\right)^{n}}{\Delta t}\right] \qquad (3.20)$$

where

$$A = \Delta y\,\Delta z$$
$$V = \Delta x\,\Delta y\,\Delta z$$

Equation (3.20) becomes in the limit:

$$\frac{\partial}{\partial x}\left(\frac{k_o}{\mu_o B_o}\frac{\partial P}{\partial x}\right) = \frac{\partial}{\partial t}\left(\frac{\phi S_o}{B_o}\right) \qquad (3.21)$$

For a radial system the equivalent system is:

$$\frac{1}{r}\frac{\partial}{\partial r}\left(r\frac{k_o}{\mu_o B_o}\frac{\partial P}{\partial r}\right) = \frac{\partial}{\partial t}\left(\frac{\phi S_o}{B_o}\right) \qquad (3.22)$$

Gas: A mass balance on the gas phase must include all possible sources of gas (Fig. 3.11). For a linear system we can write:

Mass rate in − Mass rate out = Mass rate of accumulation

Each of the sources of gas as indicated in Fig. 3.11 is incorporated in the mass rate term. Thus:

$$\left[-A\left(\frac{k_g}{\mu_g B_g} + \frac{R_{so}k_o}{\mu_o B_o} + \frac{R_{sw}k_w}{\mu_w B_w}\right)\frac{\partial P}{\partial x}\right]_x - \left[-A\left(\frac{k_g}{\mu_g B_g} + \frac{R_{so}k_o}{\mu_o B_o} + \frac{R_{sw}k_w}{\mu_w B_w}\right)\frac{\partial P}{\partial x}\right]_{x+\Delta x}$$

$$\underset{\substack{\text{Free}\\\text{gas}}}{} \qquad \underset{\substack{\text{Gas}\\\text{in}\\\text{oil}}}{} \qquad \underset{\substack{\text{Gas}\\\text{in}\\\text{water}}}{}$$

$$= V\left[\frac{\phi\left(\frac{S_g}{B_g} + \frac{R_{so}S_o}{B_o} + \frac{R_{sw}S_w}{B_w}\right)^{n+1} - \left(\frac{S_g}{B_g} + \frac{R_{so}S_o}{B_o} + \frac{R_{sw}S_w}{B_w}\right)^{n}}{\Delta t}\right] \qquad (3.23)$$

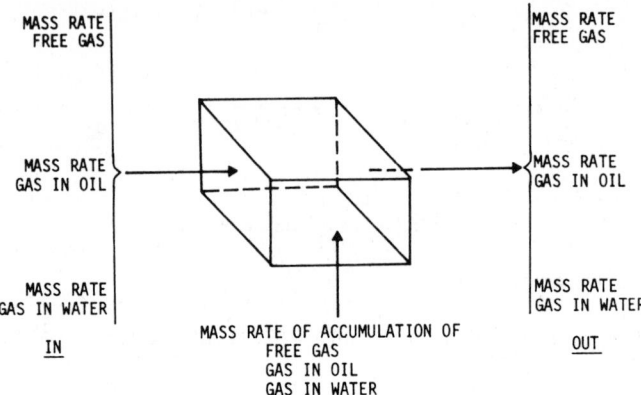

Figure 3.11: Gas mass balance on element.

which becomes in the limit:

$$\frac{\partial}{\partial x}\left[\left(\frac{k_g}{\mu_g B_g} + \frac{R_{so}k_o}{\mu_o B_o} + \frac{R_{sw}k_w}{\mu_w B_w}\right)\frac{\partial P}{\partial x}\right] = \frac{\partial}{\partial t}\left[\phi\left(\frac{S_g}{B_g} + \frac{R_{so}S_o}{B_o} + \frac{R_{sw}S_w}{B_w}\right)\right] \qquad (3.24)$$

For a radial system the following equation is obtained:

$$\frac{1}{r}\frac{\partial}{\partial r}\left[r\left(\frac{k_g}{\mu_g B_g} + \frac{R_{so}k_o}{\mu_o B_o} + \frac{R_{sw}k_w}{\mu_w B_w}\right)\frac{\partial P}{\partial r}\right] = \frac{\partial}{\partial t}\left[\phi\left(\frac{S_g}{B_g} + \frac{R_{so}S_o}{B_o} + \frac{R_{sw}k_w}{B_w}\right)\right]$$

$$(3.25)$$

Water: The water phase is essentially the same as the oil phase. For a linear system:

$$\frac{\partial}{\partial x}\left[\frac{k_w}{\mu_w B_w}\frac{\partial P}{\partial x}\right] = \frac{\partial}{\partial t}\left[\phi\frac{S_w}{B_w}\right] \qquad (3.26)$$

For a radial system:

$$\frac{1}{r}\frac{\partial}{\partial r}\left[r\frac{k_w}{\mu_w B_w}\frac{\partial P}{\partial r}\right] = \frac{\partial}{\partial t}\left[\phi\frac{S_w}{B_w}\right] \qquad (3.27)$$

Expansion in Radial Form

The generalized multiphase flow equation for the unsteady-state flow of oil, gas, and water in a porous medium is developed by combining the three single-phase flow equations into one basic equation. To do this several other observations are made. First, for all phases the following is true:

$$S_o + S_g + S_w = 1 \qquad (3.28)$$

Thus:

$$\frac{\partial}{\partial t}[S_o + S_g + S_w] = 0 \tag{3.29}$$

Pressure gradients are assumed small and the square of this term is neglected:

$$\left(\frac{\partial P}{\partial t}\right)^2 \simeq 0 \tag{3.30}$$

The derivation is as follows in radial coordinates. Multiply the oil equation (Eq. 3.22) by B_o and expand by differentiation:

$$\frac{B_o}{r}\left[r\frac{k_o}{\mu_o B_o}\frac{\partial^2 P}{\partial r^2} + r\frac{\partial P}{\partial r}\frac{k_o}{\mu_o}\left(-\frac{1}{B_o^2}\right)\frac{\partial B_o}{\partial P}\frac{\partial P}{\partial r} + \frac{1}{r}\frac{k_o}{\mu_o B_o}\frac{\partial P}{\partial r}\right]$$
$$= \phi B_o\left(\frac{1}{B_o}\frac{\partial S_o}{\partial t} + \frac{S_o}{-B_o^2}\frac{\partial B_o}{\partial P}\frac{\partial P}{\partial t}\right) \tag{3.31}$$

Thus:

$$\frac{k_o}{\mu_o}\frac{\partial^2 P}{\partial r^2} - \frac{k_o}{\mu_o B_o}\frac{\partial B_o}{\partial P}\left(\frac{\partial P}{\partial r}\right)^2 + \frac{1}{r}\frac{k_o}{\mu_o}\frac{\partial P}{\partial r} = \phi\left(\frac{\partial S_o}{\partial t} - \frac{S_o}{B_o}\frac{\partial B_o}{\partial P}\frac{\partial P}{\partial t}\right) \tag{3.32}$$

Neglecting $(\partial P/\partial r)^2$ terms, Eq. (3.32) becomes

$$\frac{k_o}{\mu_o}\frac{\partial^2 P}{\partial r^2} + \frac{1}{r}\frac{k_o}{\mu_o}\frac{\partial P}{\partial r} = \phi\left(\frac{\partial S_o}{\partial t} - \frac{S_o}{B_o}\frac{\partial B_o}{\partial P}\frac{\partial P}{\partial t}\right) \tag{3.33}$$

which is:

$$\frac{\partial}{\partial r}\left(\frac{1}{r}\frac{\partial P}{\partial r}\right)\frac{k_o}{\mu_o} = \phi\left[\frac{\partial S_o}{\partial t} - \frac{S_o}{B_o}\frac{\partial B_o}{\partial P}\frac{\partial P}{\partial t}\right] \tag{3.34}$$

The gas equation (Eq. 3.25) is multiplied by B_g and expanded as above:

$$\frac{B_g}{r}\left\{r\left(\frac{R_{so}k_o}{\mu_o B_o} + \frac{R_{sw}k_w}{\mu_w B_w} + \frac{k_g}{\mu_g B_g}\right)\frac{\partial^2 P}{\partial r^2} + r\frac{\partial P}{\partial r}\left[\frac{k_o}{\mu_o}\left(\frac{1}{B_o}\frac{\partial R_{so}}{\partial P}\frac{\partial P}{\partial r} - \frac{R_{so}}{B_o^2}\frac{\partial B_o}{\partial P}\frac{\partial P}{\partial r}\right)\right.\right.$$
$$\left. + \frac{k_w}{\mu_w}\left(\frac{1}{B_w}\frac{\partial R_{sw}}{\partial P}\frac{\partial P}{\partial r} - \frac{R_{sw}}{B_w^2}\frac{\partial B_w}{\partial P}\frac{\partial P}{\partial r}\right) - \frac{k_g}{\mu_g}\left(\frac{1}{B_g^2}\frac{\partial B_g}{\partial P}\frac{\partial P}{\partial r}\right)\right]$$
$$\left. + \frac{\partial P}{\partial r}\left(\frac{R_{so}k_o}{\mu_o B_o} + \frac{R_{sw}k_w}{\mu_w B_w} + \frac{k_g}{\mu_g B_g}\right)\right\} = \phi B_g\left(\frac{S_o}{B_o}\frac{\partial R_s}{\partial P}\frac{\partial P}{\partial t} + \frac{R_{so}}{B_o}\frac{\partial S_o}{\partial t}\right.$$
$$- \frac{R_{so}S_o}{B_o^2}\frac{\partial B_o}{\partial P}\frac{\partial P}{\partial t} + \frac{S_w}{B_w}\frac{\partial R_{sw}}{\partial P}\frac{\partial P}{\partial t} + \frac{R_{sw}}{B_w}\frac{\partial S_w}{\partial t} - \frac{R_{sw}S_w}{B_w^2}\frac{\partial B_w}{\partial P}\frac{\partial P}{\partial t}$$
$$\left. + \frac{1}{B_g}\frac{\partial S_g}{\partial t} - \frac{S_g}{B_g^2}\frac{\partial B_g}{\partial P}\frac{\partial P}{\partial t}\right) \tag{3.35}$$

Collecting terms:

$$\left(\frac{k_o}{\mu_o}\frac{R_{so}B_g}{B_o} + \frac{k_w}{\mu_w}\frac{R_{sw}B_g}{B_w} + \frac{k_g}{\mu_g}\right)\frac{\partial^2 P}{\partial r^2} + \frac{k_o}{\mu_o}\frac{B_g}{B_o}\frac{\partial R_{so}}{\partial P}\left(\frac{\partial P}{\partial r}\right)^2 + \frac{k_w}{\mu_w}\frac{B_g}{B_w}\frac{\partial R_{sw}}{\partial P}\left(\frac{\partial P}{\partial r}\right)^2$$

$$- \frac{k_o}{\mu_o}\frac{R_{so}}{B_o^2}\frac{\partial B_o}{\partial P}\left(\frac{\partial P}{\partial r}\right)^2 - \frac{k_w}{\mu_w}\frac{B_g}{B_w^2}\frac{\partial B_w}{\partial P}\left(\frac{\partial P}{\partial r}\right)^2 - \frac{k_g}{\mu_g}\frac{1}{B_g}\frac{\partial B_g}{\partial P}\left(\frac{\partial P}{\partial r}\right)^2$$

$$+ \left(\frac{k_o}{\mu_o}\frac{R_{so}B_g}{B_o} + \frac{k_w}{\mu_w}\frac{R_{sw}B_g}{B_w} + \frac{k_g}{\mu_g}\right)\frac{1}{r}\frac{\partial P}{\partial r} = \phi\left(\frac{S_o B_g}{B_o}\frac{\partial R_{so}}{\partial P} - \frac{R_{so}S_o B_g}{B_o^2}\frac{\partial B_o}{\partial P}\right.$$

$$\left. + \frac{S_w B_g}{B_w}\frac{\partial R_{sw}}{\partial P} - \frac{R_{sw}S_w B_g}{B_w^2}\frac{\partial B_w}{\partial P} - \frac{S_g}{B_g}\frac{\partial B_g}{\partial P}\right)\frac{\partial P}{\partial t}$$

$$+ \phi\left(\frac{B_g R_{so}}{B_o}\frac{\partial S_o}{\partial t} + \frac{R_{sw}B_g}{B_w}\frac{\partial S_w}{\partial t} + \frac{\partial S_g}{\partial t}\right) \tag{3.36}$$

Neglecting $(\partial P/\partial r)^2$ terms in the above equation:

$$\left(\frac{k_o}{\mu_o}\frac{R_{so}B_g}{B_o} + \frac{k_w}{\mu_w}\frac{R_{sw}B_g}{B_w} + \frac{k_g}{\mu_g}\right)\left(\frac{\partial^2 P}{\partial r^2} + \frac{1}{r}\frac{\partial P}{\partial r}\right) = \phi\left(\frac{S_o B_g}{B_o}\frac{\partial R_s}{\partial P} - \frac{R_{so}S_o B_g}{B_o^2}\frac{\partial B_o}{\partial P}\right.$$

$$\left. + \frac{S_w B_g}{B_w}\frac{\partial R_{sw}}{\partial P} - \frac{R_{sw}S_w B_g}{B_w^2}\frac{\partial B_w}{\partial P} - \frac{S_g}{B_g}\frac{\partial B_g}{\partial P}\right)\frac{\partial P}{\partial t}$$

$$+ \phi\left(\frac{R_{so}B_g}{B_o}\frac{\partial S_o}{\partial t} + \frac{R_{sw}B_g}{B_w}\frac{\partial S_w}{\partial t} + \frac{\partial S_g}{\partial t}\right) \tag{3.37}$$

The water equation (Eq. 3.27) is multiplied by B_w and expanded like the oil equation to yield:

$$\frac{k_w}{\mu_w}\frac{\partial^2 P}{\partial r^2} + \frac{k_w}{\mu_w}\frac{\partial P}{\partial r}\frac{1}{r} = \phi\left(\frac{\partial S_w}{\partial t} - \frac{S_w}{B_w}\frac{\partial B_w}{\partial P}\frac{\partial P}{\partial t}\right) \tag{3.38}$$

Combining the oil and water equations, Eq. (3.33) and Eq. (3.38), we have:

$$\left(\frac{k_o}{\mu_o} + \frac{k_w}{\mu_w}\right)\left(\frac{\partial^2 P}{\partial r^2} + \frac{1}{r}\frac{\partial P}{\partial r}\right) = \phi\left[\frac{\partial S_o}{\partial t} + \frac{\partial S_w}{\partial t} - \left(\frac{S_o}{B_o}\frac{\partial B_o}{\partial P}\frac{\partial P}{\partial t} + \frac{S_w}{B_w}\frac{\partial B_w}{\partial P}\frac{\partial P}{\partial t}\right)\right] \tag{3.39}$$

Combining Eqs. (3.37) and (3.39), we have:

$$\left(\frac{\partial^2 P}{\partial r^2} + \frac{1}{r}\frac{\partial P}{\partial r}\right)\left(\frac{k_o}{\mu_o} + \frac{k_w}{\mu_w} + \frac{k_g}{\mu_g} + \frac{k_o}{\mu_o}\frac{R_{so}B_g}{B_o} + \frac{k_w}{\mu_w}\frac{R_{sw}B_g}{B_w}\right)$$

$$= \phi\left[\left(\frac{\partial S_w}{\partial t} + \frac{\partial S_o}{\partial t} + \frac{\partial S_g}{\partial t}\right) - \frac{S_o}{B_o}\frac{\partial B_o}{\partial P}\left(1 + \frac{R_s B_g}{B_{w_o}}\right) + \frac{S_o B_g}{B_o}\frac{\partial R_s}{\partial P}\right.$$

$$\left. - \frac{S_w}{B_w}\frac{\partial B_w}{\partial P}\left(1 + \frac{R_{sw}B_g}{B_w}\right) + \frac{S_w B_g}{B_w}\frac{\partial R_{sw}}{\partial P} - \frac{S_g}{B_g}\frac{\partial B_g}{\partial P}\right]\frac{\partial P}{\partial t}$$

$$+ \frac{R_{so}B_g}{B_o}\frac{\partial S_o}{\partial t} + \frac{R_{sw}B_g}{B_w}\frac{\partial S_w}{\partial t} \tag{3.40}$$

Since

$$S_g + S_o + S_w = 1 \tag{3.28}$$

$$\frac{\partial}{\partial t}(S_g + S_o + S_w) = 0 \tag{3.29}$$

the right-hand side of Eq. (3.40) reduces to

$$\text{RHS} = \phi\left[-\frac{S_o}{B_o}\frac{\partial B_o}{\partial P}\left(1 + \frac{R_{so}B_g}{B_o}\right) + \frac{S_oB_g}{B_o}\frac{\partial R_{so}}{\partial P} - \frac{S_w}{B_w}\frac{\partial B_w}{\partial P}\left(1 + \frac{R_{sw}B_g}{B_w}\right)\right.$$
$$\left. + \frac{S_wB_g}{B_w}\frac{\partial R_{sw}}{\partial P} - \frac{S_g}{B_g}\frac{\partial B_g}{\partial P}\right]\frac{\partial P}{\partial t} + \frac{R_{so}B_g}{B_o}\frac{\partial S_o}{\partial t} + \frac{R_{sw}B_g}{B_w}\frac{\partial S_w}{\partial t} \tag{3.41}$$

Now, by substituting Eq. (3.34) and Eq. (3.38) into Eq. (3.41), the left-hand side is resolved partially in terms of $\frac{\partial}{\partial r}\left(\frac{1}{r}\frac{\partial P}{\partial r}\right)$ and saturations dependent on time:

$$\frac{1}{r}\frac{\partial}{\partial r}\left(r\frac{\partial P}{\partial r}\right)\left(\frac{k_o}{\mu_o} + \frac{k_w}{\mu_w} + \frac{k_g}{\mu_g}\right) + \left[\frac{R_{so}B_g}{B_o}\left(\frac{\partial S_o}{\partial t} - \frac{S_o}{B_o}\frac{\partial B_o}{\partial P}\frac{\partial P}{\partial t}\right)\right]\phi$$
$$+ \left[\frac{R_{sw}B_g}{B_w}\left(\frac{\partial S_w}{\partial t} - \frac{S_w}{B_w}\frac{\partial B_w}{\partial P}\frac{\partial P}{\partial t}\right)\right]\phi = \phi\left(-\frac{S_o}{B_o}\frac{\partial B_o}{\partial P} + \frac{S_oB_g}{B_o}\frac{\partial R_{so}}{\partial P}\right.$$
$$\left. - \frac{S_w}{B_w}\frac{\partial B_w}{\partial P} + \frac{S_wB_g}{B_w}\frac{\partial R_{sw}}{\partial P} - \frac{S_g}{B_g}\frac{\partial B_g}{\partial P}\right)\frac{\partial P}{\partial t} + \frac{R_{so}B_g}{B_o}\frac{S_o}{B_o}\frac{\partial B_o}{\partial P}\frac{\partial P}{\partial t}$$
$$- \frac{S_w}{B_w}\frac{\partial B_w}{\partial P}\frac{R_{sw}B_g}{B_w}\frac{\partial P}{\partial t} + \frac{R_{so}B_g}{B_o}\frac{\partial S_o}{\partial t} + \frac{R_{sw}B_g}{B_w}\frac{\partial S_w}{\partial t} \tag{3.42}$$

Collecting like terms in Eq. (3.42) and letting

$$c_t = -\frac{S_o}{B_o}\frac{\partial B_o}{\partial P} + \frac{S_oB_g}{B_o}\frac{\partial R_{so}}{\partial P} - \frac{S_w}{B_w}\frac{\partial B_w}{\partial P} + \frac{S_wB_g}{B_w}\frac{\partial R_{sw}}{\partial P} - \frac{S_g}{B_g}\frac{\partial B_g}{\partial P} \tag{3.43}$$

then Eq. (3.42) becomes

$$\frac{1}{r}\frac{\partial}{\partial r}\left(r\frac{\partial P}{\partial r}\right)\left(\frac{k}{\mu}\right)_t + \phi\frac{R_{so}B_g}{B_o}\frac{\partial S_o}{\partial t} - \phi\frac{S_oR_{so}B_g}{B_o^2}\frac{\partial B_o}{\partial P}\frac{\partial P}{\partial t} + \phi\frac{R_{sw}B_g}{B_w}\frac{\partial S_w}{\partial t}$$
$$- \phi\frac{R_{sw}B_g}{B_w}\frac{S_w}{B_w}\frac{\partial B_w}{\partial P}\frac{\partial P}{\partial t} = \phi\left(c_t\frac{\partial P}{\partial t} - \frac{S_o}{B_o}\frac{\partial B_o}{\partial P}\frac{R_{so}B_g}{B_o}\frac{\partial P}{\partial t}\right.$$
$$\left. - \frac{S_w}{B_w}\frac{\partial B_w}{\partial P}\frac{R_{sw}B_g}{B_w}\frac{\partial P}{\partial t} + \frac{R_{so}B_g}{B_o}\frac{\partial S_o}{\partial t} + \frac{R_{sw}B_g}{B_w}\frac{\partial S_w}{\partial t}\right) \tag{3.44}$$

where $\left(\frac{k}{\mu}\right)_t = \frac{k_o}{\mu_o} + \frac{k_w}{\mu_w} + \frac{k_g}{\mu_g}$ Total mobility[1]

Collecting like terms in Eq. (3.44) and simplifying the equation by cancel-
ing equal terms of opposite sign:

$$\frac{1}{r}\frac{\partial}{\partial r}\left(r\frac{\partial P}{\partial r}\right)\left(\frac{k}{\mu}\right)_t = \phi c_t \frac{\partial P}{\partial t} \tag{3.45}$$

Finally:

$$\frac{1}{r}\frac{\partial}{\partial r}\left(r\frac{\partial P}{\partial r}\right) = \frac{\phi c_t}{(k/\mu)_t}\frac{\partial P}{\partial t} \tag{3.46}$$

This equation assumes that mobilities do not vary with radius. Equation
(3.46) is the three-phase unsteady-state flow equation for oil, gas, and water
in a radial system. Solving this equation gives the values of pressures at any
radius at any time. This equation forms the basis for pressure analysis of
multiphase flow.

Expansion in One-Dimensional Forms: Given the equations for each
fluid phase in a one-dimensional system:

$$A_x\frac{\partial}{\partial x}\left(\frac{k_o}{\mu_o B_o}\frac{\partial\Phi_o}{\partial x}\right) + q_0 = V_R\frac{\partial}{\partial t}\left(\frac{\phi S_o}{B_o}\right) \quad \text{Oil} \tag{3.47}$$

$$A_x\frac{\partial}{\partial x}\left(\frac{k_w}{\mu_w B_w}\frac{\partial\Phi_w}{\partial x}\right) + q_w = V_R\frac{\partial}{\partial t}\left(\frac{\phi S_w}{B_w}\right) \quad \text{Water} \tag{3.48}$$

$$A_x\frac{\partial}{\partial x}\left(\frac{k_g}{\mu_g B_g}\frac{\partial\Phi_g}{\partial x} + \frac{R_{so}k_o}{\mu_o B_o}\frac{\partial\Phi_o}{\partial x} + \frac{R_{sw}k_w}{\mu_w B_w}\frac{\partial\Phi_w}{\partial x}\right) + q_g$$

$$= V_R\frac{\partial}{\partial t}\left[\phi\left(\frac{S_g}{B_g} + \frac{R_{so}S_o}{B_o} + \frac{R_{sw}S_w}{B_w}\right)\right] \quad \text{Gas} \tag{3.49}$$

We can combine these to obtain the equations for flow in a reservoir. In
order to do this, however, we need to express some accessory conditions.

The potential terms are defined as:[6]

$$\Phi_o = P_o + \rho_o gh \tag{3.50}$$

$$\Phi_g = P_g + \rho_g gh \tag{3.51}$$

$$\Phi_w = P_w + \rho_w gh \tag{3.52}$$

The capillary pressure terms are:[7]

$$P_{cw} = P_o - P_w \tag{3.53}$$

$$P_{cg} = P_g - P_o \tag{3.54}$$

Equations (3.47) through (3.54) can be combined using, in addition, the saturation equation (Eq. 3.29) to obtain:

$$A_x \frac{\partial}{\partial x}\left(\lambda_T \frac{\partial P_o}{\partial x}\right) + A_x \frac{\partial}{\partial x}\left(\lambda_g \frac{\partial P_{cg}}{\partial x} - \lambda_w \frac{\partial P_{cw}}{\partial x}\right) + A_x \frac{\partial}{\partial x}\left[\lambda_g \frac{\partial(\rho_g gh)}{\partial x}\right.$$
$$\left. + \lambda_o \frac{\partial(\rho_o gh)}{\partial x} + \lambda_w \frac{\partial(\rho_w gh)}{\partial x}\right] = \beta_1 \frac{\partial P_o}{\partial t} + \beta_2 \qquad (3.55)$$

where the λ-variables are mobility terms, β_1-variables are functions of PVT (pressure-volume-temperature) terms, and β_2-variables are production terms.

For two-dimensional flow, Eq. (3.55) is expanded to include the y-coordinate terms.

Solution Outline: The two basic methods to solve the simulator equations are covered in more detail in Chapter 5. A brief outline of one method is presented in Fig. 3.12 to introduce the engineer to the procedure.

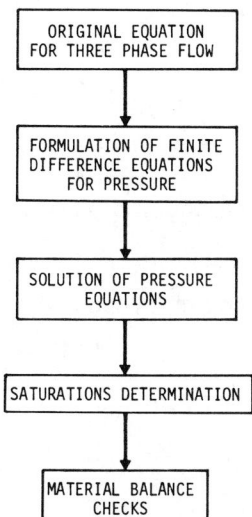

Figure 3.12: Procedure outline for solution of flow equation.

3.4 MULTICOMPONENT SYSTEMS[8,9]

In some hydrocarbon systems there is considerable mass transfer between the flowing phases. This mass transfer complicates the already complex system, since a mass balance must be made on every flowing fraction instead of on each phase. In a reservoir system there are generally several species of chemical compounds. These components vary in concentration in different phases, while each phase flows at a different rate.

Consider an element of the reservoir as shown in Fig. 3.13. There are N species of chemical compounds flowing into the reservoir element in three

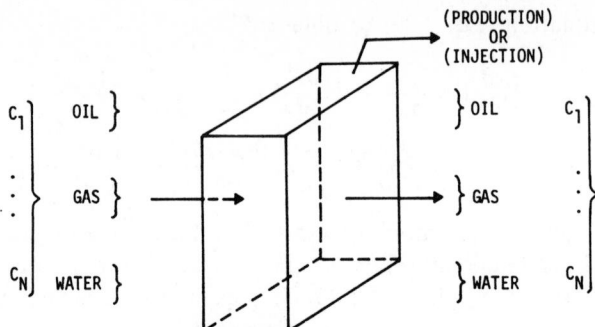

Figure 3.13: Compositional mass balance on element.

possible mobile phases. Within the element there are changes due to either or all of the following:

1. Production
2. Injection
3. Pressure change

It no longer suffices to maintain a mass balance on each phase, but each component must be conserved throughout the system.

Consider the conservation of mass applied to one component. Let

C_{oj} = mass fraction of jth component in oil

C_{gj} = mass fraction of jth component in gas

C_{wj} = mass fraction of jth component in water

Then as before we can write:

$$\frac{\partial}{\partial x}\left(\frac{k_o \rho_o}{\mu_o} C_{oj} \frac{\partial P_o}{\partial x} + \frac{k_g \rho_g}{\mu_g} C_{gj} \frac{\partial P_g}{\partial x} + \frac{k_w \rho_w}{\mu_w} C_{wj} \frac{\partial P_w}{\partial_x}\right)$$
$$= \frac{\partial}{\partial t}(\phi S_o \rho_o C_{oj} + \phi S_g \rho_g C_{gj} + \phi S_w \rho_w C_{wj}) \qquad (3.56)$$

Equation (3.56) describes the flow of a single component—e.g., CH_4—in a linear system without any sources or sinks. These complications (sources and sinks) will be discussed later. A closer look at Eq. (3.56) shows that each term on the left represents the mass flux of the jth component in each phase, which is simply derived by the following:

$$\text{Total mass flux} = \text{Volumetric rate} * \text{Density}$$
$$= q_o \rho_o$$
$$= \frac{k_o \rho_o}{\mu_o} \frac{\partial P_o}{\partial x} \qquad (3.57)$$

$$\text{Component mass flux} = C_{oj} * (\text{Total mass flux})$$

$$= C_{oj} \frac{k_o \rho_o}{\mu_o} \frac{\partial P_o}{\partial x} \tag{3.58}$$

Similarly, the accumulation term embodies the changes in each phase of the specific component:

$$\text{Mass rate of change} = \frac{\text{Mass at time } (t + \Delta t) - \text{Mass at time } t}{\Delta t}$$

A general equation for the N species under observation will be of the form:

$$\frac{\partial}{\partial x} \left(\sum_{i=1}^{3} \frac{k_i \rho_i}{\mu_i} C_{ij} \frac{\partial P_i}{\partial x} \right) = \frac{\partial}{\partial t} \left(\sum_{i=1}^{3} \phi S_i \rho_i C_{ij} \right), \qquad j = 1, \ldots, N \tag{3.59}$$

where the index i represents the phases and the index j represents the components.

At this point we must determine the number of independent variables in the system. These data are summarized below for an N-component system.

Unknowns	Number
C_{ij}	$3N$
P_i	3
S_i	3
ρ_i	3
μ_i	3
k_i	3
	$3N + 15$

Note: $C_{ij}, i = 1, 2, 3$
$\quad\quad\quad j = 1, \ldots, N \quad$ Total $= 3N$

In order to solve this system uniquely, we *must* have $3N + 15$ *independent* relationships. The relations can be differential or algebraic.

These relationships come from several sources:

1. Differential equations
2. Phase equilibria
3. PVT data
4. Relative permeability data
5. Conservation principles
6. Capillarity data

Let us develop the necessary relationships:

1. One partial differential equation can be written for each component in the system, thus providing N relationships.

2. The fluid phase saturations must always sum to unity, since the pore space is always fluid-filled:

$$S_g + S_o + S_w = 1 \qquad (3.60)$$

This is one relationship.

3. The mass fractions of each component in each fluid phase must sum to unity, since mass conservation of each component is required. Thus:

$$\sum_{j=1}^{N} C_{oj} = 1$$

$$\sum_{j=1}^{N} C_{gj} = 1 \qquad (3.61)$$

$$\sum_{j=1}^{N} C_{wj} = 1$$

This provides three relationships.

4. From the PVT data the following are readily obtained:

$$\rho_o = f(P_o, C_{oj})$$
$$\rho_g = f(P_g, C_{gj}) \qquad (3.62)$$
$$\rho_w = f(P_w, C_{wj})$$

$$\mu_o = f(P_o, C_{oj})$$
$$\mu_g = f(P_g, C_{gj}) \qquad (3.63)$$
$$\mu_w = f(P_w, C_{wj})$$

This provides six more relationships. In actual practice the density and viscosity are computed by using experimentally determined correlations which relate these parameters to concentrations and pressures. Two well-known correlations are the Alani-Kennedy[10] and Avasti-Kennedy[11] correlations for hydrocarbons.

5. Relative permeability data allows us to obtain the needed data for mobility calculations:

$$k_o = f(S_g, S_o, S_w)$$
$$k_g = f(S_g, S_o, S_w) \qquad (3.64)$$
$$k_w = f(S_g, S_o, S_w)$$

This provides three more relationships.

6. Phase equilibria: The equilibrium constant which can be devised from thermodynamic principles governs the distribution of a component between its liquid and gaseous states. For example,

$$\frac{C_{gj}}{C_{oj}} = K_{jgo} \tag{3.65}$$

This states that the ratio of the mass fraction of component j in the gas to the mass fraction of component j in oil is a constant. This constant, called the equilibrium constant, is a function of several variables:

$$\frac{C_{gj}}{C_{oj}} = K_{jgo} = f(T, P, C_{ij}) \tag{3.66}$$

Also:

$$\frac{C_{gj}}{C_{wj}} = K_{jgw} = f(T, P, C_{ij}) \tag{3.66}$$

from which:

$$\frac{C_{jo}}{C_{jw}} = \frac{K_{jgw}}{K_{jgo}} = K_{gow}$$

Equation (3.66) provides $2N$ independent relationships when written for each component in the system. The last equation does *not* produce an independent relationship, since it is derived from the others.

7. Capillary pressure provides the remaining relationship:

$$\begin{aligned} P_g - P_o &= P_{c_{go}} = f(S_g, S_o, S_w) \\ P_o - P_w &= P_{c_{ow}} = f(S_g, S_o, S_w) \end{aligned} \tag{3.67}$$

The origin of the relationships are summarized below:

Relationship	Unknowns	Equation(s)
Differential equations	N	(3.59)
Phase equilibria	$2N$	(3.66)
PVT data	6	(3.62), (3.63)
Relative permeability	3	(3.64)
Σ Mass fractions	3	(3.61)
Σ Saturations	1	(3.60)
Capillary	2	(3.67)
	$3N + 15$	

We therefore have $3N + 15$ independent unknowns and $3N + 15$ independent relationships which can be used to solve the system.

In practice, several simplifying assumptions are usually made to make the formidable problem more amenable to solution. These are the following:

1. Capillary pressure between oil and gas is generally neglected.
2. Several components are usually grouped together—e.g., a system containing the following seven components will be grouped as shown:

$$
\begin{aligned}
&C_1 \quad \text{Component 1} \\
&\left.\begin{aligned} &C_2 \\ &C_3 \\ &C_4 \\ &C_5 \\ &C_6 \end{aligned}\right\} \quad \text{Component 2} \\
&C_7 \quad \text{Component 3}
\end{aligned}
$$

This entails the development of a set of consistent PVT data, equilibrium data, and other pertinent data for the system as defined.

3. The mass fraction of components present in the water is so small that the C_{wj}-terms are all zero. This means that oil and gas are the only phases in which mass transfer occurs. The conservation equation for the water present is still needed, as discussed earlier in Chapter 2.

Sources and Sinks

The basic equation derived for the linear compositional model did not include sources and/or sinks as shown in Eq. (3.59). These can be included by the addition of a term representing the source or sink:

$$
\frac{\partial}{\partial x}\left(\sum_{i=1}^{3} \frac{k_i \rho_i}{\mu_i} C_{ij} \frac{\partial P_i}{\partial x}\right) - \sum_{i=1}^{3} q_i \alpha_{ij}\, \delta(x) = \frac{\partial}{\partial t}\left(\sum_{i=1}^{3} \phi S_i \rho_i C_{ij}\right) \tag{3.68}
$$

where

q_i = mass injection rate of phase i in suitable units

α_{ij} = mass fraction of jth component in ith phase

$\delta(x)$ = Dirac delta function which is defined as follows:

Production or injection in cell at x: $\delta(x) = 1$

No production or injection in cell at x: $\delta(x) = 0$

The locations of these wells are shown in Fig. 3.14.

The solution of the compositional reservoir system is by far the most difficult problem in reservoir simulation.

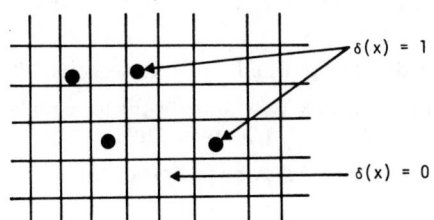

Figure 3.14: Well locations.

Solution Outline

The solution of the compositional model is by virtue of its complexity an iterative one. The processes indicated in Fig. 3.15 are but a superficial outline and are essentially the technique followed by Tsutsumi and Dixon.[12]

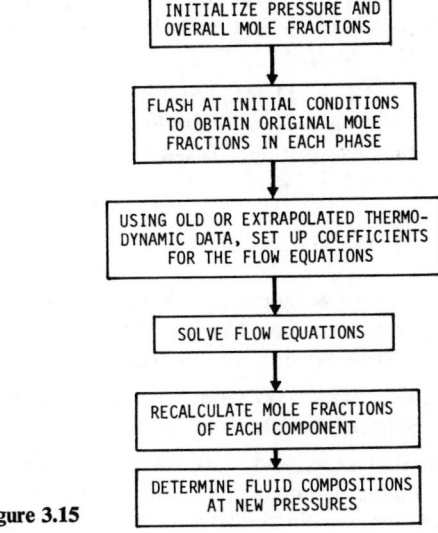

Figure 3.15

REFERENCES

1. J. C. MARTIN, "Simplified Equations of Flow in Gas Drive Reservoirs and Theoretical Foundation of Multiphase Pressure Buildup Analysis," *Trans. AIME* (1959), **216**, 309–11.

2. R. E. COLLINS, *Flow of Fluids Through Porous Materials* (New York: Reinhold, 1961).

3. M. MUSKAT, *The Flow of Homogeneous Fluids Through Porous Media* (New York: McGraw-Hill, 1937).

4. M. Muskat, *Physical Principles of Oil Production* (New York: McGraw-Hill, 1949).

5. W. Hurst, "Unsteady Flow of Fluids in Oil Reservoirs," *Physics* (Jan. 1934), 5.

6. M. King Hubbert, "Darcy's Law and the Field Equations of the Flow of Underground Fluids," *Trans. AIME* (1956), **207**, 222–39.

7. M. C. Leverett, "Capillary Behavior in Porous Solids," *Trans. AIME* (1941), **142**.

8. I. F., Roebuck, G. E. Henderson, J. Douglas, Jr., and W. T. Ford, "The Compositional Reservoir Simulator: The Linear Model," *Trans. AIME* (1969), **246**, 115.

9. W. Abel, R. F. Jackson, and R. A. Wattenbarger, "Simulation of a Partial Pressure Maintenance Gas Cycling Project with a Compositional Model, Carson Creek Field, Alberta," *J. Pet. Tech.* (Jan. 1970), 38–46.

10. G. H. Alani, and H. T. Kennedy, "Volumes of Liquid Hydrocarbons at High Temperatures and Pressures," *Trans. AIME* (1960), **219**, 288–92.

11. S. M. Avasti, and H. T. Kennedy, "The Prediction of Volumes, Compressibilities, and Thermal Expansion Coefficients of Hydrocarbon Mixtures," *Soc. Pet. Eng. J.* (1968), **8**, 98–106.

12. G. Tsutsumi, and T. N. Dixon, "Mathematical Simulation of Two Phase Flow with Interphase Mass Transfer in Petroleum Reservoirs," SPE 4074, 47th Annual Meeting, San Antonio, Texas, Oct. 8–11, 1972.

BIBLIOGRAPHY

Abel, W., R. F. Jackson, and R. A. Wattenbarger, "Simulation of a Partial Pressure Maintenance Gas Cycling Project With a Compositional Model, Carson Creek Field, Alberta," *J. Pet. Tech.* (Jan. 1970), 38–46.

Bondor, P. L., G. J. Hirasaki, and M. J. Tham, "Mathematical Simulation of Polymer Flooding in Complex Reservoirs," *Soc. Pet. Eng. J.* (Oct. 1972), 369–82.

Cook, R. E., R. H. Jacoby, and A. B. Ramesh, "A Beta-type Reservoir Simulator for Approximating Compositional Effects During Gas Injection," SPE 4272, Third Symposium on Numerical Simulation of Reservoir Performance, Houston, Texas, Jan. 10–12, 1973.

Culham, W. E., S. M. Fauouq Ali, and C. D. Stahl, "Experimental and Numerical Simulation of Two-phase Flow With Interphase Mass Transfer in One and Two Dimensions," *Soc. Pet. Eng. J.* (Sept. 1969), 323–37; *Trans. AIME*, **246**.

Davidson, L. B., F. G. Miller, and T. D. Mueller, "A Mathematical Model of Reservoir Response During the Cyclic Injection of Steam," *Soc. Pet. Eng. J.* (June 1967), 174–88; *Trans. AIME*, **240**.

Deans, H. A., "A Mathematical Model for Dispersion in the Direction of Flow in Porous Media," *Soc. Pet. Eng. J.* (March 1963), 49–52.

Eilerts, C. K., "Integration of Partial Differential Equation for Transient Linear

Flow of Gas-condensate Fluids in Porous Structures," *Soc. Pet. Eng. J.* (Dec. 1964), 291–306.

EILERTS, C. K., and E. F. SUMNER, "Integration of Partial Differential Equations for Multicomponent, Two-phase Transient Radial Flow," *Soc. Pet. Eng. J.* (June 1967), 125–35; *Trans. AIME,* **240.**

EILERTS, C. K., E. F. SUMNER, and N. L. POTTS, "Integration of Partial Differential Equation for Transient Radial Flow of Gas-condensate Fluids in Porous Structures," *Soc. Pet. Eng. J.* (June 1964), 141–52.

GILCHRIST, R. E., R. R. HARVEY, and M. R. DEAN, "Three-component Analysis by Dispersivity in Fluid-flow Analogs," SPE of AIME (1962), II-194.

GRIEVES, R. G., and G. THODOS, "The Cricondentherm and Cricondenbar Pressures of Multicomponent Hydrocarbon Mixtures," SPE of AIME (1964), II-240.

HIATT, W. N., "Mathematical Basis of Two-phase, Incompressible, Vertical Flow Through Porous Media and Its Implications in the Study of Gravity-drainage-type Petroleum Reservoirs," *Soc. Pet. Eng. J.* (Sept. 1968), 225–30; *Trans. AIME,* **243.**

HIGGINS, R. V., and A. J. LEIGHTON, "A Computer Method to Calculate Two-phase Flow in Any Irregularly Bounded Porous Medium," *J. Pet. Tech.* (June 1962), 679–83; *Trans. AIME,* **225.**

HIGGINS, R. V., and A. J. LEIGHTON, "Computer Prediction of Water Drive of Oil and Gas Mixtures Through Irregularly Bounded Porous Media—Three-phase Flow," *J. Pet. Tech.* (Sept. 1962), 1048–54; *Trans. AIME,* **225.**

HUANG, E.T.S., "A Sensitivity Study of Reservoir Performance Using a Compositional Reservoir Simulator," *Soc. Pet. Eng. J.* (Feb. 1972), 3–12.

KNIAZEFF, V. J., and S. A. NAVILLE, "Two-phase Flow of Volatile Hydrocarbons," *Soc. Pet. Eng. J.* (March 1965), 37–44; *Trans. AIME,* **234.**

LAIRD, A.D.K., and J. A. PUTNAM, "Three Component Saturation in Porous Media by X-ray Techniques," SPE of AIME (1959), **216.**

MODINE, A. D., and K. H. COATS, "Multidimensional Displacement of a Viscous Phase by a Non-viscous Phase," *Soc. Pet. Eng. J.* (Dec. 1968), 325–30.

NIELSEN, R. L., and M. R. TEK, "Evaluation of Scale-up Laws for Two-phase Flow Through Porous Media," *Soc. Pet. Eng. J.* (June 1963), 164–76; *Trans. AIME,* **228.**

NOLEN, J. S., "Numerical Simulation of Compositional Phenomena in Petroleum Reservoirs," SPE 4274, Third Symposium on Numerical Simulation of Reservoir Performance, Houston, Texas, Jan. 10–12, 1973.

PERRINE, R. L., and G. M. GAY, "Unstable Miscible Flow in Heterogeneous Systems," *Soc. Pet. Eng. J.* (Sept. 1966), 228–38; *Trans. AIME,* **237.**

PRICE, H. S., and D. A. T. DONOHUE, "Isothermal Displacement Processes with Interphase Mass Transfer," *Soc. Pet. Eng. J.* (June 1967), 205–20; *Trans. AIME,* **240.**

RAMESH, A. B., and T. N. DIXON, "Numerical Simulation of Carbonated Waterflooding in a Heterogeneous Reservoir," SPE 4075, Third Symposium on Numerical Simulation of Reservoir Performance, Houston, Texas, Jan. 10–12, 1973.

ROEBUCK, I. F., JR., G. E. HENDERSON, J. DOUGLAS, JR., and W. T. FORD, "The Compositional Reservoir Simulator: Case 1—The Linear Model," *Soc. Pet. Eng. J.* (March 1969), 115–30; *Trans. AIME*, **246**.

ROWE, A. M., JR. and I. H. SILBERBERG, "Prediction of the Phase Behavior Generated by the Enriched-gas-drive Process," *Soc. Pet. Eng. J.* (June 1965), 160–66.

SHELDON, J. W., B. ZONDEK, and W. T. CARDWELL, JR., "One-dimensional, Incompressible, Non-capillary, Two-phase Fluid Flow in a Porous Medium," *Trans. AIME* (1959), **216**, 290–96.

SMIALEK, R. J., and G. THODOS, "Prediction of K-values for Ternary Hydrocarbon Systems: Conditions up to the Critical Point," SPE of AIME (1964), II-329.

STROUD, L., W. E. DEVANCY, and J. E. MILLER, "Multiple Liquid Phases in a Natural-gas System," SPE of AIME, 1961, II-137.

THOMPSON, FRED R., and RICHARD A. THACHUK, "Compositional Simulation of a Gas Cycling Project, Bonnie Glen D-3A Pool, Alberta, Canada," Third Symposium on Numerical Simulation of Reservoir Performance, Houston, Texas, Jan. 10–12, 1973.

TSUTSUMI, GINJIRO, and THOMAS N. DIXON, "Mathematical Simulation of Two-phase Flow with Interphase Mass Transfer in Petroleum Reservoirs," SPE 4074, 47th Annual Meeting, San Antonio, Texas, Oct. 8–11, 1972.

VAN-QUY, N., P. SIMANDOUX, and J. CORTEVILLE, "A Numerical Study of Diphasic Multi-component Flow," *Soc. Pet. Eng. J.* (April 1972), 171–84; *Trans. AIME*, **253**.

WOERTZ, BYRON B., "Vapor-liquid Equilibrium Ratios (K-values) of Light Hydrocarbons at Reservoir Conditions," SPE 3100, 45th Annual Meeting, Houston, Tex., Oct. 4–7, 1970.

YAMAMOTO, R. H., W. T. FORD, and A. BOUBEGUIRA, "Compositional Reservoir Simulator for Fissured Systems—The Single Block Model," *Soc. Pet. Eng. J.* (June 1971), 113–28.

4 Setting Up the Finite-Difference Model

4.1 INTRODUCTION

The equations which govern the flow of fluids in porous media were derived earlier in the text. These equations are nonlinear partial differential equations which relate the pressure and saturation changes with time throughout the medium. These equations are extremely complex, and their applications are complicated by the presence of specialized boundary conditions.

The solution of these equations by analytical means is generally impossible except for the most trivial cases. The solutions, when they do exist, give a *continuous* distribution of the dependent parameters (pressure or saturation), as shown in Fig. 4.1.

The numerical solution of these equations is generally the *only* way that a solution can be obtained in most applications. The numerical solution

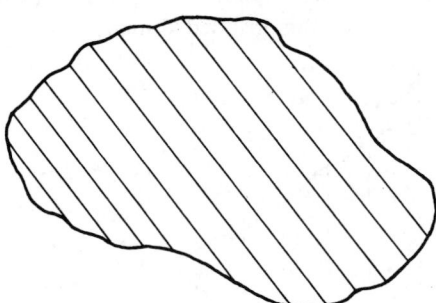

Figure 4.1: Continuous system.

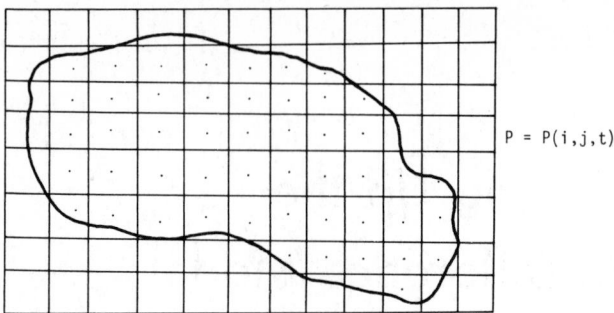

Figure 4.2: Discrete system.

produces answers at *discrete* points within the system, as shown in Fig. 4.2, and the location of the points can be quite arbitrarily determined. The transformation of the continuous differential equation to a discrete form is made by the use of *finite differences*. In this process both space and time are discretized.

Discretization Process

The solution to the systems of flow equations commonly encountered in reservoir engineering work involves the determination of some dependent parameters in space and time. As mentioned above, the solution is obtained at discrete points in space and time. The spatial domain is broken up into a number of cells, grids, or blocks by superimposing some type of a grid. This grid is usually rectangular in form but not necessarily so. Figure 4.2 illustrates the use of a two-dimensional grid. The time domain is also discretized into a number of time steps, during each of which the problem is solved to obtain new values of the dependent parameter. The size of these steps depends on the particular problem being solved, and generally the smaller the time step the more accurate is the solution. An example of time discretization is shown in Fig. 4.3. The finite-difference equations are formulated to solve for the dependent parameters over this gridded domain.

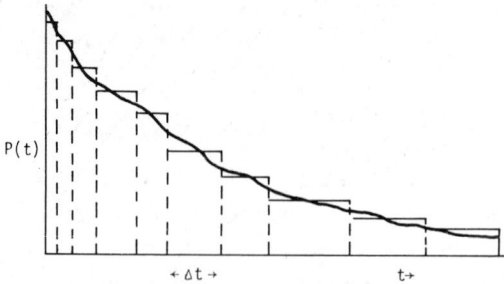

Figure 4.3: Time discretization.

Finite Differences[1,2]

The partial differential equation is replaced by its finite-difference equivalent. The finite-difference equations can be derived by making a *Taylor series* expansion of the function at a given point and then solving for the required derivative.

Consider the following Taylor series expansions:

$$P(x + \Delta x) = P(x) + \Delta x\, P'(x) + \tfrac{1}{2}\Delta x^2\, P''(x) + \tfrac{1}{6}\Delta x^3\, P'''(x) \qquad (4.1)$$
Forward difference

$$P(x - \Delta x) = P(x) - \Delta x\, P'(x) + \tfrac{1}{2}\Delta x^2\, P''(x) - \tfrac{1}{6}\Delta x^3\, P'''(x) \qquad (4.2)$$
Backward difference

where

$$P' = \frac{\partial P}{\partial x}$$

$$P'' = \frac{\partial^2 P}{\partial x^2}, \quad \text{etc.}$$

First Derivative (Fig. 4.4)

Equation (4.1) or (4.2) could be solved for the first or second derivatives as required—e.g.:

$$P'(x) = \frac{P(x + \Delta x) - P(x)}{\Delta x} + 0(\Delta x) \qquad (4.3)$$

$$P'(x) = \frac{P(x) - P(x - \Delta x)}{\Delta x} + 0(\Delta x) \qquad (4.4)$$

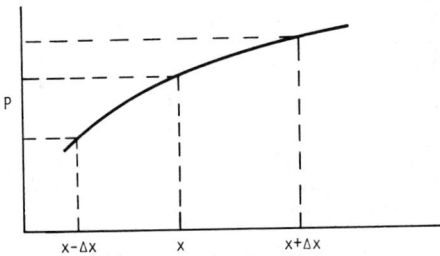

FIRST DERIVATIVE:

FORWARD: $\dfrac{\partial P}{\partial x} = \dfrac{P(x+\Delta x) - P(x)}{\Delta x}$

BACKWARD: $\dfrac{\partial P}{\partial x} = \dfrac{P(x) - P(x-\Delta x)}{\Delta x}$

CENTRAL: $\dfrac{\partial P}{\partial x} = \dfrac{P(x+\Delta x) - P(x-\Delta x)}{\Delta x}$

Figure 4.4: First derivatives.

These are the *forward* and *backward differences* respectively for the first derivative. A *central difference* can be obtained by subtracting Eqs. (4.1) and (4.2):

$$P'(x) = \frac{P(x + \Delta x) - P(x - \Delta x)}{2\,\Delta x} + 0(\Delta x^2) \qquad (4.5)$$

Note that the errors associated with these approximations are different; the forward and backward schemes have errors of the order of Δx, while the error in the central form is of the order of Δx^2. This error associated with the finite-difference form of the partial differential equation is called the *truncation* error.

Second Derivative (Fig. 4.5)

Consider the *addition* of Eqs. (4.1) and (4.2):

$$P(x + \Delta x) + P(x - \Delta x) = 2P(x) + \Delta x^2\, P''(x) + 0(\Delta x^4) \qquad (4.6)$$

Solving for $P''(x)$:

$$P'' = \frac{P(x + \Delta x) - 2P(x) + P(x - \Delta x)}{\Delta x^2} + 0(\Delta x^2) \qquad (4.7)$$

Therefore, the error associated with the second derivative is of the order of Δx^2.

FINITE DIFFERENCE FORMULATION

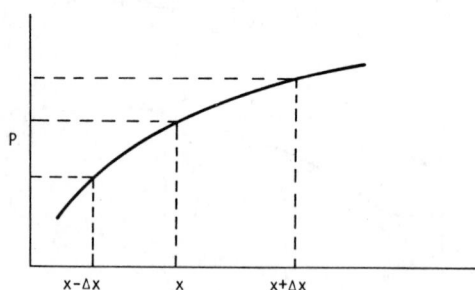

Figure 4.5: Second derivatives.

Summary

For a discrete set of numbered points:

$$\frac{\partial P}{\partial x} = \frac{P_{i+1} - P_i}{\Delta x}, \qquad \frac{\partial P}{\partial x} = \frac{P_i - P_{i-1}}{\Delta x}$$

$$\frac{\partial P}{\partial x} = \frac{P_{i+1} - P_{i-1}}{2\,\Delta x}, \qquad \frac{\partial^2 P}{\partial x^2} = \frac{P_{i+1} - 2P_i + P_{i-1}}{\Delta x^2} \qquad (4.8)$$

4.2 FINITE-DIFFERENCE SCHEMES[3]—CONCEPT OF EXPLICIT AND IMPLICIT FORMS

Introduction

Let us consider a simple process which is time-dependent—i.e., the solution varies as a function of time. Consider the temperature distribution in a one-dimensional rod. The governing equation for the temperature behavior will be of the following type with the appropriate boundary conditions:

$$\frac{\partial^2 T}{\partial x^2} = \frac{1}{\alpha}\frac{\partial T}{\partial t},$$

By solving this system we can obtain $T(x,t)$, which allows us to determine the temperature distribution at any location x at any time t. In analytical treatments we obtain a solution which is continuous in time and space; in numerical solutions, however, the solution process allows us to have temperature values at fixed locations of x and at discrete points in time. Our solution now can be visualized as a series of still photographs at particular instances of time. This is indicated graphically in Fig. 4.6.

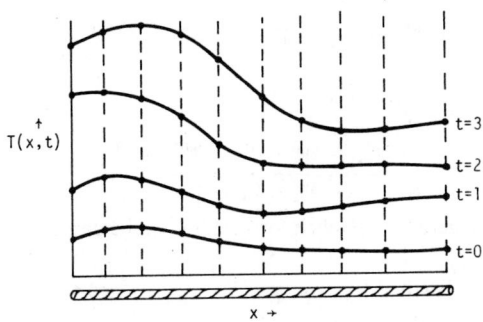

Figure 4.6: Temperature distribution in a rod.

The time values 0, 1, 2, 3, . . . correspond to different time levels which get progressively larger as the solution progresses. The information at a given time level—e.g., $t = 1$—is used to compute that at a higher level—e.g., $t = 2$. There are two basic ways in going from the old time-level values to the new time-level values. The new values can be calculated *individually* for each location in space x—a process which begins at $x = 0$ and terminates at $x = L$ for a given time value. This method of calculating the new values one at a time is the *explicit* scheme. The explicit method involves the sequential solution of *one equation with one unknown*. On the other hand, all the new values between $x = 0$ and $x = L$ can be calculated simultaneously for a given time value. This method of calculating the new values simultaneously is the *implicit* method. The implicit scheme involves the solution of an $N \times N$ *system of simultaneous linear equations*.

Let us examine the two basic formulations in some detail:

Explicit Formulation

In the explicit scheme we solve for one unknown at a time as indicated in Fig. 4.7. Consider the two-dimensional equation:

$$\frac{\partial^2 P}{\partial x^2} + \frac{\partial^2 P}{\partial y^2} = \frac{\partial P}{\partial t} \tag{4.9}$$

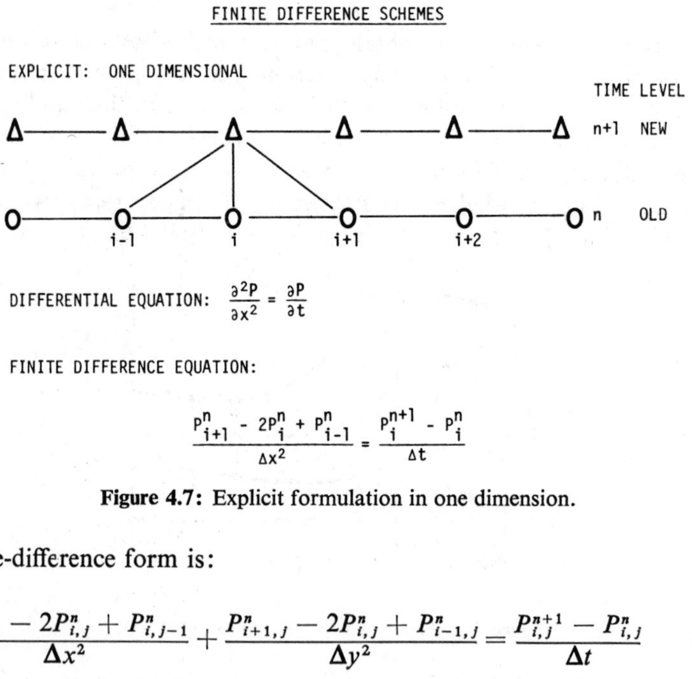

FINITE DIFFERENCE SCHEMES

EXPLICIT: ONE DIMENSIONAL

DIFFERENTIAL EQUATION: $\frac{\partial^2 P}{\partial x^2} = \frac{\partial P}{\partial t}$

FINITE DIFFERENCE EQUATION:

$$\frac{P_{i+1}^n - 2P_i^n + P_{i-1}^n}{\Delta x^2} = \frac{P_i^{n+1} - P_i^n}{\Delta t}$$

Figure 4.7: Explicit formulation in one dimension.

A finite-difference form is:

$$\frac{P_{i,j+1}^n - 2P_{i,j}^n + P_{i,j-1}^n}{\Delta x^2} + \frac{P_{i+1,j}^n - 2P_{i,j}^n + P_{i-1,j}^n}{\Delta y^2} = \frac{P_{i,j}^{n+1} - P_{i,j}^n}{\Delta t} \tag{4.10}$$

where

$i, j =$ cell location in grid

$n =$ old time level

$n + 1 =$ new time level

Equation (4.10) has only one unknown value—i.e., the new pressure at time $(n + 1)$. This value involves the time derivative. This equation can be rearranged to obtain the new pressure *explicitly* in terms of the other adjacent pressures:

$$P_{i,j}^{n+1} = P_{i,j}^n + \Delta t\left(\frac{P_{i,j+1}^n - 2P_{i,j}^n + P_{i,j-1}^n}{\Delta x^2} + \frac{P_{i+1,j}^n - 2P_{i,j}^n + P_{i-1,j}^n}{\Delta y^2}\right)$$

(4.11)

Note that every value on the right-hand side is now known and there is *one* equation with *one* unknown. All values at the new time are solved for by moving through all the (i, j) locations in the model in some systematic manner. Figure 4.8 shows the cell arrangement in a two-dimensional grid.

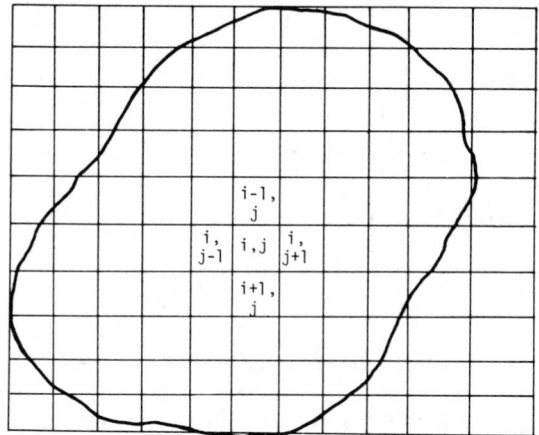

Figure 4.8: Cell arrangement in two dimensions.

Equation (4.11) can be simplified to:

$$P_{i,j}^{n+1} = P_{i,j}^n + \alpha(P_{i,j+1}^n - 2P_{i,j}^n + P_{i,j-1}^n) + \beta(P_{i+1,j}^n - 2P_{i,j}^n + P_{i-1,j}^n)$$

(4.12)

where

$$\alpha = \frac{\Delta t}{\Delta x^2}, \qquad \beta = \frac{\Delta t}{\Delta y^2}$$

Discussion of Explicit Formulation: Explicit methods are not generally used in reservoir simulation because of the usually severe restrictions on the time step size. The programming effort needed to build a simulator based on the explicit scheme is much less than any other; the running time, however, is quite substantial on these programs.

The stability criteria are discussed in Sec. 4.4.

Implicit Formulation

In the implicit scheme we solve for all unknown values simultaneously. (See Fig. 4.9.) Consider the following partial differential equation:

$$\frac{\partial^2 P}{\partial x^2} = \frac{\partial P}{\partial t} \tag{4.13}$$

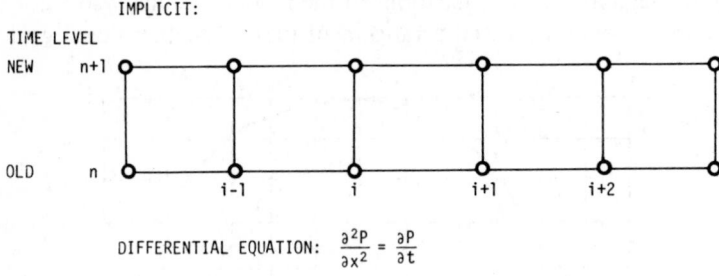

Figure 4.9: Implicit formulation in one dimension.

A finite-difference analog of this equation is as follows:

$$\frac{P_{i-1}^n - 2P_i^n + P_{i+1}^n}{\Delta x^2} = \frac{P_i^{n+1} - P_i^n}{\Delta t} \tag{4.14}$$

This equation has only one unknown, P_i^{n+1}; however, we can set up Eq. (4.14) to solve for all three P_i-values in the equation as follows:

$$\frac{P_{i-1}^{n+1} - 2P_i^{n+1} + P_{i+1}^{n+1}}{\Delta x^2} = \frac{P_i^{n+1} - P_i^n}{\Delta t} \tag{4.15}$$

This equation has all unknown pressures at the new time level (Fig. 4.10). Simplifying Eq. (4.15):

$$P_{i-1}^{n+1} - 2P_i^{n+1} + P_{i+1}^{n+1} = \frac{\Delta x^2}{\Delta t}(P_i^{n+1} - P_i^n) \tag{4.16}$$

Collecting terms of similar kind:

$$P_{i-1}^{n+1} - \left(2 + \frac{\Delta x^2}{\Delta t}\right)P_i^{n+1} + P_{i+1}^{n+1} = \frac{\Delta x^2}{\Delta t}P_i^n \tag{4.17}$$

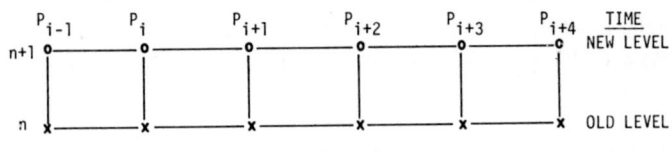

Figure 4.10

Note that this is one equation with three unknowns, in which point i is linked or coupled to points $(i + 1)$, $(i - 1)$. Equation (4.17) is of the general type:

$$a_i P_{i-1} + b_i P_i + c_i P_{i+1} = d_i \qquad (4.18)$$

where the coefficients a_i, b_i, and c_i are related to the geometry of the system and its physical properties and d_i contains known terms.

Writing Eq. (4.18) for N cells in a linear grid results in one equation for each cell; the total results are N equations with N unknowns.

EXAMPLE:

Cell

1	$a_1 P_0 - b_1 P_1 + c_1 P_2$	$= d_1$
2	$a_2 P_1 - b_2 P_2 + c_2 P_3$	$= d_2$
3	$a_3 P_2 - b_3 P_3 + c_3 P_4$	$= d_3$
4	$a_4 P_3 - b_4 P_4 + c_4 P_5$	$= d_4$

(4.19)

$$N \qquad a_n P_{n-1} - b_n P_n + C_n P_{n+1} = d_N$$

The cells with 0 and $n + 1$ subscripts are generally fictitious cells. They do not form part of the model and are deleted by use of the appropriate boundary conditions.

Note that this matrix has a characteristic form: there are three diagonal elements, and all the off-diagonal elements are zero. This matrix is called a *tridiagonal matrix*. The set of simultaneous equations (Eq. 4.19) can be written in matrix notation:

$$AP = d \qquad (4.20)$$

where

$$\begin{bmatrix} & & \\ a_i & b_i & c_i \\ & & \end{bmatrix} * \begin{bmatrix} P_i \end{bmatrix} = \begin{bmatrix} d_i \end{bmatrix} \qquad (4.21)$$

This system is solved for the unknown pressures P by using the Thomas algorithm, which is a modified form of Gaussian elimination. This algorithm will be discussed later.

The example so far considered was a simple one-dimensional model. For a two-dimensional system the development is identical, but the results are a little different.

Consider the following partial differential equation in two dimensions:

$$\frac{\partial^2 P}{\partial x^2} + \frac{\partial^2 P}{\partial y^2} = \frac{\partial P}{\partial t} \tag{4.22}$$

This equation describes the pressure response in a two-dimensional system.

Writing the fully implicit finite-difference formulation on the grid in Fig. 4-11:

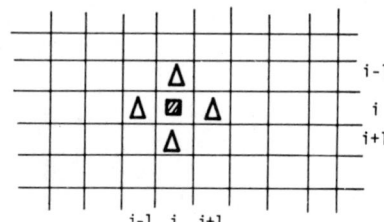

Figure 4.11: Two dimensional grid.

$$\frac{P_{i,j-1}^{n+1} - 2P_{i,j}^{n+1} + P_{i,j+1}^{n+1}}{\Delta x^2} + \frac{P_{i+1,j}^{n+1} - 2P_{i,j}^{n+1} + P_{i-1,j}^{n+1}}{\Delta y^2} = \frac{P_{i,j}^{n+1} - P_{i,j}^n}{\Delta t} \tag{4.23}$$

Note that all pressures are at the new time level and therefore unknown. There are five unknowns in the above equation.

For the sake of simplicity, let us assume $\Delta x = \Delta y$. Then, collecting terms and simplifying Eq. (4.23) on the left yields:

$$P_{i,j-1}^{n+1} + P_{i+1,j}^{n+1} - 4P_{i,j}^{n+1} + P_{i-1,j}^{n+1} + P_{i,j+1}^{n+1} = \frac{\Delta x^2}{\Delta t}(P_{i,j}^{n+1} - P_{i,j}^n) \tag{4.24}$$

which reduces to

$$P_{i,j-1}^{n+1} + P_{i+1,j}^{n+1} - \left(4 + \frac{\Delta x^2}{\Delta t}\right)P_{i,j}^{n+1} + P_{i-1,j}^{n+1} + P_{i,j+1}^{n+1} = -\frac{\Delta x^2}{\Delta t}P_{i,j}^n \tag{4.25}$$

This equation is of the general type:

$$e_i P_{i,j-1}^{n+1} + a_i P_{i+1,j}^{n+1} + b_i P_{i,j}^{n+1} + c_i P_{i-1,j}^{n+1} + f_i P_{i,j+1}^{n+1} = d_i \tag{4.26}$$

where the coefficients e_i, a_i, b_i, c_i, f_i, and d_i are similar to those defined for the one-dimensional system.

Writing this type equation for all N cells in a model produces N equations with N unknowns. This system is now a five-diagonal system as shown below:

$$AP = d \qquad (4.27)$$

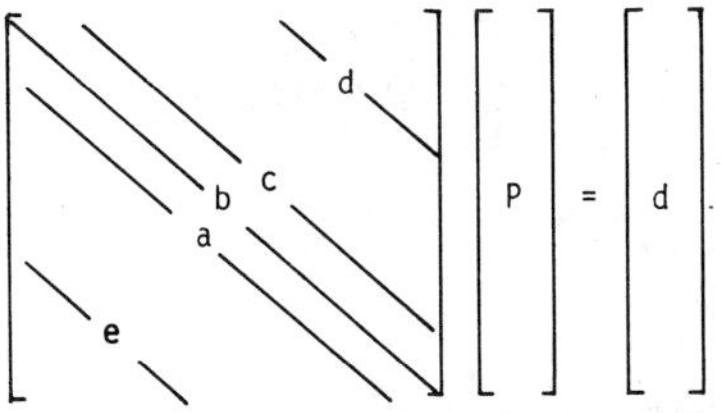

$$(4.28)$$

There is no efficient algorithm to solve this system, and special algorithms for the two-dimensional grid will be discussed later. These algorithms are designed to reduce the work requirement to obtain the solution vector P.

The implicit formulations are unconditionally stable for all values of $\Delta t/\Delta x^2$. This criterion is proved later in Sec. 4.4 on stability.

Crank-Nicholson Scheme[4]

This involves a combination of old and new time step values of the dependent variable (see Fig. 4.12).

Consider:

$$\frac{\partial^2 u}{\partial x^2} = \frac{\partial u}{\partial t} \qquad (4.29)$$

FINITE DIFFERENCE SCHEMES

MIXED (CRANK-NICHOLSON):

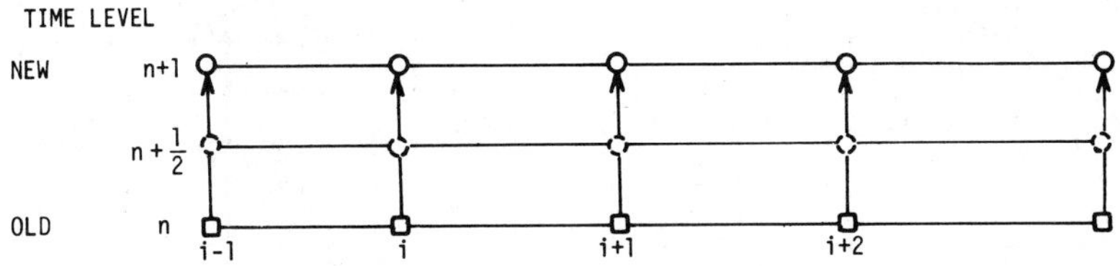

DIFFERENTIAL EQUATION: $\dfrac{\partial^2 P}{\partial x^2} = \dfrac{\partial P}{\partial t}$

FINITE DIFFERENCE EQUATION:

$$\frac{[P_{i+1}^n - 2P_i^n + P_{i-1}^n]}{\Delta x^2} + \frac{(1-\ )[P_{i+1}^{n+1} - 2P_i^{n+1} + P_{i-1}^{n+1}]}{\Delta x^2} = \frac{P_i^{n+1} - P_i^n}{\Delta t}$$

Figure 4.12: Mixed formulations.

A difference formulation involving both old time and new time levels can be set up as follows:

$$\theta\left(\frac{u_{i+1}^{n+1} - 2u_i^{n+1} + u_{i-1}^{n+1}}{\Delta x^2}\right) + (1 - \theta)\left(\frac{u_{i+1}^n - 2u_i^n + u_{i-1}^n}{\Delta x^2}\right) = \frac{u_i^{n+1} - u_i^n}{\Delta t} \quad (4.30)$$

This is a *general weighted average approximation*, where

$$0 < \theta \leq 1$$

If

$\theta = 0$, then the scheme is explicit

$\theta = \frac{1}{2}$, then the scheme is Crank-Nicholson

$\theta = 1$, then the scheme is fully implicit

4.3 GRID DEFINITIONS

The finite-difference equations must be implemented over some discrete network; over this network the values of the dependent parameters are calculated.

Two types of grids are generally used:

1. *Block-centered* (Fig. 4.13)—The dependent parameters are calculated at the center of the cell or block. There are no points on the boundary.

2. *Lattice* (Fig. 4.14)—The dependent parameters are calculated at the intersection of the grid lines. There are several points on the boundary.

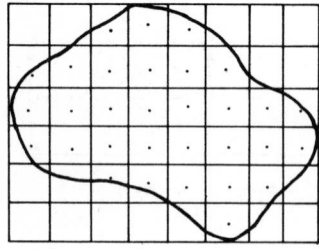

Figure 4.13: Block centered grid.

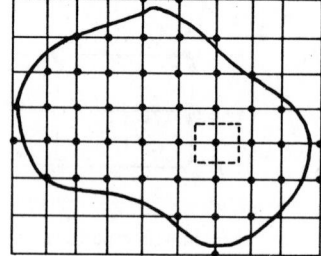

Figure 4.14: Lattice grid.

The different grid configurations are suited to different boundary conditions. The block-centered grid is generally used with a *Neumann-type boundary condition*, which specifies flow across the boundary (see Fig. 4.15). Flow across the boundary can be represented by a source term in the boundary cell; under these conditions the equations for the boundary cells are modified to include the source term. Here the gradient is $\partial P / \partial x = K$, which indicates some nonzero flux across the boundary.

The lattice-type grid is generally used when a Dirichlet-type boundary condition is specified in the problem. In this condition the function is specified on the boundary as indicated in Fig. 4.16.

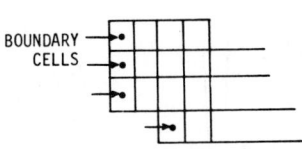

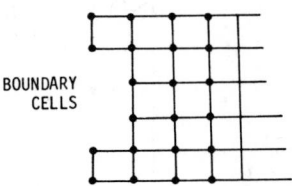

Figure 4.15: Neumann boundary.

Figure 4.16: P is specified: $P(i, j, t) = P_e$.

Irregular Grids

Irregular grids have nonuniform spacings in the x- and y-directions. These grids are used to increase the definition in regions where better control is needed. Most reservoir grids actually used are irregular, as shown in Fig. 4.17.

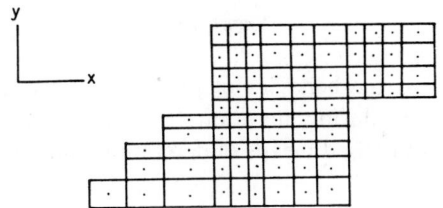

Figure 4.17: Nonuniform grids.

4.4 STABILITY CRITERIA[5-7]

The system of equations which is solved to determine the dependent parameters in a simulation study must provide a *stable* solution to be of any use. A scheme is unstable if the error continues to increase uncontrollably.

Let ϵ^n be the error between the true solution and the computed solution at any time n. Then,

$$\frac{\epsilon^{n+1}}{\epsilon^n} \leq 1 \quad \text{The system is stable}$$

$$\frac{\epsilon^{n+1}}{\epsilon^n} > 1 \quad \text{The system is unstable}$$

STABILITY

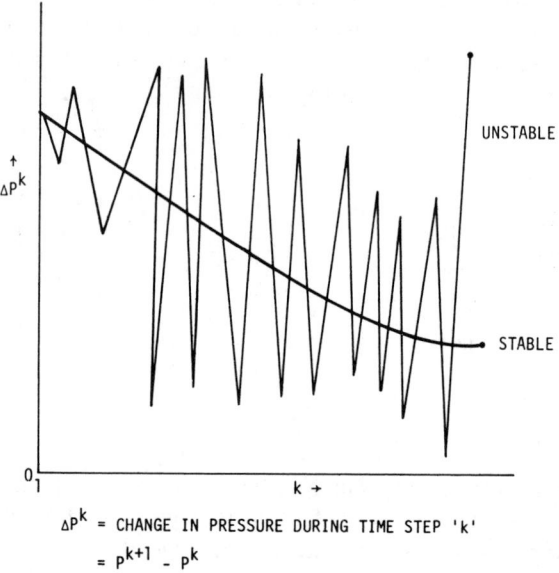

ΔP^k = CHANGE IN PRESSURE DURING TIME STEP 'k'

   = $P^{k+1} - P^k$

Figure 4.18

Figure 4.18 shows the behavior of pressure changes for both the stable and unstable systems during iteration for the correct pressure solution. There are two generally used means of determining the stability of a solution scheme:

1. Von Neumann analysis (Fourier analysis)
2. Matrix methods

Von Neumann Analysis[8] (Fourier Analysis)

The method first proposed by von Neumann is sometimes referred to as the Fourier analysis method because of the use of Fourier series in the representation of the error terms or of the solution. The initial error in the finite-difference approximation is expressed as a finite Fourier series. The growth of this error is then analyzed as the solution progresses. Since the error term and the dependent parameter both satisfy the partial differential equation, we can write the representation for either the error itself or the solution term. The stability of the solution scheme depends on the error term remaining controllable and bounded throughout the domain of the solution.

A brief development of this stability determination method will be reviewed here. The engineer is referred to the original article for in-depth study.

The Fourier series expansion of a function is usually expressed in terms of sines or cosines:

$$f(x) = \sum a_n \cos \frac{n\pi x}{L} \tag{4.31}$$

However, to facilitate the arithmetic, it is somewhat less cumbersome to use the complex exponential form:

$$f(x) = \sum A_n e^{in\pi x/L} \tag{4.32}$$

where $i = \sqrt{-1}$. The solution to the partial differential equation in finite-difference form produces a discrete set of values denoted by $u_{i,j}$ at points i, j. The Fourier series form must then be written for these discrete points. Consider the spatial domain divided into N increments of width h, then using an integer counter p the error at a point can be written E_p, where $p = 0, 1, 2, \ldots, N$. The equation in complex exponential form is now:

$$E_p = \sum_{n=0}^{N} A_n e^{in\pi ph/L} \tag{4.33}$$

or

$$E_p = \sum_{n=0}^{N} A_n e^{i\beta_n ph} \tag{4.34}$$

where $\beta = n\pi/L$, with L the total length.

Since the errors from all terms are additive, we can dispense with the summation sign in Eq. (4.34) and analyze only a single term, thus:

$$E_p = A_n e^{i\beta_n ph} \tag{4.35}$$

As time increases, we need to investigate the growth of this error; furthermore, the error term must reduce to the initial value term at time $= 0$. We need a function that will allow us to fulfill both criteria. The following form is assumed:

$$E_{p,t} = e^{i\beta_n ph} e^{\alpha t} \tag{4.36}$$

where α is a complex constant. The time domain is divided up into q increments of k time units; Eq. (4.36) is therefore rewritten:

$$E_{p,q} = e^{i\beta_n ph} e^{\alpha qk} \tag{4.37}$$

Note that the constant A_n has been dropped from the equation. Equation

(4.37) can be compacted to:

$$E_{p,q} = e^{i\beta_n ph} \zeta^q \qquad (4.38)$$

where $\zeta = e^{\alpha k}$. It is obvious from Eq. (4.38) that the error term at any point p in space and q in time will not increase as long as

$$|\zeta| \le 1 \qquad (4.39)$$

The quantity ζ is an expression which usually contains the parameters h, as well as k in some relation; in this case, for Eq. (4.39) to be satisfied we have to determine what values of k and h would satisfy the inequality relation.

Consider the following example from Smith, in which the parabolic equation is:

$$\frac{\partial u}{\partial t} = \frac{\partial^2 u}{\partial x^2} \qquad (4.40)$$

The finite-difference equation in fully implicit form is:

$$\frac{u_{p,q+1} - u_{p,q}}{k} = \frac{u_{p-1,q+1} - 2u_{p,q+1} + u_{p+1,q+1}}{h^2} \qquad (4.41)$$

Substituting Eq. (4.38) into Eq. (4.41) for the error at each term in the finite-difference form:

$$e^{i\beta ph}\zeta^{q+1} - e^{i\beta ph}\zeta^q = \frac{k}{h^2}[e^{i\beta(p-1)h}\zeta^{q+1} - 2e^{i\beta ph}\zeta^{q+1} + e^{i\beta(p+1)h}\zeta^{q+1}] \qquad (4.42)$$

Dividing through by $e^{i\beta ph}\zeta^q$ gives:

$$\zeta - 1 = \frac{k}{h^2}(\zeta e^{-i\beta h} - 2\zeta + \zeta e^{i\beta h}) \qquad (4.43)$$

This is simplified by noting:

$$\cos(\beta h) = \frac{e^{-i\beta h} + e^{i\beta h}}{2} \qquad (4.44)$$

Then:

$$\zeta - 1 = \frac{k\zeta}{h^2}(2\cos\beta h - 2) \qquad (4.45)$$

$$\zeta - 1 = -\frac{k}{h^2}4\zeta \sin^2\frac{\beta h}{2} \qquad (4.46)$$

Thus:

$$\zeta = \frac{1}{1 + 4(k/h^2)\sin^2(\beta h/2)} \qquad (4.47)$$

Since the term $4(k/h^2) \sin^2 (\beta h/2)$ is always positive for all positive values of k/h^2, the value $|\xi| \leq 1$, as required. The fully implicit form is therefore unconditionally stable.

Matrix Methods[8]

In general, the matrix method involves an analysis of the error propagation by the use of matrix algebra. Essentially, the process begins by defining the error associated with the solution of the system of simultaneous linear equations and relates this error to the continued multiplication of a given coefficient matrix A; e.g., at the $n + 1$ step the error is:

$$e^{n+1} = Ae^n = A(Ae^{n-1}) \ldots \tag{4.48}$$

Thus,

$$e^{n+1} = A^{n+1}e^0 \tag{4.49}$$

Then the matrix A must possess certain properties for the error e^{n+1} to remain bounded. The behavior of the matrix A is analyzed in terms of its eigenvalues and eigenvectors. This is possible because of the definition of an eigenvalue. For any vector V,

$$AV = \lambda V \tag{4.50}$$

defines the eigenvalue λ and the eigenvector V. Thus, the error equation (Eq. 4.49) can be written:

$$e^{n+1} = A^{n+1}e^0 = \lambda^{n+1}e^0 \tag{4.51}$$

Thus, for stability $e^{n+1} \longrightarrow 0$ as $n + 1$ increases:

$$|\lambda| \leq 1 \tag{4.52}$$

Since there are N eigenvalues for an $N \times N$ matrix, we need only to examine the largest value of the eigenvalues. Thus, Eq. (4.52) can be written:

$$|\lambda_{max}| \leq 1$$

The largest eigenvalue is called the spectral radius of the matrix.

Consider the stability treatment for the case of a parabolic equation in two dimensions:

$$\frac{\partial^2 u}{\partial x^2} + \frac{\partial^2 u}{\partial y^2} = \frac{\partial u}{\partial t} \tag{4.53}$$

Writing the fully implicit formulation for this system in two dimensions leads to a set of simultaneous linear equations.

Given the system of simultaneous equations obtained by writing the finite-difference equation for every point in the mesh:

$$Au = b \qquad (4.54)$$

In matrix form:

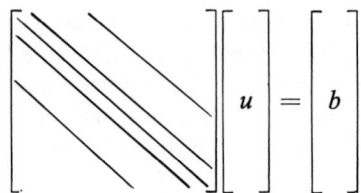

The matrix system is normalized with respect to each diagonal element a_{ii}. Then A can be decomposed into a lower and an upper triangular matrix as follows:

$$(I - H - K)u = b \qquad (4.55)$$

where

$$-H = \begin{bmatrix} 0 & & & \\ & 0 & & \\ & & \ddots & \\ & & & 0 \end{bmatrix}, \qquad -K = \begin{bmatrix} 0 & & & \\ & 0 & & \\ & & \ddots & \\ & & & 0 \end{bmatrix}$$

and I is the identity matrix:

$$I = \begin{bmatrix} 1 & & & \\ & 1 & & \\ & & \ddots & \\ & & & 1 \end{bmatrix}$$

Equation (4.55) can be written:

$$Iu = u = (H + K)u + b$$

Then

$$u^* = (H + K)u^* + b \qquad (4.56)$$

where * denotes a true value.

As shown later in Chapter 6, the line successive overrelaxation (LSOR) scheme can be used to solve Eq. 4.53. For the LSOR scheme the generalized finite-difference scheme used in the model can be written in the following recursive form:

$$Au^{n+1} = Bu^n + Cu^{n+1} + b \qquad (4.57)$$

where

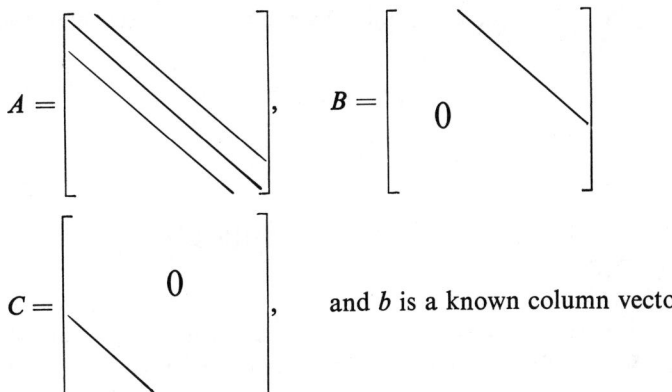

$$A = \begin{bmatrix} \end{bmatrix}, \quad B = \begin{bmatrix} 0 \end{bmatrix}$$

$$C = \begin{bmatrix} 0 \end{bmatrix}, \quad \text{and } b \text{ is a known column vector}$$

The error at any given iteration is defined as:

$$e^n = u^* - u^n \qquad (4.58)$$

where u^n is the nth approximation to the true value.

Solving Eq. (4.57) for u^{n+1}:

$$u^{n+1} = A^{-1}Bu^n + A^{-1}Cu^{n+1} + A^{-1}b \qquad (4.59)$$

Subtracting Eq. (4.59) from Eq. (4.56), the *error-propagation* term becomes:

$$e^{n+1} = (H - A^{-1}C)(u^* - u^{n+1}) + (K - A^{-1}B)(u^* - u^n) \qquad (4.60)$$

since the column vector b is a constant.

Then:

$$e^{n+1} = (H - A^{-1}C)e^{n+1} + (K - A^{-1}B)e^n \qquad (4.61)$$

Solving for e^{n+1}:

$$e^{n+1} = [I - (H - A^{-1}C)][K - A^{-1}B]e^n \qquad (4.62)$$

$$\begin{matrix} \vdots & & \vdots & & \vdots \\ \vdots & & \vdots & & \vdots \end{matrix}$$

$$= \{[I - (H - A^{-1}C)][K - A^{-1}B]\}^n e^0 \qquad (4.63)$$

The matrix $\{[I - (H - A^{-1}C)][K - A^{-1}B]\}$, must have eigenvalues less than unity for the system to converge; i.e., for the N-eigenvalues of this system, $|\lambda_m| < 1$, where λ_m is the largest eigenvalue and generally called the spectral radius. The value of λ_m can be determined by evaluating the matrix of Eq. (4.63) and determining its maximum eigenvalue by some appropriate method.

As the iterations increase, the error term decreases and eventually approaches zero—i.e.,

$$e^{n+1} \longrightarrow 0$$

$$n \longrightarrow \infty$$

4.5 CASE STUDY: EXPLICIT AND IMPLICIT CONCEPTS

The movement of a solvent-type substance through a porous medium is governed by the following type of differential equation in one dimension:

$$c\frac{\partial^2 \rho}{\partial x^2} - v\frac{\partial \rho}{\partial x} = \frac{\partial \rho}{\partial t} \tag{4.64}$$

where

$c = $ diffusion constant

$v = $ velocity term

$\rho = $ concentration term

The above equation can be solved given the necessary boundary and initial conditions.

Solution: Using explicit formulation, Eq. (4.64) can be approximated by the finite-difference equivalent and simplified to yield:

$$\rho_i^{n+1} = \left(\frac{c\,\Delta t}{\Delta x^2} - \frac{v\,\Delta t}{2\,\Delta x}\right)\rho_{i+1}^n + \left(1 - \frac{2c\,\Delta t}{\Delta x^2}\right)\rho_i^n + \left(\frac{c\,\Delta t}{\Delta x^2} + \frac{v\,\Delta t}{2\,\Delta x}\right)\rho_{i-1}^n \tag{4.65}$$

A von Neumann–type analysis indicates that this formulation is stable only if the inequality below is satisfied:

$$\tfrac{1}{2}v^2\,\Delta t \leq 1 \tag{4.66}$$

Using Implicit Formulation: One implicit formulation using centered time difference in the finite-difference formulation produces an equation of the following type for each cell in the model:

$$-\left(\frac{c\,\Delta t}{2\,\Delta x^2} + \frac{v\,\Delta t}{4\,\Delta x}\right)p_{i-1}^{n+1} + \left(1 + \frac{2c\,\Delta t}{2\,\Delta x^2}\right)p_i^{n+1} - \left(\frac{c\,\Delta t}{2\,\Delta x^2} - \frac{v\,\Delta t}{4\,\Delta x}\right)p_{i+1}^{n+1} =$$

$$\left(\frac{c\,\Delta t}{2\,\Delta x^2} + \frac{v\,\Delta t}{4\,\Delta x}\right)p_{i-1}^{n} + \left(1 - \frac{2c\,\Delta t}{2\,\Delta x^2}\right)p_i^{n} + \left(\frac{c\,\Delta t}{2\,\Delta x^2} - \frac{v\,\Delta t}{4\,\Delta x}\right)p_{i+1}^{n} \qquad (4.67)$$

which is of the type:

$$a_i p_{i-1}^{n+1} + b_i p_i^{n+1} + c_i p_{i+1}^{n+1} = d_i \qquad (4.68)$$

This formulation is unconditionally stable, as indicated by a similar analysis.

The explicit and implicit formulations above are tested using the following boundary and initial conditions. At time $= 0$:

$$p(x) = 0$$

For time 0:

Outlet end: $\left.\dfrac{\partial p}{\partial x}\right|_{x=1} = -p(x)$

Inlet end: A slug is injected as indicated in Fig. 4.19.

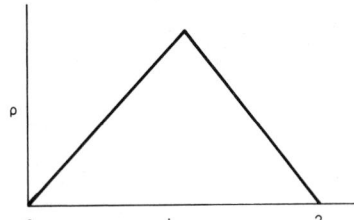

Figure 4.19: Inlet end boundary condition.

RESERVOIR SIMULATION CLASS   DR.H.B.CRICHLOW

PROJECT ENGINEER .. FORREST F. CRAIG, III   UNIV. OF OKLA.

EXPLICIT   SOLUTION OF POLYMER CONCENTRATION EQUATION

GRID INTERVAL= 0.10   TIME STEP = 0.0045
DIFFUSION CONST= 1.00   VELOCITY = 1.0000
TOTAL TIME = 3.00   NO OF POINTS= 11

POLYMER CONCENTRATION IN X DIRECTION

TIME	1	2	3	4	5	6	7	8	9	10	11	DPDX
0.090	0.090	0.064	0.044	0.029	0.018	0.011	0.006	0.004	0.002	0.001	0.001	-0.001
0.180	0.180	0.143	0.112	0.087	0.066	0.049	0.036	0.026	0.019	0.015	0.013	-0.013
0.270	0.270	0.226	0.188	0.154	0.126	0.101	0.082	0.066	0.053	0.044	0.039	-0.039
0.360	0.360	0.311	0.267	0.228	0.193	0.163	0.137	0.116	0.099	0.085	0.076	-0.076
0.450	0.450	0.397	0.348	0.304	0.265	0.230	0.199	0.173	0.151	0.133	0.119	-0.119
0.540	0.540	0.484	0.431	0.383	0.340	0.301	0.266	0.235	0.208	0.186	0.167	-0.167
0.630	0.630	0.571	0.515	0.464	0.417	0.373	0.334	0.299	0.268	0.241	0.217	-0.217
0.720	0.720	0.658	0.600	0.545	0.495	0.448	0.405	0.365	0.330	0.298	0.269	-0.269
0.810	0.810	0.746	0.685	0.627	0.573	0.523	0.476	0.433	0.393	0.356	0.323	-0.323
0.900	0.900	0.833	0.770	0.709	0.652	0.599	0.548	0.501	0.457	0.415	0.377	-0.377
0.990	0.920	0.921	0.855	0.792	0.732	0.675	0.621	0.570	0.521	0.475	0.431	-0.431
1.080	0.830	0.898	0.867	0.828	0.783	0.735	0.685	0.634	0.584	0.534	0.485	-0.485
1.170	0.740	0.828	0.817	0.817	0.772	0.740	0.703	0.662	0.618	0.570	0.520	-0.520
1.260	0.650	0.751	0.753	0.747	0.734	0.714	0.688	0.656	0.618	0.574	0.525	-0.525
1.350	0.560	0.669	0.681	0.684	0.681	0.670	0.652	0.627	0.594	0.555	0.508	-0.508
1.440	0.470	0.585	0.604	0.614	0.618	0.614	0.602	0.583	0.556	0.521	0.478	-0.478
1.530	0.380	0.500	0.523	0.540	0.548	0.550	0.544	0.530	0.508	0.477	0.438	-0.438
1.620	0.290	0.414	0.441	0.462	0.475	0.481	0.480	0.471	0.453	0.428	0.393	-0.393
1.710	0.200	0.327	0.358	0.382	0.400	0.410	0.413	0.408	0.395	0.374	0.344	-0.344
1.800	0.110	0.240	0.274	0.301	0.322	0.336	0.343	0.343	0.334	0.318	0.293	-0.293
1.890	0.020	0.153	0.189	0.220	0.244	0.262	0.273	0.276	0.272	0.260	0.240	-0.240
1.980	0.000	0.065	0.104	0.138	0.165	0.186	0.201	0.208	0.209	0.201	0.187	-0.187
2.070	0.000	0.024	0.049	0.074	0.096	0.116	0.131	0.141	0.145	0.142	0.133	-0.133
2.160	0.000	0.009	0.030	0.045	0.060	0.073	0.083	0.091	0.094	0.093	0.087	-0.087
2.250	0.000	0.006	0.019	0.029	0.038	0.046	0.053	0.058	0.060	0.059	0.055	-0.055
2.340	0.000	0.004	0.012	0.018	0.024	0.030	0.034	0.037	0.038	0.038	0.035	-0.035
2.430	0.000	0.002	0.008	0.012	0.015	0.019	0.022	0.023	0.024	0.024	0.022	-0.022
2.520	0.000	0.002	0.005	0.007	0.010	0.012	0.014	0.015	0.016	0.015	0.014	-0.014
2.610	0.000	0.001	0.003	0.005	0.006	0.008	0.009	0.010	0.010	0.010	0.009	-0.009
2.700	0.000	0.001	0.002	0.003	0.005	0.005	0.006	0.006	0.006	0.006	0.006	-0.006
2.790	0.000	0.000	0.001	0.002	0.003	0.003	0.004	0.004	0.004	0.004	0.004	-0.004
2.880	0.000	0.000	0.001	0.001	0.002	0.002	0.002	0.002	0.003	0.003	0.002	-0.002
2.970	0.000	0.000	0.001	0.001	0.001	0.001	0.001	0.002	0.002	0.002	0.002	-0.002

RESERVOIR SIMULATION CLASS   DR.H.B.CRICHLOW

PROJECT ENGINEER .. FORREST F. CRAIG, III   UNIV. OF OKLA.

EXPLICIT   SOLUTION OF POLYMER CONCENTRATION EQUATION

```
GRID INTERVAL=   0.10      TIME STEP = 0.0060
DIFFUSION CONST=  1.00     VELOCITY = 1.0000
TOTAL TIME  =     3.00     NO OF POINTS=  11
```

POLYMER CONCENTRATION IN X DIRECTION

```
TIME       1        2        3        4        5        6        7        8        9        10        11      CPCX

                  0.062    0.047    0.024    0.022    0.007    0.010    0.001    0.003   -0.000     0.002    -0.002
0.090   0.09C
0.180   0.18C     0.237   -C.074    C.355   -0.269    0.428   -0.370    0.445   -0.409    0.460    -0.469     0.469
0.270   0.27C    -8.490   18.063  -26.927   36.123  -44.318   52.386  -59.717   67.166  -74.572   82.613   -82.613
0.360   0.36C1058.717*****3397.308******5893.520*****8425.113*********************************************************
0.450   0.450****************************************************************************************************************
0.540   0.540*************************************************************************************************************
0.630   0.630*************************************************************************************************************
0.720   0.720*************************************************************************************************************
0.810   0.810*************************************************************************************************************
0.900   0.900*************************************************************************************************************
0.990   0.990*************************************************************************************************************
1.080   0.830*************************************************************************************************************
1.170   0.740*************************************************************************************************************
1.260   0.650*************************************************************************************************************
1.350   0.560*************************************************************************************************************
1.440   0.470*************************************************************************************************************
1.530   0.380*************************************************************************************************************
1.620   0.290*************************************************************************************************************
1.710   0.200*************************************************************************************************************
1.800   0.110*************************************************************************************************************
1.890   0.020*************************************************************************************************************
1.980   0.000*************************************************************************************************************
2.070   0.000*************************************************************************************************************
2.160   0.000*************************************************************************************************************
2.250   0.000*************************************************************************************************************
2.340   0.000*************************************************************************************************************
2.430   0.000*************************************************************************************************************
2.520   0.000*************************************************************************************************************
2.610   0.000*************************************************************************************************************
2.700   0.000*************************************************************************************************************
2.790   0.000*************************************************************************************************************
2.880   0.000*************************************************************************************************************
2.970   0.000*************************************************************************************************************
```

$\Delta t > \Delta t_{max}$ . . unstable

PROJECT ENGINEER .. FORREST F. CRAIG, III UNIV. OF OKLA.

IMPLICIT   SOLUTION OF POLYMER CONCENTRATION EQUATION

GRID INTERVAL=   0.10   TIME STEP = 0.0030
DIFFUSION CONST= 1.00   VELOCITY = 1.0000
TOTAL TIME =   3.00   NO OF POINTS= 11

POLYMER CONCENTRATION IN X DIRECTION

TIME	1	2	3	4	5	6	7	8	9	10	11	DPDX
0.090	0.090	0.096	0.066	0.044	0.029	0.018	0.011	0.006	0.004	0.002	0.002	-0.002
0.180	0.180	0.216	0.169	0.131	0.100	0.075	0.055	0.041	0.030	0.024	0.020	-0.020
0.270	0.270	0.340	0.283	0.233	0.190	0.154	0.124	0.100	0.082	0.068	0.060	-0.060
0.360	0.360	0.467	0.401	0.343	0.291	0.246	0.207	0.175	0.149	0.129	0.115	-0.115
0.450	0.450	0.596	0.524	0.458	0.399	0.346	0.300	0.261	0.228	0.201	0.180	-0.180
0.540	0.540	0.726	0.648	0.576	0.511	0.452	0.399	0.353	0.313	0.279	0.251	-0.251
0.630	0.630	0.857	0.774	0.697	0.626	0.561	0.502	0.450	0.403	0.362	0.327	-0.327
0.720	0.720	0.988	0.900	0.819	0.743	0.673	0.608	0.549	0.496	0.448	0.405	-0.405
0.810	0.810	1.119	1.028	0.942	0.861	0.785	0.715	0.650	0.590	0.535	0.485	-0.485
0.900	0.900	1.251	1.155	1.065	0.980	0.899	0.823	0.752	0.686	0.624	0.566	-0.566
0.990	0.990	1.382	1.283	1.189	1.099	1.013	0.932	0.855	0.783	0.713	0.647	-0.647
1.080	0.920	1.346	1.299	1.240	1.173	1.101	1.026	0.950	0.875	0.800	0.727	-0.727
1.170	0.830	1.241	1.224	1.196	1.157	1.108	1.053	0.991	0.924	0.853	0.778	-0.778
1.260	0.740	1.125	1.128	1.119	1.099	1.069	1.030	0.981	0.924	0.859	0.785	-0.785
1.350	0.650	1.003	1.020	1.025	1.019	1.003	0.976	0.938	0.889	0.830	0.760	-0.760
1.440	0.560	0.877	0.904	0.920	0.925	0.919	0.901	0.872	0.832	0.779	0.714	-0.714
1.530	0.470	0.749	0.784	0.808	0.821	0.823	0.814	0.793	0.760	0.714	0.656	-0.656
1.620	0.380	0.620	0.661	0.692	0.712	0.721	0.719	0.705	0.679	0.640	0.588	-0.588
1.710	0.290	0.490	0.536	0.572	0.599	0.614	0.618	0.611	0.592	0.560	0.515	-0.515
1.800	0.200	0.359	0.410	0.451	0.483	0.504	0.514	0.513	0.501	0.476	0.439	-0.439
1.890	0.110	0.228	0.283	0.329	0.366	0.392	0.408	0.413	0.407	0.389	0.360	-0.360
1.980	0.020	0.097	0.156	0.206	0.247	0.279	0.300	0.312	0.312	0.302	0.280	-0.280
2.070	0.000	0.036	0.074	0.111	0.145	0.175	0.197	0.212	0.218	0.213	0.199	-0.199
2.160	0.000	0.022	0.045	0.069	0.091	0.110	0.126	0.137	0.142	0.140	0.131	-0.131
2.250	0.000	0.014	0.029	0.044	0.058	0.070	0.081	0.088	0.091	0.090	0.084	-0.084
2.340	0.000	0.009	0.018	0.028	0.037	0.045	0.052	0.056	0.058	0.058	0.054	-0.054
2.430	0.000	0.006	0.012	0.018	0.024	0.029	0.033	0.036	0.037	0.037	0.034	-0.034
2.520	0.000	0.004	0.008	0.011	0.015	0.018	0.021	0.023	0.024	0.024	0.022	-0.022
2.610	0.000	0.002	0.005	0.007	0.010	0.012	0.014	0.015	0.015	0.015	0.014	-0.014
2.700	0.000	0.002	0.003	0.005	0.006	0.008	0.009	0.009	0.010	0.010	0.009	-0.009
2.790	0.00C	0.001	0.002	0.003	0.004	0.005	0.006	0.006	0.006	0.006	0.006	-0.006
2.880	0.000	0.001	0.001	0.002	0.003	0.003	0.004	0.004	0.004	0.004	0.004	-0.004
2.970	0.000	0.000	0.001	0.001	0.002	0.002	0.002	0.002	0.003	0.003	0.002	-0.002

RESERVOIR SIMULATION CLASS   DR.H.B.CRICHLOW

PROJECT ENGINEER .. FORREST F. CRAIG, III   UNIV. OF OKLA.

IMPLICIT   SOLUTION OF POLYMER CONCENTRATION EQUATION

```
GRID INTERVAL=    0.10        TIME STEP =  0.0900
DIFFUSION CONST=  1.00        VELOCITY =   1.0000
TOTAL TIME =      3.00        NO OF POINTS=    11
```

POLYMER CONCENTRATION IN X DIRECTION

TIME	1	2	3	4	5	6	7	8	9	10	11	DPDX
0.090	0.090	0.118	0.078	0.051	0.034	0.022	0.015	0.010	0.007	0.005	0.004	-0.004
0.180	0.180	0.231	0.188	0.147	0.112	0.084	0.063	0.047	0.035	0.028	0.024	-0.024
0.270	0.270	0.363	0.300	0.248	0.204	0.167	0.135	0.109	0.090	0.075	0.066	-0.066
0.360	0.360	0.485	0.422	0.361	0.307	0.260	0.220	0.187	0.160	0.139	0.123	-0.123
0.450	0.450	0.619	0.542	0.476	0.416	0.362	0.315	0.274	0.240	0.212	0.190	-0.190
0.540	0.540	0.745	0.669	0.595	0.528	0.468	0.415	0.368	0.327	0.292	0.262	-0.262
0.630	0.630	0.879	0.793	0.716	0.644	0.579	0.519	0.465	0.417	0.375	0.339	-0.339
0.720	0.720	1.008	0.921	0.838	0.761	0.690	0.625	0.565	0.510	0.461	0.417	-0.417
0.810	0.810	1.141	1.047	0.961	0.880	0.803	0.732	0.666	0.605	0.549	0.497	-0.497
0.900	0.900	1.271	1.176	1.085	0.998	0.917	0.841	0.769	0.701	0.638	0.579	-0.579
0.990	0.990	1.404	1.303	1.209	1.118	1.032	0.950	0.872	0.798	0.728	0.660	-0.660
1.080	0.920	1.324	1.294	1.242	1.178	1.107	1.033	0.958	0.883	0.809	0.735	-0.735
1.170	0.830	1.230	1.209	1.184	1.151	1.107	1.054	0.994	0.928	0.857	0.781	-0.781
1.260	0.740	1.101	1.114	1.107	1.089	1.061	1.024	0.978	0.923	0.858	0.756	-0.756
1.350	0.650	0.987	1.000	1.009	1.006	0.992	0.967	0.930	0.883	0.825	0.707	-0.707
1.440	0.560	0.854	0.887	0.904	0.910	0.905	0.889	0.862	0.823	0.771	0.646	-0.646
1.530	0.470	0.731	0.763	0.790	0.805	0.809	0.800	0.781	0.749	0.704	0.577	-0.577
1.620	0.380	0.597	0.643	0.673	0.694	0.705	0.704	0.691	0.666	0.628	0.504	-0.504
1.710	0.290	0.471	0.515	0.553	0.581	0.597	0.602	0.596	0.578	0.547	0.426	-0.426
1.800	0.200	0.336	0.391	0.432	0.464	0.486	0.498	0.498	0.486	0.463	0.347	-0.347
1.890	0.110	0.208	0.262	0.309	0.347	0.374	0.391	0.397	0.392	0.376	0.267	-0.267
1.980	0.020	0.074	0.136	0.186	0.228	0.260	0.283	0.295	0.297	0.287	0.188	-0.188
2.070	0.000	0.037	0.067	0.102	0.135	0.164	0.186	0.200	0.206	0.202	0.123	-0.123
2.160	0.000	0.018	0.043	0.064	0.084	0.102	0.117	0.128	0.133	0.131	0.078	-0.078
2.250	0.000	0.007	0.025	0.040	0.054	0.065	0.075	0.081	0.084	0.083	0.050	-0.050
2.340	0.000	0.003	0.015	0.026	0.034	0.041	0.047	0.052	0.054	0.053	0.031	-0.031
2.430	0.000	0.002	0.010	0.016	0.022	0.026	0.030	0.033	0.034	0.034	0.020	-0.020
2.520	0.000	0.000	0.008	0.010	0.014	0.017	0.019	0.021	0.022	0.021	0.013	-0.013
2.610	0.000	-0.000	0.004	0.007	0.009	0.011	0.012	0.013	0.014	0.014	0.008	-0.008
2.700	0.000	-0.000	0.003	0.004	0.005	0.007	0.008	0.008	0.009	0.009	0.005	-0.005
2.790	0.000	-0.000	0.001	0.003	0.004	0.004	0.005	0.005	0.006	0.005	0.003	-0.003
2.880	-0.000	-0.000	-0.001	0.002	0.002	0.003	0.003	0.003	0.004	0.003	0.002	-0.002
2.970	-0.000	-0.001	-0.002	0.001	0.001	0.002	0.002	0.002	0.002	0.002	0.001	-0.001
3.060	-0.000	-0.000	-0.001	0.001	0.001	0.001	0.001	0.001	0.001	0.001	0.001	-0.001

SUCCESSFUL END OF RUN
SUCCESSFUL END OF RUN

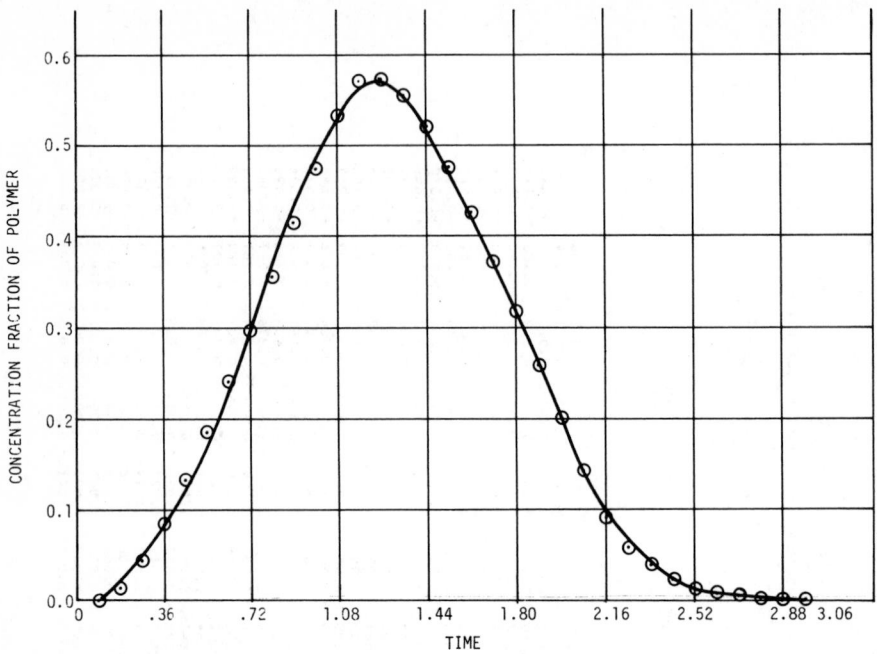

BREAKOUT CURVE FOR LINEAR MODEL, EXPLICIT METHOD - V = 1.00, Δt = 0.0030.

WORK PROBLEMS

I. *Finite-Difference Formulations*

(a) Set up the finite-difference equations for:

(1) $\dfrac{\partial \mu}{\partial t}$ (2) $\dfrac{\partial \mu}{\partial x}$

(3) $\dfrac{\partial P}{\partial x} + hP = 0$ (4) $a\dfrac{\partial P}{\partial y} + b\dfrac{\partial P}{\partial y} = C$

(5) $\dfrac{\partial^2 P}{\partial x^2}$ (6) $\dfrac{\partial^2 P}{\partial x^2} + \dfrac{\partial P}{\partial x} = 0$

(7) $\dfrac{\partial^2 P}{\partial x^2} + \dfrac{\partial^2 P}{\partial y^2} + a\dfrac{\partial P}{\partial x} + b\dfrac{\partial P}{\partial y} = 0$

(8) $\dfrac{\partial^2 P}{\partial x^2} = \dfrac{\partial^2 P}{\partial y^2}$

II. *Explicit Formulation*

(a) Set up the explicit formulation for the following equations:

(1) $\dfrac{\partial^2 P}{\partial x^2} = C\dfrac{\partial P}{\partial t}$

(2) $a\dfrac{\partial^2 P}{\partial x^2} + b\dfrac{\partial P}{\partial x} = C\dfrac{\partial P}{\partial t}$

(3) $\dfrac{\partial^2 P}{\partial x^2} + \dfrac{\partial^2 P}{\partial y^2} = \dfrac{\partial P}{\partial t}$

(4) $k_{(x)}\dfrac{\partial^2 P}{\partial x^2} = \alpha\dfrac{\partial P}{\partial t}$

(5) $k_{(x)}\dfrac{\partial^2 P}{\partial x^2} = \alpha_{(t)}\dfrac{\partial P}{\partial t}$

(6) $k_{(x)}\dfrac{\partial^2 P}{\partial x^2} + k_{(y)}\dfrac{\partial^2 P}{\partial y^2} = \alpha_{(t)}\dfrac{\partial P}{\partial t}$

III. *Implicit Formulation*

(a) Set up the implicit formulation for:

$$\dfrac{\partial^2 P}{\partial x^2} = \dfrac{\partial P}{\partial t}$$

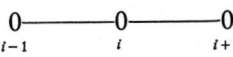

$$
\begin{array}{ccc}
0 & 0 & 0 \\
i-1 & i & i+1
\end{array}
$$

Fill in the matrix system below:

$$\begin{bmatrix} \quad \\ \quad \end{bmatrix} * \begin{bmatrix} \quad \\ \quad \end{bmatrix} = \begin{bmatrix} \quad \\ \quad \end{bmatrix}$$

(b) Set up the implicit formulation for:

$$a\dfrac{\partial^2 P}{\partial x^2} + b\dfrac{\partial^2 P}{\partial y^2} = \alpha\dfrac{\partial P}{\partial t}$$

Fill in the matrix system below:

$$\begin{bmatrix} \quad \\ \quad \end{bmatrix} * \begin{bmatrix} \quad \\ \quad \end{bmatrix} = \begin{bmatrix} \quad \\ \quad \end{bmatrix}$$

(c) Set up the implicit formulation for:

$$k_{(x)}\dfrac{\partial^2 P}{\partial x^2} + k_{(y)}\dfrac{\partial^2 P}{\partial y^2} + k_{(z)}\dfrac{\partial^2 P}{\partial z^2} = \alpha_{(t)}\dfrac{\partial P}{\partial t}$$

Fill in the matrix system below:

$$\begin{bmatrix} \quad \\ \quad \end{bmatrix} * \begin{bmatrix} \quad \\ \quad \end{bmatrix} = \begin{bmatrix} \quad \\ \quad \end{bmatrix}$$

REFERENCES

1. G. E. Forsythe, and W. R. Wasow, *Finite Difference Methods for Partial Differential Equations* (New York: Wiley, 1960).

2. R. D. Richtmyer, *Difference Methods for Initial Value Problems* (New York: Interscience, 1957).

3. J. H. Bramble, ed., *Numerical Solution of Partial Differential Equations*, Proc. of Symposium at University of Maryland, May 3–8, 1965 (New York: Academic Press, 1961).

4. J. Crank and P. Nicolson, "A Practical Method for the Numerical Evaluation of Solutions of Partial Differential Equations of the Heat Conduction Type," *Proc. Camb. Phil. Soc.* (1947), **43**, 50–67.

5. P. D. Lax, and R. D. Richtmyer, "Survey of the Stability of Linear Finite Difference Equations," *Comm. Pure Appl. Math.* (1956), **9**, 267–93.

6. J. Douglas, Jr., "The Effect of Round-off Error in the Numerical Solution of the Heat Equation," *J. ACM* (1959), **6**, 48–58.

7. E. C. DuFort, and S. P. Frankel, "Stability Conditions in the Numerical Treatment of Parabolic Differential Equations," *Math. Tables Aids Comput.* (1953), **7**, 135–52.

8. G. D. Smith, *Numerical Solution of Partial Differential Equations* (Oxford: Oxford University Press, 1965).

BIBLIOGRAPHY

Blair, P. M., and C. F. Weinaug, "Solution of Two-phase Flow Problems Using Implicit Difference Equations," *Soc. Pet. Eng. J.* (Dec. 1969), 417–24; *Trans. AIME*, **246**.

Coats, K. H., W. D. George, Chieh Chu, and B. E. Marcum, "Three-dimensional Simulation of Steamflooding," SPE 4500, 48th Annual Meeting, Las Vegas, Nev., Sept. 30–Oct. 3, 1973.

Culham, W. E., and Richard S. Varga, "Numerical Methods for Time-dependent, Nonlinear Boundary Value Problems," SPE 2806, Second Symposium on Numerical Simulation of Reservoir Performance, Dallas, Tex., Feb. 5–6, 1970.

Douglas, J., Jr., D. W. Peaceman, and H. H. Rachford, Jr., "Calculation of Unsteady-state Gas Flow Within a Square Drainage Area," *J. Pet. Tech.* (Nov. 1955), 190–95.

Fagin, R. G., and C. H. Stewart, Jr., "A New Approach to the Two-dimensional Multiphase Reservoir Simulator," *Soc. Pet. Eng. J.* (June 1966), 175–82; *Trans. AIME*, **237**.

Forsythe, G. E., and W. R. Wasow, *Finite Difference Methods for Partial Differential Equations.* (New York: Wiley, 1960).

FOURNIER, K. P., "A Numerical Method for Computing Recovery of Oil by Hot Water Injection in a Radial System," *Soc. Pet. Eng. J.* (June 1965), 131–40.

GARDER, A. O., JR., D. W. PEACEMAN, and A. L. POZZI, JR., "Numerical Calculation of Multi-dimensional Miscible Displacement by the Method of Characteristics," *Soc. Pet Eng. J.* (March 1964), 26–36.

GARRETT, J. E., and H. C. OSBORNE, "Use of Eight Flux Streams Per Cell in a Two-dimensional, Three-phase, Unsteady-state Reservoir Simulation Model," SPE 2026, First Symposium on Numerical Simulation of Reservoir Performance, Dallas, Texas, April 22–23, 1968.

GOTTFRIED, B. S., W. H. GUILINGER, and R. W. SNYDER, "Numerical Solutions of the Equations for One-dimensional Multi-phase Flow in Porous Media," *Soc. Pet. Eng. J.* (March 1966), 62–72; *Trans. AIME*, **237**.

HENSON, W. L., P. L. WEARDEN, and J. D. RICE, "A Numerical Solution to the Unsteady-state Partial-water-drive Reservoir Performance Problem," *Soc. Pet. Eng. J.* (Sept. 1961), 184–94; *Trans. AIME*, **222**.

HIRASAKI, G. J., and P. M. O'DELL, "Representation of Reservoir Geometry for Numerical Simulation," *Soc. Pet. Eng. J.* (Dec. 1970), 393–404; *Trans. AIME*, **249**.

JAVANDEL, I., and P. A. WITHERSPOON, "Application of the Finite Element Method to Transient Flow in Porous Media," *Soc. Pet. Eng. J.* (Sept. 1968), 241–52; *Trans. AIME*, **243**.

LANTZ, R. B., "Quantitative Evaluation of Numerical Diffusion (Truncation Error)," *Soc. Pet. Eng. J.* (Sept. 1971), 315–20; *Trans. AIME*, **251**.

LAUMBACH, DALLAS D., "A High Accuracy Finite-difference Technique for Treating the Convection-Diffusion Equation," SPE 3996, 47th Annual Meeting, San Antonio, Texas, Oct. 8–11, 1972.

LEE, E. H., and F. J. FAYERS, "The Use of the Method of Characteristics in Determining Boundary Conditions for Problems in Reservoir Analysis," *Trans. AIME* (1959), **216**, 284–89.

PEACEMAN, D. W., "Improved Treatment of Dispersion in Numerical Calculation of Multidimensional Miscible Displacement," *Soc. Pet. Eng. J.* (Sept. 1966), 213–16; *Trans. AIME*, **237**.

PEACEMAN, D. W., and H. H. RACHFORD, JR., "Numerical Calculation of Multidimensional Miscible Displacement," *Soc. Pet. Eng. J.* (Dec. 1962), 327–39.

PRICE, H. S., J. C. CAVENDISH, and R. S. VARGA, "Numerical Methods of Higher-order Accuracy for Diffusion-Convection Equations," *Soc. Pet. Eng. J.* (Sept. 1968), 293–303; *Trans. AIME*, **243**.

SETTARI, A., and K. AZIZ, "Use of Irregular Grid in Reservoir Simulation," *Soc. Pet. Eng. J.* (April 1972), 103–14.

SHUM, Y. M., "Use of the Finite-element Method in the Solution of Diffusion-Convection Equations," *Soc. Pet. Eng. J.* (June 1971), 139–44.

SMITH, G. D., *Numerical Solution of Partial Differential Equations* (Oxford: Oxford University Press, 1965).

SONIER, F., and P. CHAUMET, "A Fully Implicit Three-dimensional Model in Curvilinear Coordinates," SPE 4543, 48th Annual Meeting, Las Vegas, Nev., Sept. 30–Oct. 3, 1973.

STONE, H. L., "Iterative Solution of Implicit Approximations of Multidimensional Partial Differential Equation," *SIAM Numer. Anal.* (1968), **5,** 530.

TSUTSUMI, GINJIRO, and THOMAS N. DIXON, "Mathematical Simulation of Two-phase Flow with Interphase Mass Transfer in Petroleum Reservoirs," SPE 4074, 47th Annual Meeting, San Antonio, Texas, Oct. 8–11, 1972.

5 Solution of the Simulator Equations

5.1 THE SOLUTION PROCESS

Having derived the equations for the simultaneous flow of all fluid phases, we must now try to solve the system of equations for the necessary unknown parameters. The unknown values are the following:

1. Oil pressure
2. Gas pressure
3. Water pressure
4. Oil saturation
5. Water saturation
6. Gas saturation

Other quantities can be *derived* from these variables—namely,

1. Oil production rate
2. Gas production rate
3. Water production

The solution process depends to some extent on the particular system being modeled—i.e., areal sweep or coning, single-well or multi-well, etc.

There are two basic means of solving the simulator equations as shown on the flowchart in Fig. 5.1: the *implicit pressure–explicit saturation* and the *implicit pressure–implicit saturation* methods. The formulation of the equations differs for the two cases, and the methods' degrees of sophistication vary considerably. A summary of each method is given on pp. 98–109.

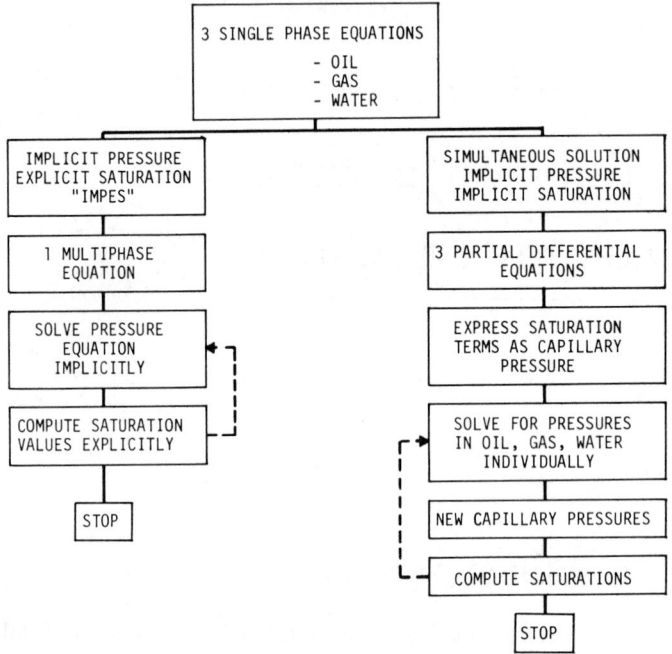

Figure 5.1: Summary of solution processes.

5.2 IMPLICIT PRESSURE–EXPLICIT SATURATION (IMPES) METHOD

Introduction

The IMPES method combines the single-phase equations into a single multiphase equation based on pressure, then solves the pressure equation implicitly for the pressure distribution. The saturation distribution is then explicitly calculated for each point. The IMPES process is formulated as follows.

We start with each single-phase equation in a single dimension:

$$A_x \frac{\partial}{\partial x}\left(\frac{k_o}{\mu_o B_o}\frac{\partial \Phi_o}{\partial x}\right) + q_o = V_R \frac{\partial}{\partial t}\left(\phi \frac{S_o}{B_o}\right) \tag{5.1}$$

$$A_x \frac{\partial}{\partial x}\left(\frac{k_w}{\mu_w B_w}\frac{\partial \Phi_o}{\partial x}\right) + q_w = V_R \frac{\partial}{\partial t}\left(\phi \frac{S_w}{B_w}\right) \tag{5.2}$$

$$A_x \frac{\partial}{\partial x}\left[\left(\frac{k_g}{\mu_g B_g} + \frac{R_{s_o}k_o}{\mu_o B_o} + \frac{R_{s_w}k_w}{\mu_w B_w}\right)\frac{\partial \Phi_g}{\partial x}\right] + q_g$$

$$= V_R \frac{\partial}{\partial t}\left[\phi\left(\frac{S_g}{B_g} + \frac{R_{s_o}S_o}{B_o} + \frac{R_{s_w}S_w}{B_w}\right)\right] \tag{5.3}$$

Equations (5.1), (5.2), and (5.3) are combined to yield a single equation relating the behavior of all phases in the reservoir. In order to make this transformation, the following additional definitions are required:
Potential terms:

$$\text{Oil} \qquad \Phi_o = P_o + \rho_o gh \qquad\qquad (5.4)$$

$$\text{Gas} \qquad \Phi_g = P_g + \rho_g gh \qquad\qquad (5.5)$$

$$\text{Water} \qquad \Phi_w = P_w + \rho_w gh \qquad\qquad (5.6)$$

Capillary terms:

$$\text{Water/oil} \qquad P_{cw} = P_o - P_w \qquad\qquad (5.7)$$

$$\text{Gas/oil} \qquad P_{cg} = P_g - P_o \qquad\qquad (5.8)$$

The simplification of Eqs. (5.1), (5.2), and (5.3) is carried out by first expanding the gas Eq. (5.3) and substituting the oil and water equations into this equation for the appropriate terms. The resulting equation contains only oil pressure, capillary pressure, and hydrostatic head terms.
One form of this equation is:

$$A_x \frac{\partial}{\partial x}\left(\lambda_T \frac{\partial P_o}{\partial x}\right) + A_x \frac{\partial}{\partial x}\left(\lambda_g \frac{\partial P_{cg}}{\partial x} - \lambda_w \frac{\partial P_{cw}}{\partial x}\right)$$

$$+ A_x \frac{\partial}{\partial x}\left[\lambda_g \frac{\partial(\rho_g gh)}{\partial x} + \lambda_o \frac{\partial(\rho_o gh)}{\partial x} + \lambda_w \frac{\partial(\rho_w gh)}{\partial x}\right] = B_1 \frac{\partial P_o}{\partial t} + B_2 \qquad (5.9)$$

where B_1 and B_2 are collections of terms that include saturation, PVT terms, and production terms, respectively. This equation is normally solved to obtain the oil pressures at each point in the reservoir, and subsequently the phase saturations and productions can be determined. Inspection of Eq. (5.9), however, indicates that there are pressure- and saturation-dependent terms. The variable λ is the mobility, which is a function of saturation and pressure:

$$\lambda_i = \frac{k_i}{\mu_i B_i} \qquad i = \text{o, g, w} \qquad\qquad (5.10)$$

A problem then arises: How can we evaluate the pressures if the solution of the equation depends on the calculation of mobilities which themselves depend on these pressures? There are two ways out of this dilemma. First, we can let all pressure- and saturation-dependent terms lag behind the pressure terms. This is done by evaluating the quantities λ, P_{og}, and P_{ow} at the previous saturations and pressures. Implied in this approach is the belief (or hope) that the saturation and pressure values are not changing very rapidly. In some cases this is justified, an example being in a large areal model study with well-distributed production rates. This approach can be illustrated qualitatively by:

$$\text{(Mobility, capillary data)}^n \text{(Pressures)}^{n+1} = \text{RHS}^{n+1} \qquad (5.11)$$

The mobility, pressure, and capillary pressure terms are evaluated at the time step level n, while the pressures and the right-hand side (RHS) Eq. 5.9 are formulated at time step level $(n + 1)$. These $(n + 1)$-level quantities are then solved for, since they comprise the unknowns. This approach is noniterative and completed in one "pass."

The second formulation uses an iterative scheme to update the pressure-, saturation-, and capillary-dependent quantities. The approach is illustrated qualitiatively by:

$$\text{(Mobility, capillary data)}^{n+1,k} \text{(Pressures)}^{n+1,k+1} = \text{RHS}^{n+1,k+1} \qquad (5.12)$$

The pressure, saturation, and capillary data at any given iteration level k in a given time step $(n + 1)$ are assumed known from the most recent computational value. At the start of a new time step, this most recent value is obviously the old time step value; however, during a given time step the most current value is the last iteration. As the pressures and saturations are solved, the updating of these coefficients continues using the new values of pressure and saturation. The iteration process terminates when the convergence criteria are satisfied. This formulation is more rigorous mathematically and is physically more appropriate, but from a practical standpoint the differences between the two approaches are significant only for a special class of problems, and the increased computational workload of the iterative method should only be borne in those situations where it is warranted. These include coning studies, cross-sectional studies, and other special-purpose cases.

Having decided on the level at which the terms in Eq. (5.9) will be evaluated, the equation is partitioned such that all known terms—i.e., those at level n—are grouped together and the resulting unknown terms—i.e., level $(n + 1)$—are together. The equation is still a nonlinear partial differential equation and as such is not amenable to solution by any known analytical techniques. Finite-difference techniques as detailed in Chapter 4 must be used to formulate and solve the partial differential equation over a rectilinear grid.

Finite-Difference Analog

The finite-difference formulation of Eq. (5.9) can be carried out by substituting for the differential equation the appropriate difference equation. First, let us assume that the noniterative approach will be used for the pressure, saturation, and capillary terms, in which case we can then use the time levels as follows:

$$A_x \frac{\partial}{\partial x}\left(\lambda_T^n \frac{\partial P_o^{n+1}}{\partial x}\right) + A_x \frac{\partial}{\partial x}\left(\lambda_g \frac{\partial P_{cg}}{\partial x} - \lambda_w \frac{\partial P_{cw}}{\partial x}\right)^n$$

$$+ A_x \frac{\partial}{\partial x}\left[\lambda_g^n \frac{\partial(\rho_g g h)}{\partial x} + \lambda_o^n \frac{\partial(\rho_o g h)}{\partial x} + \lambda_w^n \frac{\partial(\rho_w g h)}{\partial x}\right]$$

$$= B_1^n \frac{\partial P_o^{n+1}}{\partial t} + B_2^{n+1} \tag{5.13}$$

Each term above can be differenced separately; the most important term is the first term with the oil pressure at level $(n + 1)$. Since all the other terms are considered known, they can be grouped together and collectively put on the right-hand side.

The finite-difference form of the first term on the left of Eq. (5.13) is:

$$A_x \frac{\partial}{\partial x}\left(\lambda_T^n \frac{\partial P_o^{n+1}}{\partial x}\right) = A_x \left[\frac{\lambda_{T_{i+1/2}}\left(\dfrac{P_{o_{i+1}} - P_{o_i}}{\left(\dfrac{\Delta x_i + \Delta x_{i+1}}{2}\right)}\right) - \lambda_{T_{i+1/2}}\left(\dfrac{P_{o_i} - P_{o_{i-1}}}{\left(\dfrac{\Delta x_i + \Delta x_{i-1}}{2}\right)}\right)}{\Delta x_i}\right] \tag{5.14}$$

Equation (5.14) is based on the area A_x being a constant. If the area varies, the value A_x is inserted into the brackets thus:

$$\frac{\partial}{\partial x}\left(A_x \lambda_T^n \frac{\partial P_o^{n+1}}{\partial x}\right) = \frac{1}{\Delta x_i}\left[A_{x_{i+1/2}}\lambda_{T_{i+1/2}}\left(\dfrac{P_{o_{i+1}} - P_{o_i}}{\left(\dfrac{\Delta x_i + \Delta x_{i+1}}{2}\right)}\right)\right.$$

$$\left. - A_{x_{i-1/2}}\lambda_{T_{i-1/2}}\left(\dfrac{P_{o_i} - P_{o_{i-1}}}{\left(\dfrac{\Delta x_i + \Delta x_{i-1}}{2}\right)}\right)\right] \tag{5.15}$$

The mobility term $\lambda_{T_{i\pm1/2}}$ is evaluated between the contiguous cells through which flow is occurring. This term is generally selected to reflect the mobility at the upstream direction. Equations (5.14) and (5.15) can be simplified rather easily to obtain terms that comprise the mobility and the geometry and the pressure terms remain; for example, simplifying the left-hand side and equating it to the right-hand side, we obtain:

$$\chi_{i+1/2}(P_{o_{i+1}} - P_{o_i})^{n+1} - \chi_{i-1/2}(P_{o_i} - P_{o_{i-1}})^{n+1} = \frac{\partial P_{o_i}^{n+1}}{\partial t} + C^n \tag{5.16}$$

In Eq. (5.16) the χ-terms on the left contain mobility and rock geometry terms and the C^n-term on the right contains all the known quantities evaluated earlier at time level n. The time derivative of pressure is differenced as follows:

$$\frac{\partial P_{o_i}}{\partial t} = \frac{P_{o_i}^{n+1} - P_{o_i}^n}{\Delta t^n} \tag{5.17}$$

Notice that this now contains a new level pressure value. Equation (5.16) becomes:

$$\chi_{i+1/2}(P_{o_{i+1}} - P_{o_i})^{n+1} - \chi_{i-1/2}(P_{o_i} - P_{o_{i-1}})^{n+1} = \frac{P_{o_i}^{n+1} - P_{o_i}^{n}}{\Delta t_n} + C^n \qquad (5.18)$$

Grouping all $(n + 1)$-terms we obtain:

$$\chi_{i+1/2}P_{i+1}^{n+1} - \left(\chi_{i+1/2} + \chi_{i-1/2} + \frac{1}{\Delta t^n}\right)P_{o_i}^{n+1} + \chi_{i-1/2}P_{o_{i-1}}^{n+1} = -\frac{P_{o_i}^{n}}{\Delta t^n} + C^n$$

$$(5.19)$$

This is a single equation in pressure which is the finite-difference analog of the original partial differential equation written for one point i in the model. The same equation is written for every cell in the model and the pressures are computed at each time level $(n + 1)$. In one dimension there are three unknown pressures coupled together, in two dimensions there are five, and in three dimensions there are seven.

The finite-difference form of the pressure equation is solved using some suitable algorithm—e.g., ADIP, LSOR, SIP—to obtain the *pressure distribution*. The potential distribution is then computed.

From the potential distribution the new saturations are calculated as follows:

COMPUTATION OF SATURATION AT NEW LEVEL

$$\left[\frac{S_o}{B_o}\right]^{n+1} = \left[\frac{S_o}{B_o}\right]^{n} + \frac{\Delta t}{\phi}\left[\frac{\partial}{\partial x}\left[\frac{k_o}{\mu_o B_o}\frac{\partial \Phi_o}{\partial x}\right]\right]$$

$$= \frac{S_o}{B_o}^{n} + \Sigma_{i=1}^{4} \text{ Flow Terms}$$

$$\frac{\left(\phi\frac{S_o}{B_o}\right)^{n+1} - \left(\phi\frac{S_o}{B_o}\right)^{n}}{\Delta t} = \frac{\partial}{\partial t}\left(\phi\frac{S_o}{B_o}\right) = \frac{\partial}{\partial x}\left(\frac{k_o}{\mu_o B_o}\frac{\partial \Phi_o}{\partial x}\right) \qquad (5.20)$$

from which, by rearranging, the new saturation can be solved:

$$\left(\frac{S_o}{B_o}\right)^{n+1} = \left(\frac{S_o}{B_o}\right)^{n} + \frac{\Delta t}{\phi}\left[\frac{\partial}{\partial x}\left(\frac{k_o}{\mu_o B_o}\frac{\partial \Phi_o}{\partial x}\right)\right] = S_o^n + \Sigma \text{ Flux terms} \qquad (5.21)$$

Figures 5.2 through 5.5 summarize the IMPES process.

START TIME STEP:

TIME = t   STEP 1

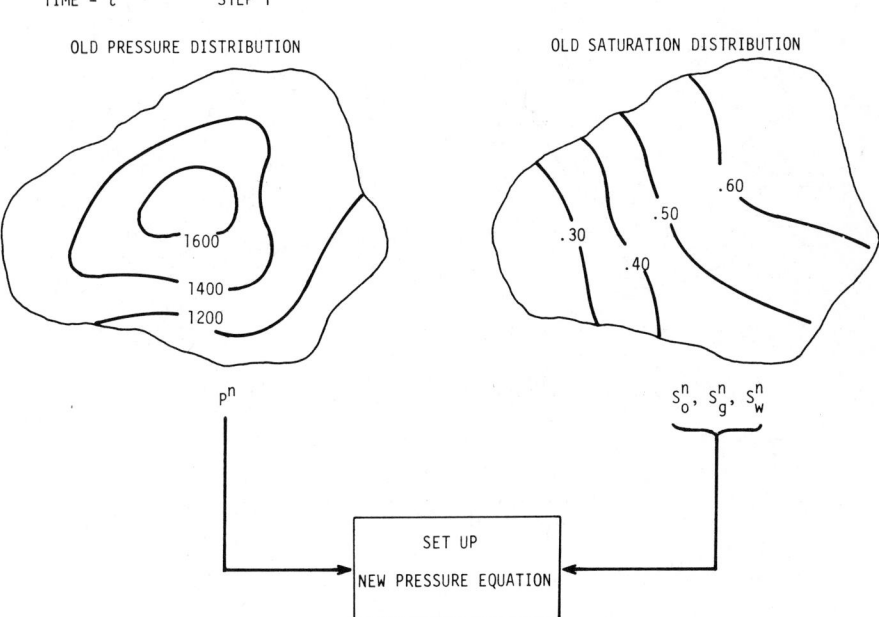

Figure 5.2: Setting up pressure equations—IMPES.

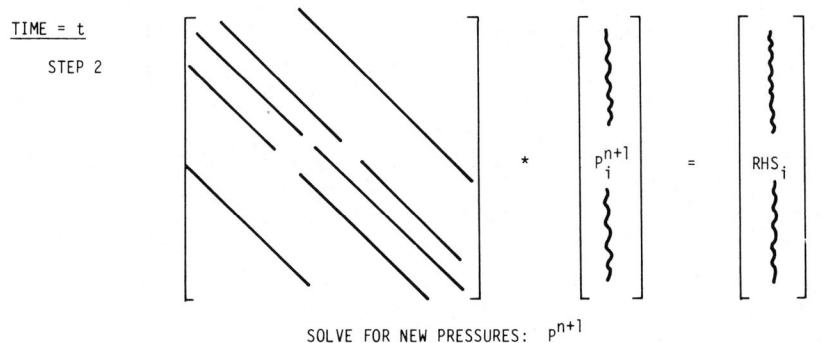

SOLVE FOR NEW PRESSURES: p^{n+1}

TIME = t + Δt

STEP 3

CALCULATE NEW SATURATIONS USING THE NEW PRESSURE GRADIENTS

$$\rightarrow S_o^{n+1}, \ S_g^{n+1}, \ S_w^{n+1}$$

Figure 5.3: Form of pressure matrix—IMPES.

103

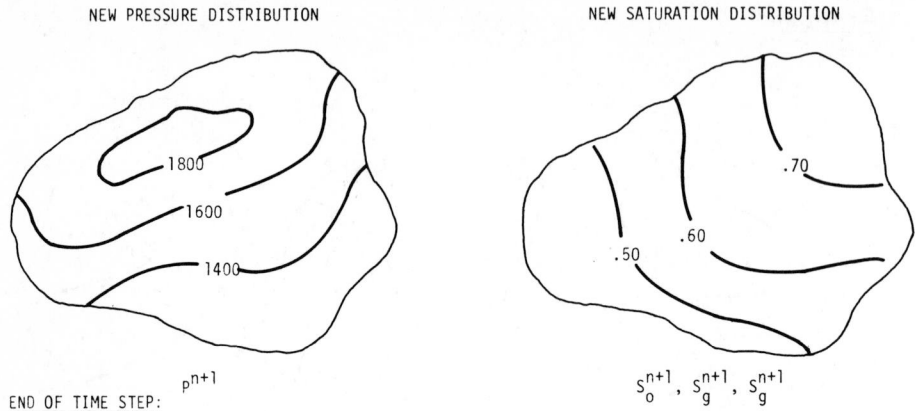

Figure 5.4: Results at end of time step—IMPES.

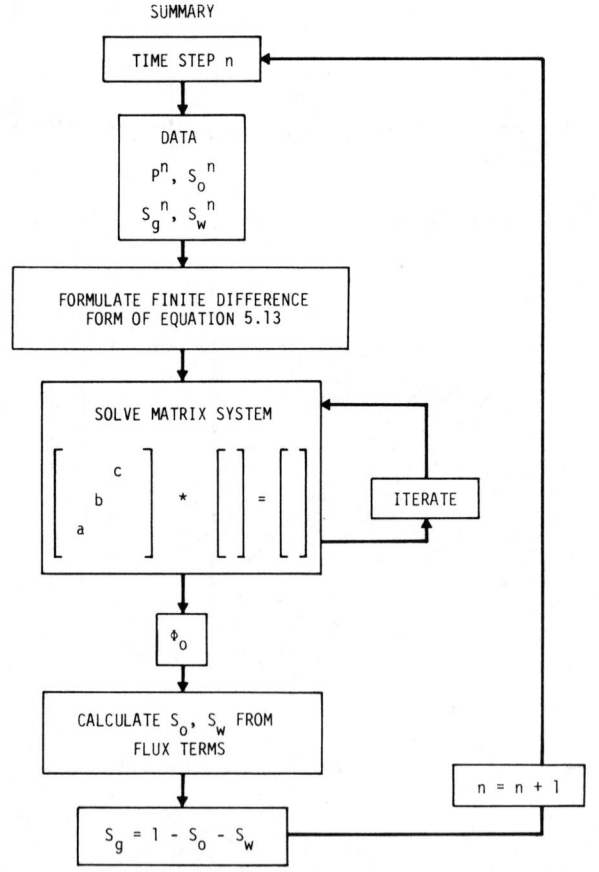

Figure 5.5: IMPES summary.

5.3 IMPLICIT PRESSURE–IMPLICIT SATURATION (SIMULTANEOUS SOLUTION) METHOD

This procedure involves the simultaneous solution of the partial differential equations for flow of oil, water, and gas to obtain the pressures in each phase. The saturations of each phase are calculated implicitly using capillary pressure relations.

This method is very complex and computationally involved, and it is suited to a special class of problems.

An outline of this method is shown below to familiarize the student with the procedure. To understand it more fully, consultation of a more advanced treatise is recommended. The following formulation is patterned after that by Douglas et al.[1]

Consider incompressible two-phase flow in a single dimension:

$$\frac{\partial}{\partial x}\left(\frac{k_o}{\mu_o}\frac{\partial \Phi_o}{\partial x}\right) = \phi\frac{\partial S_o}{\partial t} \tag{5.22}$$

$$\frac{\partial}{\partial x}\left(\frac{k_w}{\mu_w}\frac{\partial \Phi_w}{\partial x}\right) = \phi\frac{\partial S_w}{\partial t} \tag{5.23}$$

where $\Phi = p + \rho gh = $ flow potential
$h = $ height above a horizontal reference plane
$g = $ gravitational acceleration
$\rho = $ density of oil or water
$p = $ phase pressure of oil or water

Then

$$\Phi_o = P_o + \rho_o gh$$
$$\Phi_w = P_w + \rho_w gh$$

Note that Eqs. (5.22) and (5.23) are simply the one-dimensional partial differential equations governing flow of the particular phase. Further, we know that since only oil and water exist in our system,

$$S_o + S_w = 1 \tag{5.24}$$

Then:

$$S_o = 1 - S_w$$

Thus:

$$\frac{\partial S_o}{\partial t} = \frac{\partial(1 - S_w)}{\partial t} = -\frac{\partial S_w}{\partial t} \tag{5.25}$$

Some other concepts must be introduced before we move along. First, the capillary pressure at any point in the system is defined mathematically as:

$$P_c = P_o - P_w \tag{5.26}$$

which can also be written

$$P_c = \Phi_o - \Phi_w + \Delta\rho gh \tag{5.27}$$

We would like to express the saturation change in terms of the capillary pressure term and finally in terms of flow potentials. To do this we invoke the chain rule in calculus:

$$\frac{\partial S}{\partial t} = \frac{\partial S}{\partial P_c} \frac{\partial P_c}{\partial t} \tag{5.28}$$

which is written:

$$\frac{\partial S}{\partial t} = S' \frac{\partial P_c}{\partial t} \tag{5.29}$$

where $S' = $ slope of the saturation capillary pressure curve as indicated in Fig. 5.6.

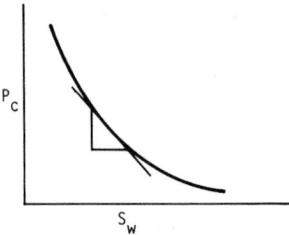

Figure 5.6: Capillary pressure saturation relation.

If we use the definition of P_c from Eq. (5.27), Eq. (5.28) becomes:

$$\frac{\partial S}{\partial t} = S' \frac{\partial P_c}{\partial t} = S' \left(\frac{\partial \Phi_o}{\partial t} - \frac{\partial \Phi_w}{\partial t} \right) \tag{5.30}$$

since

$$\frac{\partial(\Delta\rho gh)}{\partial t} = 0$$

Equations (5.22) and (5.23) can be rewritten:

$$\frac{\partial}{\partial x} \left(\frac{k_o}{\mu_o} \frac{\partial \Phi_o}{\partial x} \right) = -\phi S' \left(\frac{\partial \Phi_o}{\partial t} - \frac{\partial \Phi_w}{\partial t} \right) \tag{5.31}$$

$$\frac{\partial}{\partial x} \left(\frac{k_o}{\partial_o} \frac{\partial \Phi_w}{\partial x} \right) = \phi S' \left(\frac{\partial \phi_o}{\partial t} - \frac{\partial \Phi_w}{\partial t} \right) \tag{5.32}$$

Now S', the derivative $\partial S/\partial P_c$, can be written in terms of the potentials:

$$S' = \frac{\partial S}{\partial P_c} = \frac{S^{n+1} - S^n}{\Phi_o^{n+1} - \Phi_w^n} \tag{5.33}$$

Note that this actually takes the derivative at a time level n which lags behind the computation step. This procedure could be modified to calculate S' at $(n + \frac{1}{2})$-level. Equations (5.31), (5.32), and (5.33) are expanded by their finite-difference equivalents to set up the model over a set of discrete points.

The oil phase is shown below, where the partial differential equations are written for each cell i using a fully implicit formulation:

$$\frac{1}{\Delta x}\left[\left(\frac{k_o}{\mu_o}\right)_{i+1/2}\left(\frac{\Phi_{o_{i+1}}^{n+1} - \Phi_{o_i}^{n+1}}{\Delta x}\right) - \left(\frac{k_o}{\mu_o}\right)_{i-1/2}\left(\frac{\Phi_{o_i}^{n+1} - \Phi_{o_{i-1}}^{n+1}}{\Delta x}\right)\right]$$
$$= -\frac{\Phi S'}{\Delta t}[(\Phi_{o_i}^{n+1} - \Phi_{o_i}^n) - (\Phi_{w_i}^{n+1} - \Phi_{w_i}^n)] \tag{5.34}$$

Similarly for the water phase:

$$\frac{1}{\Delta x}\left[\left(\frac{k_w}{\mu_w}\right)_{i+1/2}\left(\frac{\Phi_{w_{i+1}}^{n+1} - \Phi_{w_i}^{n+1}}{\Delta x}\right) - \left(\frac{k_w}{\mu_w}\right)_{i-1/2}\left(\frac{\Phi_{w_i}^{n+1} - \Phi_{w_{i-1}}^{n+1}}{\Delta x}\right)\right]$$
$$= +\frac{\Phi S'}{\Delta t}[(\Phi_{o_i}^{n+1} - \Phi_{o_i}^n) - (\Phi_{w_i}^{n+1} - \Phi_{w_i}^n)] \tag{5.35}$$

Therefore, at a given cell there are *two* unknowns Φ_o^{n+1}, Φ_w^{n+1} at new time levels.

Consider Eqs. (5.34) and (5.35) written in the typical form where all the unknowns are on the left and all the knowns are on the right. Then, for the oil equation (Eq. 5.34) written for the finite-difference grid in Fig. 5.7:

$$e\Phi_{o_{i-1}}^{n+1} + f\Phi_{o_i}^{n+1} + g\Phi_{o_{i+1}}^{n+1} + h\Phi_{w_i}^{n+1} = D_{o_i} \tag{5.36}$$

where e, f, g, and h are collections of terms; these include Δx, k_o, μ_o, etc. Note that this gives three terms for oil and one for water. Similarly, the water equation (Eq. 5.35) produces:

$$a\Phi_{w_{i-1}}^{n+1} + b\Phi_{w_i}^{n+1} + c\Phi_{w_{i+1}}^{n+1} + d\Phi_{o_i}^{n+1} = D_{w_i} \tag{5.37}$$

Note that there are three water terms and one oil term.

i-1   i   i+1   i+2

Figure 5.7: One-dimensional grid.

Equations (5.36) and (5.37) written for each cell in the grid give a set of simultaneous equations within which the unknowns are Φ_o, Φ_w at a time level.

Consider a three-cell system, as shown in Fig. 5.8. Writing Eqs. (5.36) and (5.37) for each cell gives six equations with six unknowns as shown in Fig. 5.9.

Since cell 0 is fictitious, and so is cell 4, we can (by using the appropriate boundary conditions not shown here) obtain a matrix form as in Fig. 5.10.

Note the very sharp similarity to the regular tridiagonal solution used

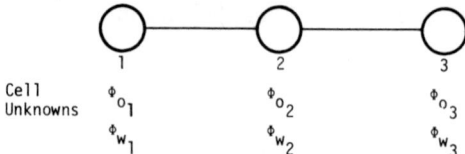

Cell
Unknowns Φ_{o_1} Φ_{o_2} Φ_{o_3}
           Φ_{w_1} Φ_{w_2} Φ_{w_3}

Figure 5.8: Three-cell system in one dimension.

CELL 1 $\begin{cases} a_1\Phi_{w_0} + b_1\Phi_{w_1} + c_1\Phi_{w_1} + d_1\Phi_{o_1} \quad \cdots\cdots\cdots\cdots\cdots\cdots\cdots\cdots\cdots\cdots = D_{w_1} \\[2ex] e_1\Phi_{o_0} + f_1\Phi_{o_1} + g_1\Phi_{o_2} + h_1\Phi_{w_1} \quad \cdots\cdots\cdots\cdots\cdots\cdots\cdots\cdots\cdots\cdots = D_{o_1} \end{cases}$

CELL 2 $----\begin{cases} a_2\Phi_{w_1} + b_2\Phi_{w_2} + c_2\Phi_{w_3} + d_2\Phi_{o_2} \quad \cdots\cdots\cdots\cdots\cdots\cdots\cdots = D_{w_2} \\[2ex] e_2\Phi_{o_1} + f_2\Phi_{o_2} + g_2\Phi_{o_3} + h_2\Phi_{w_2} \quad \cdots\cdots\cdots\cdots\cdots\cdots\cdots = D_{o_2} \end{cases}$

CELL 3 $------\begin{cases} a_3\Phi_{w_2} + b_3\Phi_{w_3} + c_3\Phi_{w_4} + d_3\Phi_{o_3} \quad \cdots\cdots\cdots\cdots\cdots = D_{w_3} \\[2ex] e_3\Phi_{o_2} + f_3\Phi_{o_3} + g_3\Phi_{o_4} + h_3\Phi_{w_3} \quad \cdots\cdots\cdots\cdots\cdots = D_{o_3} \end{cases}$

Figure 5.9: Equation system for three-cell model.

$$\begin{bmatrix} b_1 & d_1 & c_1 & 0 & & \\ h_1 & f_1 & 0 & g_1 & & \\ a_2 & 0 & b_2 & d_2 & c_2 & 0 \\ 0 & e_2 & h_2 & f_2 & 0 & g_2 \\ & & a_3 & 0 & b_3 & d_3 \\ & & 0 & e_3 & h_3 & f_3 \end{bmatrix} \star \begin{bmatrix} w_1 \\ o_1 \\ w_2 \\ o_2 \\ w_3 \\ o_3 \end{bmatrix} = \begin{bmatrix} D_{w_1} \\ D_{o_1} \\ D_{w_2} \\ D_{o_2} \\ D_{w_3} \\ D_{o_3} \end{bmatrix}$$

Figure 5.10: Matrix representation.

in the IMPES method—except where the IMPES method has numbers as entries in each diagonal element, this simultaneous method has matrices of the size 2×2. A solution method similar to the Thomas[2] algorithm is used except:

Matrix inversion occurs everywhere that division had been used.

Care must be utilized on pre- and post-multiplying.

The solution of the above matrix system gives the values of Φ_{o_i} and Φ_{w_i}, and from there the capillary pressure is calculated using Eq. (5.38):

$$P_{c_i} = \Phi_{o_i} - \Phi_{w_i} + \Delta \rho g h_i \tag{5.38}$$

From this capillary pressure data at cell i the saturation of the water could be calculated and that of the oil deduced (see Fig. 5.11). A flow chart of the simultaneous solution process is summarized in Fig. 5.12.

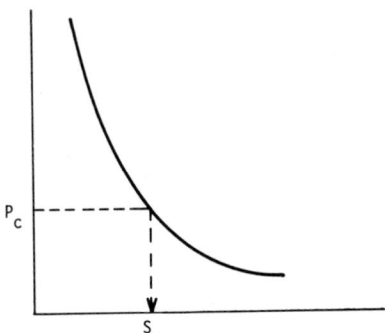

Figure 5.11: Saturation determination from capillary pressure.

5.4 UPSTREAM AND DOWNSTREAM RELATIVE PERMEABILITIES

In a reservoir simulator, flow is computed between elements or blocks of a rectangular grid. Flow between these elements is a function of two parameters, as can be seen from the following flow equation:

$$q_{1 \to 2} = k \frac{k_{ro} A}{\mu} \frac{(P_1 - P_2)}{\Delta x} \tag{5.39}$$

$$q = \lambda \, \Delta P \tag{5.40}$$

where λ is a mobility term and ΔP is the pressure difference. These terms *must be nonzero* for any fluid movement.

SUMMARY

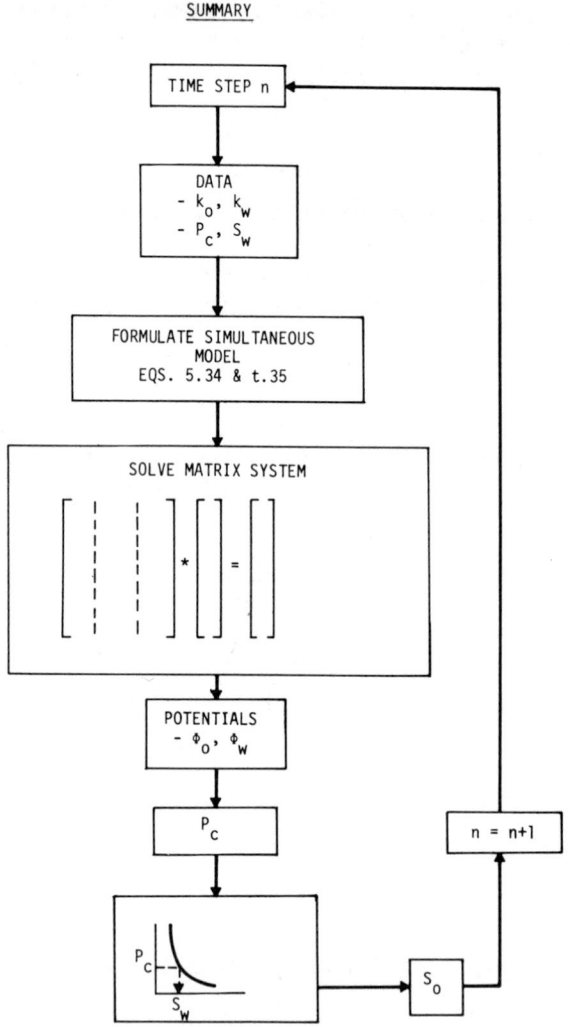

Figure 5.12: Simultaneous solution summary.

Consider the adjacent cells i and $(i + 1)$ of the reservoir model shown in Fig. 5.13. Assuming there is an arbitrary pressure distribution such that $P_i > P_{i+1}$, the P_i is called the *upstream cell* and P_{i+1} the *downstream cell*. Flow of any fluid would then be from cell i to cell $(i + 1)$. However, we are still faced with the problem of determining what mobility should be used in computing flow.

i	i+1

Figure 5.13: Cell configuration.

There are several possible mobilities:

1. *Upstream*—where the fluid mobility is based on the relative permeability and viscosity data computed from fluid saturations and pressures in cell i (the upstream cell).
2. *Downstream*—where the fluid mobility is based on the relative permeability and viscosity data computed from fluid saturations and pressures in cell $(i + 1)$ (the downstream cell).
3. *Average*—a weighted average in which the mobility is a function of both the i- and $(i + 1)$-cells.
4. *Harmonic*—a harmonic average value where the mobilities are averaged as series resistances.

The flow equations will be of the following type:

$$q = \lambda_i(P_i - P_{i+1}) \qquad\qquad \text{Upstream} \qquad (5.41)$$

$$q = \lambda_{i+1}(P_i - P_{i+1}) \qquad\qquad \text{Downstream} \qquad (5.42)$$

$$q = [w\lambda_i + (1 - w)\lambda_{i+1}](P_i - P_{i+1}) \quad \text{Weighted} \qquad (5.43)$$

$$q = \frac{\lambda_i\lambda_{i+1}}{(\lambda_i + \lambda_{i+1})}(P_i - P_{i+1}) \qquad \text{Harmonic} \qquad (5.44)$$

where w is a weighting parameter. Note that when $w = 1$, the third equation reverts to the upstream, and when $w = 0$, the third equation reverts to the downstream value.

In practice the downstream mobility is rarely used. The reason is illustrated below for two cells (see Fig. 5.14). Consider the following:

$$P_i > P_{i+1}$$

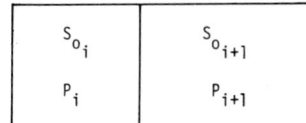

Figure 5.14: Downstream mobility.

The flow would be from P_i to P_{i+1}. Now let $S_{o_{i+1}} > S_{o_c}$ and $S_{o_i} < S_{o_c}$; e.g., the ith cell may be behind a flood front in a water drive. The downstream mobility which will be evaluated at $S_{o_{i+1}}$ will be nonzero:

$$\lambda_{i+1} > 0$$

Then from the flow equation:

$$q = \lambda_{i+1}\, \Delta P \qquad (5.45)$$

Oil will flow from cell i to cell $(i + 1)$ even though the oil saturation is less than the critical.

If the upstream permeability were used, then

$$q = \lambda_i \Delta P \tag{5.46}$$

However, λ_i is computed from S_{o_i}, which is less than the critical:

$$\lambda_i = 0 \tag{5.47}$$

$$q = 0 \tag{5.48}$$

Thus, no flow occurs from cell i to $(i + 1)$ as we should expect in the physical system.

The harmonic average leads to erroneous flow rates when there exist sharp changes in saturation. Consider an extreme case (Fig. 5.15):

$$\left. \begin{array}{l} P_i > P_{i+1} \\ k_{ro} = 0.6 \\ kr_{o_{i+1}} = 0 \end{array} \right\} \; S_{o_i} \gg S_{o_{i+1}} \tag{5.49}$$

Then the flow equation will be:

$$\begin{aligned} q &= \frac{\lambda_i \lambda_{i+1}}{\lambda_i + \lambda_{i+1}} (P_i - P_{i+1}) \\ &= \frac{0.6 * 0}{0.6} (P_i - P_{i+1}) \\ &= 0 \end{aligned} \tag{5.50}$$

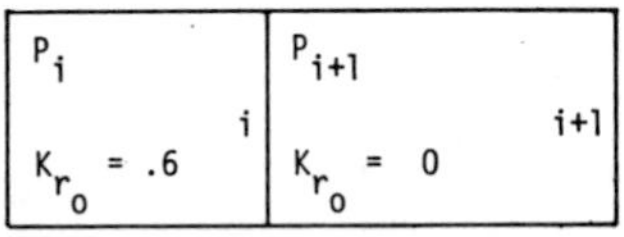

Figure 5.15: Harmonic average mobility.

This is unrealistic, since the flow will be zero, and in the physical system there will be considerable flow from i to $(i + 1)$.

The upstream weighting is inherently more reliable, and this has been shown in two ways:

1. Comparison with known analytical solutions for simple cases using Buckley Leverett approach

2. Comparison with experimental data—the upstream weighting seems to produce more physically consistent results.

Examples of the upstream and midpoint weighting are shown in Figs. 5.16 and 5.17.

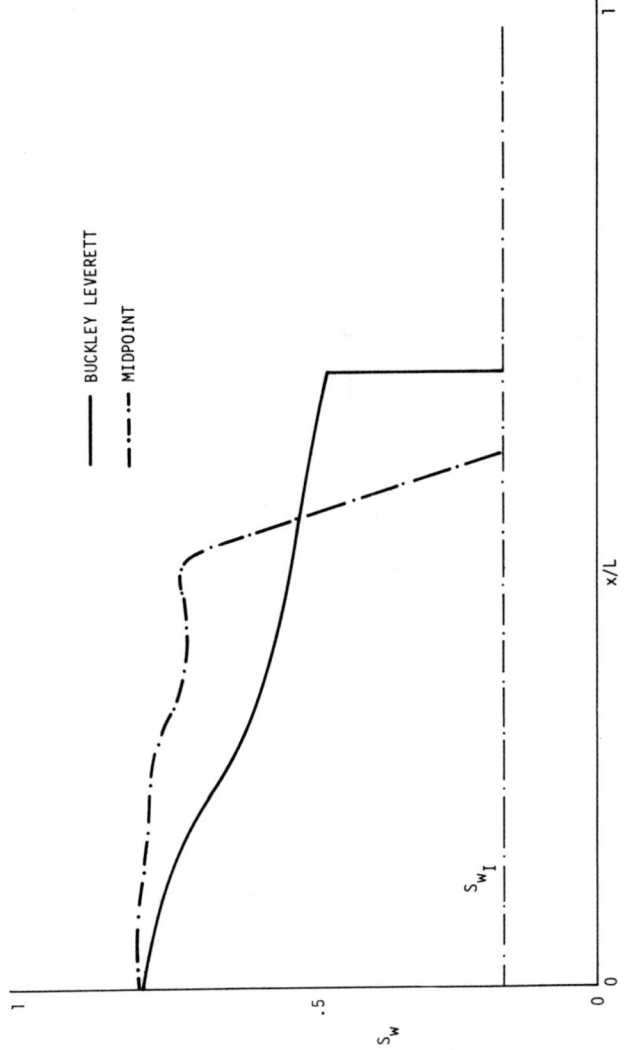

Figure 5.16: One-dimensional water oil displacement.

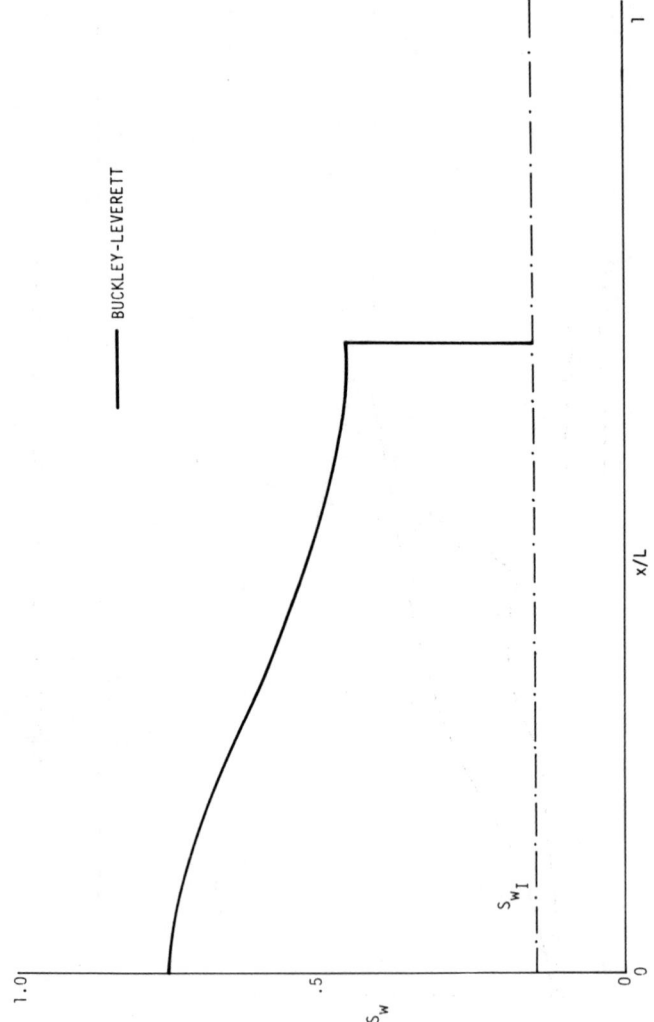

Figure 5.17: One-dimensional water oil displacement.

Extrapolated Relative Permeability Values

We have just concluded that upstream weighting of relative permeability values produces the most physically consistent results. Garret[3] pointed out that this could cause excessive numerical dispersion of flood fronts and that there is a marked sensitivity to the computed areal displacement due to grid orientation. Todd et al.[4] have developed an extrapolated or two-point scheme which alleviates some of the dispersion and displacement sensitivity. The method is based on the extrapolation of a relative permeability value at the interblock location based on the *two* upstream values if they exist. In the case where these two values are not available, modifications are made to the general equation. The approach is as follows.

Consider that portion of a simulator grid system shown in Fig. 5.18. For flow in the positive x-direction in Fig. 5.18 the mobility terms are needed at the $(i + \frac{1}{2})$ and $(i - \frac{1}{2})$ locations. These values can be obtained by applying a linear extrapolation of the k_r-term using the two upstream points as follows. The equation for linear extrapolation of a function is:

$$y_{n+1} = y_n + \frac{\partial y}{\partial x}\bigg|_n \Delta x_n \tag{5.51}$$

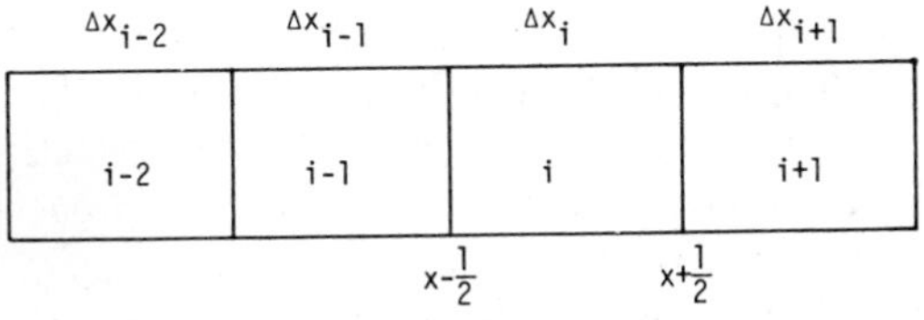

Figure 5.18: Averaging relative permeability.

This is illustrated in Fig. 5.19. The interblock value then becomes:

$$k_{r_{i+1/2}} = k_{r_i} + \frac{k_{r_i} - k_{r_{i-1}}}{\frac{\Delta x_i + \Delta x_{i-1}}{2}} * \frac{\Delta x_i}{2} \tag{5.52}$$

$$= k_{r_i} + \left(\frac{k_{r_i} - k_{r_{i-1}}}{\Delta x_i + \Delta x_{i-1}}\right) \Delta x_i \tag{5.53}$$

Similarly for the other mobility term:

$$k_{r_{i-1/2}} = k_{r_{i-1}} + \left(\frac{k_{r_{i-1}} - k_{r_{i-2}}}{\Delta x_{i-1} + \Delta x_{i-2}}\right) \Delta x_{i-1} \tag{5.54}$$

A general equation for this extrapolation is as follows:

$$k_{r_{i\pm1/2}} = k_{r_{up}} + \left(\frac{k_{r_{up}} - k_{r_{i\pm j}}}{\sum\limits_{j=1}^{2} \Delta x_{up}}\right) \Delta x_{up} \tag{5.55}$$

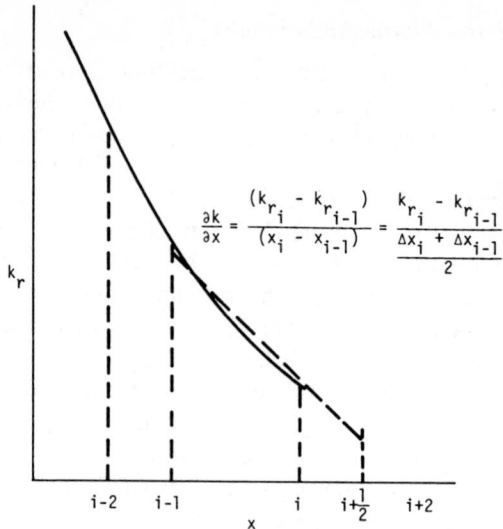

$$\frac{\partial k}{\partial x} = \frac{(k_{r_i} - k_{r_{i-1}})}{(x_i - x_{i-1})} = \frac{k_{r_i} - k_{r_{i-1}}}{\dfrac{\Delta x_i + \Delta x_{i-1}}{2}}$$

Figure 5.19: Linear extrapolation of interblock relative permeability term.

where the subscript "up" refers to the upstream cells at all times. Remember, *upstream* refers to direction of flow and bears no relation to i-index direction. The i-index could be with the flow or against the flow direction.

There are obvious complications to this simple approach. It is possible to formulate a set of circumstances wherein the use of Eq. (5.55) produces a negative value for $k_{r_{i\pm1/2}}$. This can occur in cells of equal Δx-dimensions with a very sharp displacing front just entering a given block i from block $(i - 1)$. In this situation the mobility must be limited to prevent the physically unrealistic condition of negative mobility; therefore,

$$k_{r_{i\pm1/2}} \geq 0 \tag{5.56}$$

The physical makeup of the reservoir could also produce complications in utilizing Eq. (5.55). For example, consider what happens at a boundary cell when the two upstream cells no longer exist (Fig. 5.20). Near sources and

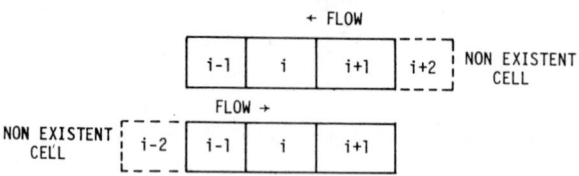

Figure 5.20: Boundary cells.

sinks—i.e., injectors and producers (Fig. 5.21)—and also at fluid contacts (Fig. 5.22), the general equations break down. The model must revert to the regular upstream formulations for these cases. The use of the extrapolated relative permeability values by reducing numerical dispersion could allow the engineer to use fewer cells in his model, which reduces the running time. The extrapolated technique is more computationally complex, but the computing "overhead" involved in the screening of cells in this method is less than the incremental computing time expended if a larger-number-celled model is used.

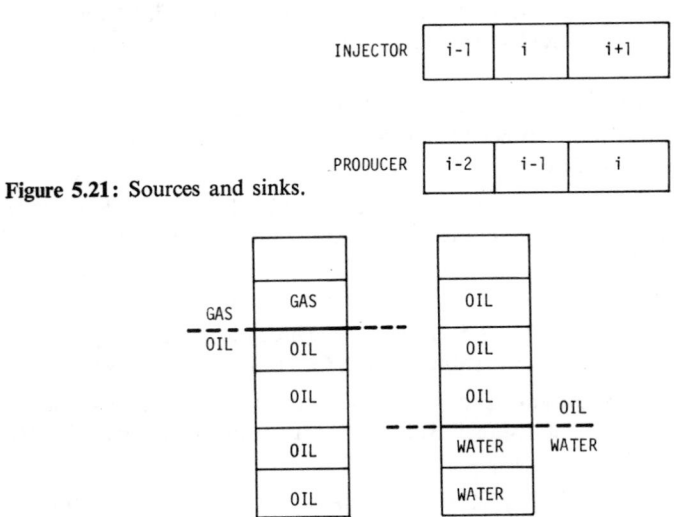

Figure 5.21: Sources and sinks.

Figure 5.22: Fluid contacts.

REFERENCES

1. J. Douglas, Jr., D. W. Peaceman, and H. H., Rachford, Jr., "A Method for Calculating Multi-dimensional Immiscible Displacement," *Trans. AIME* (1959), **216**, 297–306.

2. G. H., Bruce, D. W., Peaceman, and H. H. Rachford, Jr., "Calculations of Unsteady-state Gas Flow through Porous Media," *Trans. AIME* (1953), **198**, 79.

3. J. E. Garrett, and H. C. Osborne, "Use of Eight Flux Streams Per Cell in a Two-dimensional, Three-phase, Unsteady-state Reservoir Simulation Model," SPE 2026, First Symposium on Numerical Simulation of Reservoir Performance, Dallas, Texas, April 22–23, 1968.

4. M. R. Todd, P. M. O'Dell, and R. J. Hirasaki, "Methods of Increased Accuracy in Numerical Reservoir Simulators," *Trans. AIME* (1972), **253**.

BIBLIOGRAPHY

BLAIR, P. M. and C. F. WEINAUG, "Solution of Two-phase Flow Problems Using Implicit Difference Equations," *Soc. Pet. Eng. J.* (Dec. 1969), 417–24; *Trans. AIME*, **246**.

BREITENBACH, E. A., D. H. THURNAU, and H. K. VAN POOLLEN, "Solution of the Immiscible Fluid Flow Simulation Equations," *Soc. Pet. Eng. J.* (June 1969), 155–69.

CHAPPELEAR, J. E. and W. L. P. ROGERS, "Some Practical Considerations in the Construction of a Semi-implicit Simulator," SPE 4276, Third Symposium on Numerical Simulation of Reservoir Performance, Houston, Texas, Jan. 10–12, 1973.

CHAUDHARI, N. M., "An Improved Numerical Technique for Solving Multidimensional Miscible Displacement Equations," *Soc. Pet. Eng. J.* (Sept. 1971), 277–84; *Trans. AIME*, **251**.

COATS, K. H., "A Treatment of the Gas Percolation Problem in Simulation of Three-dimensional, Three-phase Flow in Reservoirs," *Soc. Pet. Eng. J.* (Dec. 1968), 413–19; *Trans. AIME*, **243**.

COATS, K. H., J. R. DEMPSEY, and J. H. HENDERSON, "The Use of Vertical Equilibrium in Two-dimensional Simulation of Three-dimensional Reservoir Performance," *Soc. Pet. Eng. J.* (March 1971), 63–71; *Trans. AIME*, **251**.

COATS, K. H., W. D. GEORGE, CHIEH CHU, and B. E. MARCUM, "Three-dimensional Simulation of Steamflooding," SPE 4500, 48th Annual Meeting, Las Vegas, Nev., Sept. 30–Oct. 3, 1973.

COATS, K. H., R. L. NIELSEN, M. H. TERHUNE, and A. G. WEBER, "Simulation of Three-dimensional, Two-phase Flow in Oil and Gas Reservoirs," *Soc. Pet. Eng. J.* (Dec. 1967), 377–88; *Trans. AIME*, **240**.

DOUGLAS, J., JR., D. W. PEACEMAN, and H. H. RACHFORD, JR., "A Method for Calculating Multi-dimensional Immiscible Displacement," *Trans. AIME* (1959) **216**, 297–306.

FAGIN, R. G. and C. H. STEWART, JR., "A New Approach to the Two-dimensional Multiphase Reservoir Simulator," *Soc. Pet. Eng. J.* (June, 1966), 175–82; *Trans. AIME*, **237**.

FAROUQ ALI, S. M. and C. D. STAHL, "Computer Models for Simulating Alcohol Displacement in Porous Media," *Soc. Pet. Eng. J.* (March 1965), 89–99.

GARRETT, J. E. and H. C. OSBORNE, "Use of Eight Flux Streams Per Cell in a Two-dimensional, Three-phase, Unsteady-state Reservoir Simulation Model," SPE 2026, First Symposium on Numerical Simulation of Reservoir Performance, Dallas, Texas, April 22–23, 1968.

GIVENS, J. W., "A Practical Two-dimensional Model for Simulating Dry Gas Reservoirs with Bottom Water Drive," *J. Pet. Tech.* (Nov. 1968), 1229–33.

LANTZ, R. B., "Rigorous Calculation of Miscible Displacement Using Immiscible Reservoir Simulators," *Soc. Pet. Eng. J.* (June 1970), 192–202; *Trans. AIME,* **249**.

LETKEMAN, J. P. and R. L. RIDINGS, "A Numerical Coning Model," *Soc. Pet. Eng. J.* (Dec. 1970), 418–42; *Trans. AIME,* **249**.

MacDONALD, R. C. and K. H. COATS, "Methods for Numerical Simulation of Water and Gas Coning," *Soc. Pet. Eng. J.* (Dec. 1970), 425–36; *Trans. AIME,* **249**.

MORSE, R. A. and R. L. WHITING, "A Numerical Model Study of Gravitational Effects and Production Rate on Solution Gas Drive Performance of Oil Reservoirs," *J. Pet. Tech.* (May 1970), 625–36; *Trans. AIME,* **249**.

PEERY, J. H. and E. H. HERRON, JR., "Three-phase Reservoir Simulation," *J. Pet. Tech.* (Feb. 1969), 211–20; *Trans. AIME,* **246**.

RACHFORD, H. H., JR., "Numerical Calculation of Immiscible Displacement by a Moving Reference Point," *Soc. Pet. Eng. J.* (June 1966), 87–101; *Trans. AIME,* **237**.

RACHFORD, H. H., JR., R. D. TAYLOR, J., DOUGLAS, JR., and P. M. DYKE, "Application of Numerical Methods to Predict Recovery From Thin Oil Columns," *Trans. AIME* (1958), **213,** 193–201.

RICHARDSON, J. G. and R. J. BLACKWELL, "Use of Simple Mathematical Models for Predicting Reservoir Behavior," *J. Pet. Tech.* (Sept. 1971), 1145–54; *Trans. AIME,* **251**.

SHEFFIELD, M., "Three-phase Fluid Flow Including Gravitational, Viscous and Capillary Forces," *Soc. Pet. Eng. J.* (June 1969), 232–46; *Trans. AIME,* **246**.

SHELDON, J. W. and E. W. DOUGHERTY, "A Numerical Method for Computing the Dynamical Behavior of Fluid-Fluid Interfaces in Permeable Media," *Soc. Pet. Eng. J.* (June 1964), 158–70.

SHELDON, J. W. and F. J. FAYERS, "The Motion of an Interface Between Two Fluids in a Slightly Dipping Porous Medium," *Soc. Pet. Eng. J.* (Sept. 1962), 275–82; *Trans. AIME,* **225**.

SNYDER, L. J., "Two-phase Reservoir Flow Calculations," *Soc. Pet. Eng. J.* (June 1969), 170–82.

SONIER F. and P. CHAUMET, "A Fully Implicit Three-dimensional Model in Curvilinear Coordinates," SPE 4543, 48th Annual Meeting, Las Vegas, Nev., Sept. 30–Oct. 3, 1973.

SPILLETTE, A. G., J. G. HILLESTAD, and H. L. STONE, "A High-stability Sequential Solution Approach to Reservoir Simulation," SPE 4542, 48th Annual Meeting, Las Vegas, Nev., Sept. 30–Oct. 3, 1973.

TSUTSUMI, GINJIRO and THOMAS N. DIXON, "Mathematical Simulation of Two-phase Flow with Interphase Mass Transfer in Petroleum Reservoirs," SPE 4074, 47th Annual Meeting, San Antonio, Texas, Oct. 8–11, 1972.

WEINSTEIN, H. G., H. L. STONE, and T. V. KWAN, "Simultaneous Solution of Multiphase Reservoir Flow Equations," *Soc. Pet. Eng. J.* (June, 1970), 99–110.

WELGE, H. J. and A. G. WEBER, "Use of Two-dimensional Methods for Calculating Well Coning Behavior," *Soc. Pet. Eng. J.* (Dec. 1964), 345–55.

WEST, W. J., W. W. GARVIN, and J. W. SHELDON, "Solution of the Equations of Unsteady State Two-phase Flow in Oil Reservoirs," *Trans. AIME* (1954), **201,** 217–29.

6 Solving the Matrix of Simultaneous Equations

6.1 INTRODUCTION

The efficiency of a simulator as a working and economical tool hinges upon the ability of its algorithms to solve the pressure equations efficiently. The guts of the model is the algorithm which performs the pressure solution, and no matter how elaborate the other algorithms within the model are, they are primarily just "window dressing" if the pressure solution is inadequate. The calculations which are normally done before or after the pressure calculations are rarely calculated more than once, but the pressure values are usually iterated upon several times during a single time step until some convergence criterion is satisfied. For this reason it is mandatory that the minimum work be done during the pressure solution.

The correct pressure solution is a necessary prerequisite to a good material balance check, because in order for a model to maintain stability, it should conserve mass at all times. There are many published schemes for solving the systems of equations obtained in the formulation of the simulator equations. As mentioned in Chapter 5, the implicit pressure–explicit saturation method produces a diagonal matrix which must be inverted to obtain the new pressure values. The simultaneous solution produces a block diagonal system which is more difficult to handle computationally, but the end result is the same: the oil and water potentials are calculated at each cell block. In order to make a decision on an effective method, let us first formulate a set of criteria with which we can compare the various methods. In this chapter we shall discuss the various methods within the following reference frame:

1. *Computational speed*—the actual number of machine operations required to solve the system.

2. *Accuracy and stability*—a measure of the correctness of the solution.

3. *Programming ease*—a measure of the time and money required to develop a model under a given formulation.

4. *Areas of applicability*—some techniques work well under selected conditions but are impossible to use or inadequate in other applications.

6.2 CONCEPTS IN SIMULTANEOUS LINEAR EQUATIONS

Before we start detailing the various methods available for solving the pressure equations, let us review some concepts in simultaneous linear equations. There are two basic means of solving a system of equations:

1. Direct process
2. Iterative process

Direct processes are those in which the solution to the system of equations is obtained upon the completion of a fixed number of operations. Stopping the operations before this point will not produce any usable data. The direct process is indicated in Fig. 6.1. Some examples of direct processes are the following:

1. Matrix inversion
2. Cramer's rule
3. Gaussian elimination
4. Gauss-Jordan method
5. Matrix decomposition

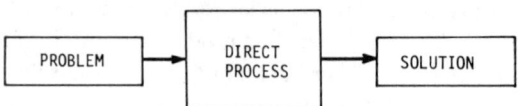

Figure 6.1: Direct process.

Iteration processes are cyclic in nature, and the solution process involves a systematic computation of better and more exact approximations to the solution at each iteration. The process involves the selection of a starting set of values of the unknown, generally called the starting vector, and this initial vector is operated on to produce a better refinement, as illustrated in Fig. 6.2. Processes *A* and *B* involve estimations and refinements of the solution vector. The error at each step is hopefully reduced, and the new value of the unknown P_n approaches the correct solution P^* as we keep iterating. Figure 6.3 illustrates an ideal convergence pattern.

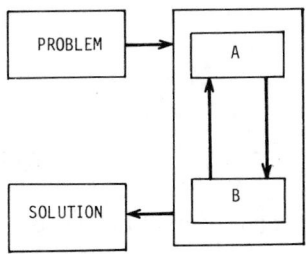

Figure 6.2: Iterative processes.

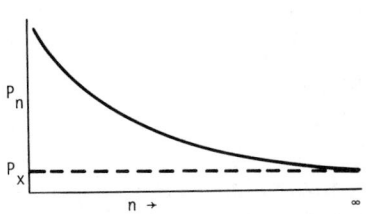

Figure 6.3: Convergence of iterative schemes.

Examples of iterative processes are:

1. Jacobi
2. Relaxation
3. Gauss-Seidel

Direct Processes

Direct processes, although they are conceptually easy, are rarely used without extensive modification to solve the pressure systems in simulation because of the large amount of computational labor involved in making even a small study and also the possibility of round-off error creeping into the solution during the backward-substitution phase of calculations. All is not lost, however, and some applications of direct methods have recently been made by engineers on simulation studies which exploit the special nature of the coefficient matrix—i.e., its sparseness or preponderance of zeros and the internal architecture of the computer that help obtain reasonably efficient and competitive algorithms. These algorithms are explored later in this chapter. Continuing with the review of basic concepts, let us formulate some of the more familiar methods.

1. Matrix Inversion: This is a straightforward process which involves the determination of the inverse matrix of the coefficient matrix and by a premultiplication obtains the solution. The process is as follows. Given:

$$AP = D \qquad (6.1)$$

and having determined the inverse of A, A^{-1},

$$A^{-1}AP = A^{-1}D \qquad (6.2)$$

Since the product below is the identity matrix,

$$A^{-1}A = I \qquad (6.3)$$

by definition:

$$P = A^{-1}D \qquad (6.4)$$

Symbolically:

$$[A][P] = [D] \qquad (6.5)$$

By premultiplying by A^{-1}, the left-hand side becomes simply the vector P, and the right-hand side becomes a new vector:

$$[I][P] = [A^{-1}][D] \qquad (6.6)$$

The solution is read off from the results of the multiplication on the right-hand side. The resulting system is:

$$\begin{bmatrix} 1 & & & & & \\ & 1 & & & & \\ & & 1 & & & \\ & & & \cdot & & \\ & & & & \cdot & \\ & & & & 1 & \\ & & & & & 1 \end{bmatrix} [P] = [D'] \qquad (6.7)$$

where D' is the new constant vector. The algorithm for determining the inverse matrix directly is developed as follows.

Given a general matrix A with elements a_{ij}, the inverse of A is:

$$A^{-1} = \frac{1}{|A|} \begin{bmatrix} A_{11} & A_{12} & \cdots & A_{1N} \\ \vdots & & & \\ \vdots & & & \\ A_{N1} & \cdots & \cdots & A_{NN} \end{bmatrix} \qquad (6.8)$$

where A_{ij} is the cofactor of element a_{ij} and A is the determinant of the original matrix A. The cofactor is defined by:

$$A_{ij} = (-1)^{i+j}M_{ij} \qquad (6.9)$$

where M_{ij} is the minor associated with the element a_{ij}. The minor is obtained by deleting the ith row and jth column of the matrix and evaluating the determinant of this reduced matrix:

$$M_{ij} = \begin{bmatrix} a_{11} & a_{12} & \cdots & \vdots & \cdots & a_{1n} \\ a_{21} & & \cdot & \vdots & \cdot & \\ \cdot & & & \vdots & & \\ \cdot & \cdots\cdots & \vdots & \cdots\cdots & \\ \cdot & & & \vdots & & \\ a_{n1} & & \cdot & \vdots & \cdot & a_{nn} \end{bmatrix} \begin{matrix} \\ \\ \\ \text{--}i \\ \\ \\ \end{matrix} \qquad (6.10)$$

$$j$$

This method is very cumbersome, and, as we shall see later, a more direct method based on elimination is preferable.

2. Cramer's Rule: This is an extremely simple but computationally disastrous method for finding the solution to the system $AP = D$. In words, the value of the jth element of the unknown vector is equal to the ratio of the determinant obtained by interchanging the right-hand-side vector of constants with the jth column of the original matrix A and the determinant of A:

$$P_i = \frac{|[\vdots D \vdots]|}{|[A]|} = \frac{\det_j}{\det_A} \qquad (6.11)$$

Since this method involves the evaluation of n determinants, it is severely limited in its use to the most trivial problems.

3. Gaussian Elimination: This is a systematic technique which involves a two-step algorithm. In the initial pass the matrix is converted to an upper triangular matrix, the operations being carried out on the right-hand-side vector of constants at the same time. In the final pass the solution is obtained by backward substitution for the unknowns. Given:

$$AP = D \qquad (6.12)$$

the constant vector D is appended to the coefficient matrix as follows:

$$\begin{bmatrix} a_{11} & a_{12} & \cdots & a_{1n} & \vdots & d_1 \\ a_{21} & \cdots & \cdots & & \vdots & d_2 \\ \cdot & & & & \vdots & \\ \cdot & & & & \vdots & \\ \cdot & & & & \vdots & \\ a_{n1} & \cdots & \cdots & a_{nn} & \vdots & d_n \end{bmatrix} \qquad (6.13)$$

This is reduced to the following:

$$\begin{bmatrix} c_{11} & c_{12} & \cdots & c_{1n} & \vdots & d'_1 \\ & c_{22} & \cdots & c_{2n} & \vdots & d'_2 \\ & & & & \vdots & \\ & \mathbf{0} & & & \vdots & \\ & & & c_{nn} & \vdots & d'_n \end{bmatrix} \qquad (6.14)$$

Equation (6.14) can be written:

$$\begin{bmatrix} c_{11} & c_{12} & \cdots & c_{1n} \\ & c_{22} & \cdots & c_{2n} \\ & & & \vdots \\ & & & \\ & & & c_{nn} \end{bmatrix} \begin{bmatrix} P_1 \\ \cdot \\ \cdot \\ P_i \\ \cdot \\ \cdot \\ P_n \end{bmatrix} = \begin{bmatrix} d'_1 \\ d'_2 \\ \cdot \\ \cdot \\ \cdot \\ d'_n \end{bmatrix} \qquad (6.15)$$

The value of P_n can be read off directly as:

$$P_n = \frac{d'_n}{c_{nn}} \qquad (6.16)$$

The $n-1, n-2, \ldots, 2, 1$ values can then be calculated simply by back substitution. The algorithm for this process is:

Forward reduction:

$$c_{ij} = a_{ij} - \frac{a_{ik}}{a_{kk}} a_{kj} \qquad (6.17)$$

where

$$j = k, \ldots, n+1$$
$$i = k+1, \ldots, n$$
$$k = 1, 2, \ldots, n-1$$

The right-hand column of constants is augmented to the original matrix to make it an $[n \times (n+1)]$ matrix. This allows the same operations to be carried out on the vector of constants.

Back substitution:

$$P_n = \frac{a_{n,n+1}}{c_{nn}} \qquad (6.18)$$

$$P_i = \frac{c_{i,n+1} - \sum\limits_{j=i+1}^{n} a_{ij} P_j}{a_{ii}} \quad \text{for } i = n-1, n-2, \ldots, 1 \qquad (6.19)$$

Note that for programming considerations the matrix A can be reduced "in place"—i.e., we do not need another matrix called c_{ij} to hold the reduced values. Furthermore, in reservoir simulation work our matrices are banded, and a very efficient algorithm could be generated which would take advantage of these special structures. The Thomas algorithm is such a procedure.

4. Gauss-Jordan Method: This is an elimination-type method which combines the two steps in the Gaussian algorithm by reducing the original matrix completely to the identity matrix. There is no back substitution, and the results can literally be read off from the final matrix. The Gauss-Jordan process can be visualized from the following.

Start with:

$$\overbrace{[\quad A}^{n} \mid \overbrace{b_i}^{1} \mid \overbrace{I \quad]}^{n} \qquad (6.20)$$

Original matrix   Identity
of coefficients   matrix

Right-hand-side
constants

Complete a set of row operations on the above $(n, 2n + 1)$ system and the following results:

$$[\quad I \quad \mid P_i \mid \quad A^{-1} \quad] \qquad (6.21)$$

Identity   Inverse of
matrix A

Solution
vector

The same set of matrix operations which reduces the original matrix to the identity matrix transforms the right-hand-side vector to the solution vector and the original identity matrix to the inverse of the original matrix. In actual practice, the engineer will usually not need the inverse of A, and there is therefore no need to augment A with the identity matrix at the start. The algorithm for the Gauss-Jordan scheme is as follows.

Normalize each element in a row:

$$a_{kj} = \frac{a_{jk}}{a_{kk}} \qquad j = k + 1, k + 2, \ldots, n + 1 \qquad (6.22a)$$

The reduction of off-diagonal elements produces:

$$a_{ij} = a_{ij} - a_{ik}a_{kj} \quad \text{for } i = k + 1, k + 2, \ldots, n \text{ and} \qquad (6.22b)$$
$$j = n + 1, n, n - 1, \ldots, k + 1 \text{ and}$$
$$k = 1, 2, \ldots n - 1 \text{ for both}$$

Note that in the algorithm the reduction of off-diagonal elements takes place by beginning at the extreme right-hand side—i.e., at the $(n + 1)$ element—of a given row. This procedure ensures that we do not operate on those columns which have already been reduced because we know the elements are already zero.

5. Matrix Decomposition: This process involves the transformation of a matrix into other matrices which are generally easier to operate on and then using these transformed matrices to obtain the solution. One method attributed to Crout[1] performs a decomposition of the matrix into a lower triangular and an upper triangular matrix. This decomposition is followed by a back substitution which computes the answer in two successive substitution steps. The process of decomposition is as follows.

Given:

$$AP = b \qquad\qquad (6.23)$$

then

$$A = LU \qquad\qquad (6.24)$$

where L is

$$\begin{bmatrix} l_{11} & & & \\ l_{21} & l_{22} & & \\ l_{31} & l_{32} & l_{33} & \\ \cdot & & & \\ \cdot & & & \\ \cdot & & & \\ l_{n1} & l_{n2} & \cdots & l_{nn} \end{bmatrix} \qquad\qquad (6.25)$$

and U is

$$\begin{bmatrix} 1 & u_{12} & u_{13} & \cdots & u_{1n} \\ & 1 & u_{23} & & \\ & & 1 & & \\ & & & \cdot & \\ & & & \cdot & \\ & & & \cdot & \\ & & & & 1 \end{bmatrix} \qquad\qquad (6.26)$$

From these two triangular matrices we can compute the solution vector P because, by Eq. (6.24),

$$A = LU \qquad\qquad (6.24)$$

Then after decomposition:

$$LUP = b \qquad\qquad (6.27)$$

Let us call the product UP the vector y. Then from Eq. (6.27):

$$Ly = b \tag{6.28}$$

and

$$UP = y \tag{6.29}$$

Solving Eqs. (6.28) and (6.29) successively for y and b respectively:

$$y = L^{-1}b \tag{6.30}$$

and

$$P = U^{-1}y \tag{6.31}$$

The algorithm for this process is as follows.

The elements of the lower are l_{ij} and $l_{ij} = 0$ if $i < j$, and the elements of the upper are u_{ij} with $u_{ii} = 1$ and $u_{ij} = 0$ if $i > j$. Then:

$$l_{im} = a_{im} - \sum_{k=1}^{m-1} l_{ik}u_{km} \quad \text{for} \quad \begin{aligned} &i = m, m+1, \dots, n \\ &m = 1, 2, \dots, n \end{aligned} \tag{6.32a}$$

$$u_{mj} = \frac{1}{l_{mm}}\left(a_{mj} - \sum_{k=1}^{m-1} l_{mk}u_{kj}\right) \quad \text{for} \quad \begin{aligned} &j = m+1, m+2, \dots, n \\ &m = 1, 2, \dots, n \end{aligned} \tag{6.32b}$$

Back substitution:

$$y_i = \frac{b_i - \sum_{k=1}^{i-1} l_{ik}y_k}{l_{ii}} \quad \text{for } i = 1, 2, \dots, n \tag{6.33}$$

and

$$P_i = y_i - \sum_{k=i+1}^{n} u_{ik}P_k \quad \text{for } i = n, n-1, \dots, 1 \tag{6.34}$$

Note that in the lower decomposition process in Eq. (6.32) the evaluation of the first summation produces:

$$l_{11} = a_{11} - \sum_{k=1}^{0} l_{11}u_{11} \tag{6.35}$$

The summation

$$\sum_{k=1}^{0} l_{11}u_{11} = 0 \tag{6.36}$$

is defined as zero; thus, the first element of the lower is a_{11}, and this calculation will be bypassed in the program.

Iterative Processes

Consider a system of equations in matrix form

$$AP = b \tag{6.37}$$

We can rewrite each individual row in such a way as to solve for the unknown quantity P as follows:

$$P_1 = \frac{1}{a_{11}}[b_1 - (a_{12}P_2 + a_{13}P_3 + \ldots + a_{1n}P_n)]$$

$$P_2 = \frac{1}{a_{22}}[b_2 - (a_{21}P_1 + a_{23}P_3 + \ldots + a_{2n}P_n)] \tag{6.38}$$

$$\cdot$$
$$\cdot$$
$$\cdot$$

$$P_n = \frac{1}{a_{nn}}[b_n - (a_{n1}P_1 + a_{n2}P_2 + \ldots + a_{n,n-1}P_n)]$$

In essence the vector P is calculated by using the right-hand-side column vector and the coefficient matrix and the other unknown elements of a given row. Knowing an original estimate of all the unknown quantities, we can sequentially plug into Eq. (6.38) and obtain a new approximation to the solution vector. This general approach can be speeded up to achieve varying rates of convergence. The following methods indicate how a progression is made from the slowest to the fastest techniques.

1. Jacobi Method: This is one of the earliest iterative methods. The system of Eq. (6.37) is solved iteratively by applying Eq. (6.38) sequentially. The solution of each row for the given unknown involves the old approximation of the unknown vector P. The algorithm is:

$$P_i^k = \frac{1}{a_{ii}}\left(b_i - \sum_{\substack{j=1 \\ j\neq i}}^{n} a_{ij}P_j^{k-1}\right) \quad \text{for } i = 1, 2, \ldots, n \tag{6.39}$$

where P_i^k is the kth approximation to the ith value of the unknown P. Note that on the right-hand side the old values P_i^{k-1} are used.

2. Gauss-Seidel Method: This improves on the Jacobi method by using the *current* approximation to the unknown vector. It is obvious that since the unknown P_i is calculated in increasing values of i, at any given value i greater than 1 some of the elements in a row of Eq. (6.38) will be new values of P_i and some old values of P_i. For example, if we are evaluating P_{17}, then the new P_1 through P_{16} will have already been computed, so they can be used in solving for the new P_{17}-value. The old P_i for i greater than 17 will be used. The algorithm is similar to the Jacobi method with a simple change on the value of P_j used on the right-hand side to reflect the level of computation:

$$P_i^k = \frac{1}{a_{ii}}\left(b_i - \sum_{\substack{j=1 \\ j\neq i}}^{n} a_{ij}P_j^{[k \text{ if } j<i \atop k-1 \text{ if } j>i]}\right) \quad \text{for } i = 1, 2, \ldots, n \qquad (6.40)$$

This method, since it uses more recent information, converges faster than the Jacobi method.

3. Relaxation Method: This method uses a technique where not only the most recent values of P_i are used in the solution process but these values are modified before being substituted into the equation to determine P_i in a given row. The modification involves the addition of a quantity to the unknown P_i before it is substituted into Eq. (6.38). The quantity added can be visualized as overcorrecting the value P_i. The algorithm is as follows:

$$P_i^k = P_i^{k-1} + \omega\frac{r_i^{k-1}}{a_{ii}} \quad \text{for } i = 1, 2, \ldots, n \qquad (6.41)$$

where the residual r_i is calculated

$$r_i^{k-1} = b_i - \sum a_{ij}P_i^{k-1} \quad \text{for } i = 1, 2, \ldots, n \qquad (6.42)$$

The factor ω in Eq. (6.41) is called the relaxation parameter, and its presence accelerates the convergence process. For any fixed value of ω between 0 and 2 the process converges for the types of equation systems we encounter in petroleum reservoir simulation. The optimum ω produces extremely rapid convergence; the selection of an optimum ω will be discussed later.

6.3 ITERATIVE METHODS IN SIMULATION PRACTICE

Iterative methods are by far the most common methods of solving the pressure equations used in simulation programs. They are generally more efficient than direct methods and can ordinarily be made to run at greater levels of accuracy. We shall discuss several basic methods.

Alternating-Direction Implicit Procedure (ADIP)

The alternating-direction implicit procedure (ADIP) can be concisely expressed as follows. The total producing time of the reservoir is divided into a number of time steps. Each time step is divided into two equal substeps. During the first substep the grid is swept in the x-direction one row at a time solving for the unknown pressures. In the second substep the system is swept in the y-direction one column at a time solving for the unknown pressures.

Consider the following pressure equation in two dimensions:

$$\frac{\partial^2 P}{\partial x^2} + \frac{\partial^2 P}{\partial y^2} = \alpha \frac{\partial P}{\partial t} \tag{6.43}$$

The fully implicit finite-difference formulation of this equation is shown below, using the indexing of Fig. 6.4.

$$\frac{P_{i+1}^{n+1} - 2P_i^{n+1} + P_{i-1}^{n+1}}{\Delta x^2} + \frac{P_{i+n}^{n+1} - 2P_i^{n+1} + P_{i-n}^{n+1}}{\Delta y^2} = \frac{\alpha(P_i^{n+1} - P_i^n)}{\Delta t} \tag{6.44}$$

This equation has five unknowns and, when written for each point in the grid below, produces an $N \times N$ system of simultaneous linear equations which are of a *pentadiagonal* form:

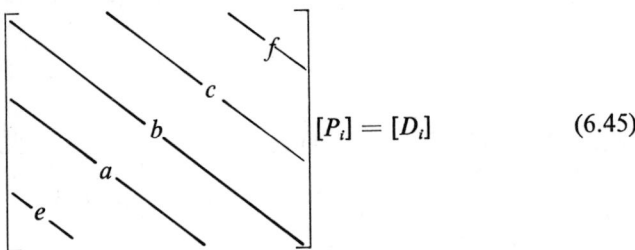

$$[P_i] = [D_i] \tag{6.45}$$

where

$$e_i P_{i-n}^{n+1} + a_i P_{i-1}^{n+1} + b_i P_i^{n+1} + c_i P_{i+1}^{n+1} + f_i P_{i+n}^{n+1} = d_i \tag{6.46}$$

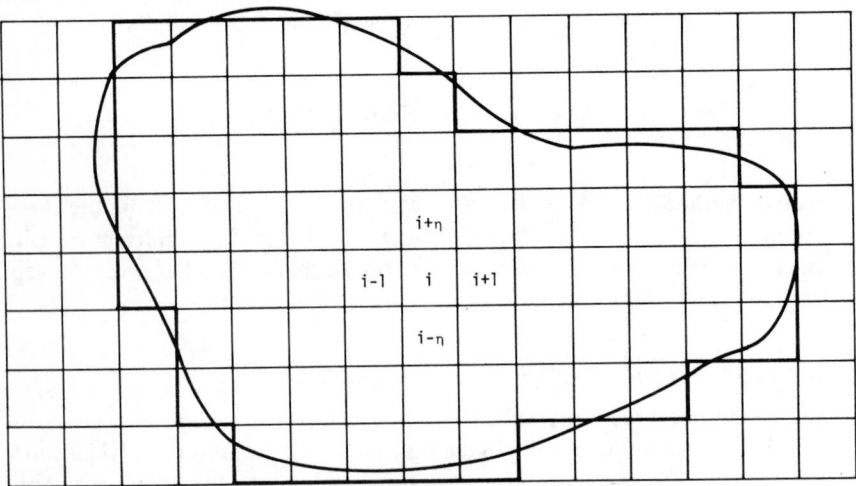

Figure 6.4: Alternating direction implicit procedure.

is the general equation for each cell. This pentadiagonal matrix could be solved for P_i^{n+1} using any of the techniques covered earlier in Sec. 6.2, but it could be very time-consuming. The ADIP technique approaches the problem as follows. In the first half of the time step, the x-direction sweep is carried out. Consider Fig. 6.5 and the general equation [Eq. (6.46)] written for

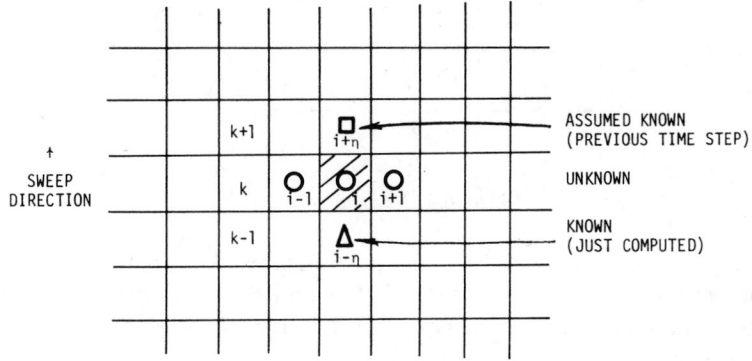

Figure 6.5: Horizontal sweep.

the cell i.. Here P_{i+n}^{n+1} is assumed known from the previous time step value, and P_{i-n}^{n+1} is *known* because it has just been calculated. (Note the direction of sweep.) There are now three unknowns instead of the original five. Equation (6.46) becomes:

$$a_i P_{i-1}^{n+(1/2)} + b_i P_i^{n+(1/2)} + c_i P_{i+1}^{n+(1/2)} = d_i - (e_i P_{i-n}^{n+(1/2)} + f_i P_{i+n}^n)$$

$$= D_i \qquad (6.47)$$

This equation, when written for each cell in the model, produces a close-band tridiagonal matrix as follows:

$$\qquad (6.48)$$

or:

$$A_H P^* = D \qquad (6.49)$$

where $P_i^* = P_i^{n+(1/2)}$ is the intermediate value of pressure P_i. (We are in essence solving a one-dimensional problem in this one-half of a time step.)

In the second half of the time step the integration process is carried out in the vertical direction (see Fig. 6.6).

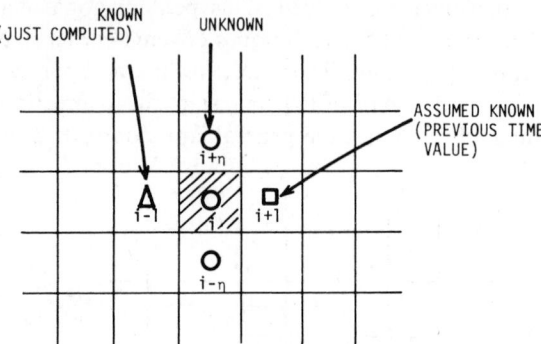

Figure 6.6: Vertical sweep.

The values of P_{i-1}^{n+1} and P_{i+1}^{n+1} are now assumed known from the most recently computed value in the case of P_{i-1}, and P_{i+1} from the value of the previous time value. The general equation (Eq. 6.46) then contains only three unknowns, as shown below:

$$e_i P_{i+n}^{n+1} + h_i P_i^{n+1} + f_i P_{i+n}^{n+1} = d_i - [a_i P_{i-1}^{n+1} + c_i P_{i+1}^{n+(1/2)}]$$
$$= H_i \qquad (6.50)$$

The values at $t + \Delta t/2$ indicated by $P_i^{n+(1/2)}$ are now transferred to the right-hand side. Equation (6.48) when written for each cell produces a wide-band tridiagonal matrix as shown below:

$$\qquad\qquad (6.51)$$

$$A_V P^{**} = H \qquad (6.52)$$

where P^{**} is the value of pressure at the end of the time step. The solution of the vertical sweep produces the new values of pressure at the new time level.

A modified form of the Thomas algorithm is used to solve this wide-band matrix.

At the end of these two half–time steps the new value for pressure at the time level $(n + 1)$ is obtained. The two-step process thus reduced the more complex two-dimensional problem to a succession of two one-dimensional processes as illustrated in Fig. 6.7.

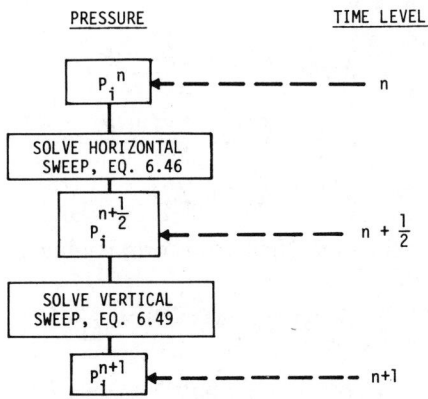

Figure 6.7: ADIP method.

Iterative Alternating-Direction Implicit Procedures[2]

An improvement on ADIP can be made by the use of a modifier called an acceleration parameter. This parameter has the same intent as the relaxation parameter mentioned in Sec. 6.2, and a judicious selection of these parameters produces rapid convergence. In implementing iterative ADIP the standard ADIP scheme is repeated several times in a given time step until there is convergence. Consider the portion of a finite-difference grid shown in Fig. 6.8. A general iterative ADIP scheme is developed as follows. A finite-difference formulation of Eq. (6.43) for the horizontal sweep is:

$$\theta(\Delta_x^2 P^* + \Delta_y^2 P^k) + (1 - \theta)\,\Delta^2 P^k = \frac{2\alpha}{\Delta t}(P^* - P^k) \qquad (6.53)$$

For the vertical sweep, it is:

$$\theta(\Delta_x^2 P^* + \Delta_y^2 P^{k+1}) + (1 - \theta)\,\Delta^2 P^k = \frac{2\alpha}{\Delta t}(P^{k+1} - P^*) \qquad (6.54)$$

where the variable θ is defined as

$$0 \le \theta \le 1 \qquad (6.55)$$

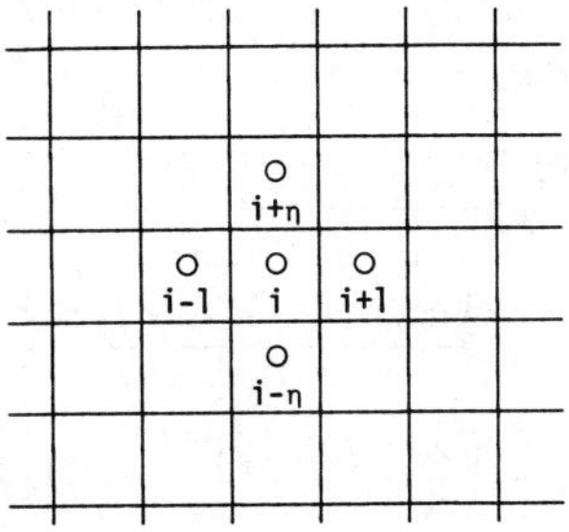

Figure 6.8: Finite-difference grid for iterative ADIP.

and

$$\Delta_x^2 P = \frac{P_{i+1} - 2P_i + P_{i-1}}{\Delta x^2} \tag{6.56}$$

$$\Delta_y^2 P = \frac{P_{i+\eta} - 2P_i + P_{i-\eta}}{\Delta y^2} \tag{6.57}$$

$$\Delta^2 P = \Delta_x^2 P + \Delta_y^2 P \tag{6.58}$$

For $\theta = \frac{1}{2}$, the normal Crank-Nicholson form is obtained:

$$\frac{1}{2}[\Delta_x^2 P^{n+(1/2)} + \Delta_y^2 P^n] + \frac{1}{2}(\Delta^2 P^n) = \frac{\alpha}{\Delta t/2}[P^{n+(1/2)} - P^n] \tag{6.59}$$

$$\frac{1}{2}[\Delta_x^2 P^{n+(1/2)} + \Delta_y^2 P^{n+1}] + \frac{1}{2}(\Delta^2 P^n) = \frac{\alpha}{\Delta t/2}[P^{n+1} - P^{n+(1/2)}] \tag{6.60}$$

where $P^{n+(1/2)}$ denotes intermediate values of P

  P^n denotes values at start of time step

  P^{n+1} denotes values at end of time step

Addition of acceleration parameters speeds the convergence, and Eq. (6.60) becomes:

$$\frac{1}{2}[\Delta_x^2 P^{n+(1/2)} + \Delta_y^2 P^n] + \frac{1}{2}(\Delta^2 P^n) - \frac{\alpha}{\Delta t/2} P^{n+(1/2)}$$

$$= H_k[P^{k+(1/2)} - P_k] - \frac{\alpha}{\Delta t/2} P^n \tag{6.61}$$

$$\frac{1}{2}[\Delta_x^2 P^{n+(1/2)} + \Delta_y^2 P^{n+1}] + \frac{1}{2}(\Delta^2 P^n) - \frac{\alpha}{\Delta t/2} P^{n+1}$$

$$= H_k[P^{k+1} - P^{k+(1/2)}] - \frac{2\alpha}{\Delta t} P^n \tag{6.62}$$

Equation (6.62), which is the finite-difference formulation of the partial differential equation for pressure, is used iteratively with a given cycle as indicated in Fig. 6.9. The successive use of a set of acceleration parameters constitutes a cycle. Several cycles are needed to converge to the pressure at the new time level.

Writing Eq. (6.62) for a given point i fully:

Horizontal sweep:

$$\frac{\frac{1}{2}[P_{i+1}^{k+(1/2)} - 2P_i^{k+(1/2)} + P_{i-1}^{k+(1/2)}]^{n+1}}{\Delta x^2} + \frac{\frac{1}{2}(P_{i+\eta}^k - 2P_i^k + P_{i-\eta}^k)^{n+1}}{\Delta y^2}$$

$$+ \frac{\frac{1}{2}(P_{i+1} - 2P_i + P_{i-1})^n}{\Delta x^2} + \frac{\frac{1}{2}(P_{i+\eta} - 2P_i + P_{i-\eta})^n}{\Delta y^2}$$

$$- \left[\frac{2\alpha}{\Delta t} P_i^{k+(1/2)}\right]^{n+1} = H_k[P_i^{k+(1/2)} - P_i^k] - \frac{2\alpha}{\Delta t} P_i^n \tag{6.63}$$

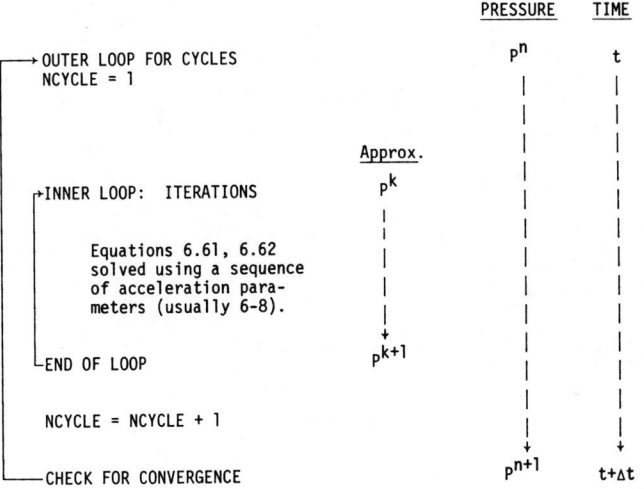

Figure 6.9: Iterative ADI process.

where

k = iteration counter
n = time step index
H_k = acceleration parameter

Vertical sweep:

$$\frac{\frac{1}{2}[P_{i+1}^{k+(1/2)} - 2P_i^{k+(1/2)} + P_{i-1}^{k+(1/2)}]^{n+1}}{\Delta x^2} + \frac{\frac{1}{2}(P_{i+\eta}^{k+1} - 2P_i^{k+1} + P_{i-\eta}^{k+1})^{n+1}}{\Delta y^2}$$

$$+ \frac{\frac{1}{2}(P_{i+1} - 2P_i + P_{i-1})^n}{\Delta x^2} + \frac{\frac{1}{2}(P_{i+\eta} - 2P_i + P_{i-\eta})^n}{\Delta y^2}$$

$$- \left(\frac{2\alpha}{\Delta t}P_i^{k+1}\right)^{n+1} = H_k[P_i^{k+1} - P_i^{k+(1/2)}] - \frac{2\alpha}{\Delta t}P_i^n \qquad (6.64)$$

Equations (6.63) and (6.64) can be written in summary form gathering all unknowns as:

$$a_i P_{i+1}^{m+(1/2)} + b_i P_i^{m+(1/2)} + c_i P_i^{m+(1/2)} = D_i \qquad (6.65)$$

$$e_i P_{i+\eta}^{m+1} + g_i P_i^{m+1} + f_i P_{i-\eta}^{m+1} = H_i \qquad (6.66)$$

Writing Eqs. (6.65) and (6.66) alternatively for each point produces two sets of tridiagonal matrices, which are solved with the Thomas algorithm. For the horizontal sweep:

$$[A][P^*] = [D] \qquad (6.67)$$

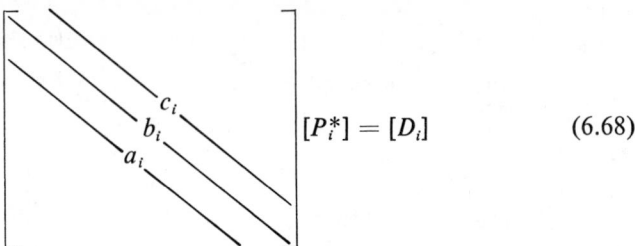

$$[P_i^*] = [D_i] \qquad (6.68)$$

For the vertical sweep:

$$[E][P^{**}] = [H] \qquad (6.69)$$

$$[P_i^{**}] = [H_i] \qquad (6.70)$$

where P^* and P^{**} are unknown approximations to the true pressure.

The selection of an optimal set of acceleration parameters is necessary to achieve the increased efficiency of this method.

Selection of Acceleration Parameters

If we use a single optimum acceleration parameter in the iterative alternating implicit technique, it has been shown by Varga[3] that ADIP has the identical asymptotic rate of convergence as the point successive overrelaxation. To achieve a faster rate of convergence a series of parameters used in sequence is employed. The determination of these parameters is generally difficult analytically but can be approached by trial-and-error techniques. The engineer usually estimates the upper and lower limits of the parameter range from the reservoir data using the following procedure of Peaceman and Rachford.[2]

Four parameters are estimated from the interblock transmissibility and the number of cells in the spatial directions of the model. These parameters are

$$M_1 = \frac{2T_x}{T_x + T_y} \frac{\pi^2}{4N_x^2} \qquad (6.71)$$

$$M_2 = \frac{2T_x}{T_x + T_y} \qquad (6.72)$$

$$M_3 = \frac{2T_y}{T_x + T_y} \frac{\pi^2}{4N_y^2} \qquad (6.73)$$

$$M_4 = \frac{2T_y}{T_x + T_y} \qquad (6.74)$$

where T_x and T_y are the x-direction and y-direction transmissibilities, respectively, and N_x and N_y are respectively the number of cells in the x- and y-directions. The lower limit of the acceleration parameter is calculated from

$$M_5 = \min(M_1, M_3) \tag{6.75}$$

and the upper limit calculated from

$$M_6 = \max(M_2, M_4) \tag{6.76}$$

The rate of convergence is usually insensitive to the upper limit, and from an examination of Eqs. (6.72) and (6.74), it can be verified that

$$\text{if} \quad T_x \gg T_y \quad \text{or} \quad T_y \gg T_x$$

$$\text{then} \quad \frac{2T_y}{T_x + T_y} = \frac{2T_x}{T_x + T_y} \simeq 2 \tag{6.77}$$

and

$$\text{if} \quad T_y \simeq T_x$$

$$\text{then} \quad \frac{2T_y}{T_x + T_y} = \frac{2T_x}{T_x + T_y} \simeq 1 \tag{6.78}$$

Several authors[5,6] have pointed out that the acceleration parameters should be spaced in a geometric series—i.e., the ratio of two consecutive values is a constant:

$$\frac{H_{k+1}}{H_k} = r \tag{6.79}$$

The value of the constant r can be ascertained from the observation for a sequence of M parameters:

$$\frac{H_M}{H_1} = r^{M-1} \tag{6.80}$$

where H_M and H_1 are the maximum and minimum parameters, respectively. Then

$$\ln r = \frac{\ln(H_M/H_1)}{M - 1} \tag{6.81}$$

Thus,

$$r = e^{[\ln(H_m/H_1)]/(m-1)} \tag{6.82}$$

Once r is known the series can be determined by Eq. (6.79). The number of parameters chosen in a cycle is a matter of judgment, and for small ranges of (H_M, H_1) four or six parameters are normally used, whereas for larger ranges of (H_M, H_1) eight to twelve parameters are used. In practice the rate of convergence is intensely sensitive to the minimum parameter value, and some degree of experimentation must be undertaken by the engineer to obtain an adequate value, since Eqs. (6.71) through (6.74) are only a guide. A difference of 0.001 in the initial value may mean the difference between convergence and divergence in a run. An estimate of minimum parameter for a two-dimensional study is shown below.

A study is done using the following data:

$$k_x = 80 \text{ md*} \qquad N_x = 40 \qquad x_{max} = 5280 \text{ ft}$$
$$k_y = 45 \text{ md} \qquad N_y = 10 \qquad y_{max} = 50 \text{ ft}$$

Calculation:

$$\text{Note: } T_x = k_x \frac{\Delta y}{\Delta x} \qquad\qquad T_y = k_y \frac{\Delta x}{\Delta y}$$

$$\Delta x = \frac{5280}{40} = 132 \text{ ft} \qquad \Delta y = \frac{50}{10} = 5 \text{ ft}$$

$$T_x = \frac{80 * 5}{132} = 3.03 \qquad T_y = \frac{45 * 132}{5} = 14.7$$

$$H_{min} = \min \left(\frac{2T_x}{T_x + T_y} \frac{\pi^2}{4N_x^2}, \frac{2T_y}{T_x + T_y} \frac{\pi^2}{4N_y^2} \right)$$

$$= \min \left(\frac{2 * 3.03 * \pi^2}{(3.03 + 14.7) * 4(40)^2}, \frac{2 * 14.7 * \pi^2}{(3.03 + 14.7) * 4(10)^2} \right)$$

$$= \min (0.000527, 0.0409)$$

$$= 0.000527$$

Using six parameters, then, the geometric multiplier is

$$\ln r = \frac{\ln (2/0.000527)}{6 - 1}$$

$$= \frac{\ln 3794}{5}$$

Thus, $r = 5.21$, and

$$H_1 = 0.000527$$
$$H_2 = 0.002745$$
$$H_3 = 0.01425$$
$$H_4 = 0.0741$$
$$H_5 = 0.3853$$
$$H_6 = 2.0$$

*Millidarcy.

The acceleration parameters must be normalized to account for possible differences in areal size of the cells and for reservoir heterogeneity. The normalizing procedure is as follows:

$$H_k = H_k \gamma_i$$

and the normalizing factor for the i cell is defined:

$$\gamma_i = \frac{\sum\limits_{n=1}^{N} K_n}{N(\Delta x * \Delta y)_i}$$

where N is the number of cells and Δx, Δy the dimensions of the i cell.

Point Relaxation

The point relaxation technique involves the application of relaxation techniques to each cell in a grid sequentially. The technique will be developed for a one-dimensional model, but its extension to more than a single dimension is obvious; the arithmetic, however, becomes slightly more involved.

Consider the Crank-Nicholson form for a one-dimensional problem:

$$\frac{\partial^2 P}{\partial x^2} = \frac{\partial P}{\partial t} \tag{6.83}$$

The finite-difference equation is

$$\frac{\frac{1}{2}(P_{i-1}^{n+1} - 2P_i^{n+1} + P_{i+1}^{n+1})}{\Delta x^2} + \frac{\frac{1}{2}(P_{i-1}^{n} - 2P_i^{n} + P_{i+1}^{n})}{\Delta x^2} = \frac{P_i^{n+1} - P_i^n}{\Delta t} \tag{6.84}$$

Solving for the quantity P_i^{n+1} on the right-hand side:

$$P_i^{n+1} = P_i^n + \frac{1}{2}\frac{\Delta t}{\Delta x^2}[(P_{i-1}^{n+1} - 2P_i^{n+1} + P_{i+1}^{n+1}) + (P_{i-1}^{n} - 2P_i^{n} + P_{i+1}^{n})]$$

$$= \frac{1}{2}\frac{\Delta t}{\Delta x^2}(P_{i-1}^{n+1} - 2P_i^{n+1} + P_{i+1}^{n+1}) + P_i^n + \frac{1}{2}\frac{\Delta t}{\Delta x^2}(P_{i-1}^{n} - 2P_i^{n} + P_{i+1}^{n})$$

$$= \frac{1}{2}\frac{\Delta t}{\Delta x^2}(P_{i-1}^{n+1} - 2P_i^{n+1} + P_{i+1}^{n+1}) + b_i \tag{6.85}$$

where b_i is a set of known quantities based on the previous time step. Equation (6.85) is obviously an implicit equation, since the unknown P_i appears on both sides.

Let $\beta = \Delta t/\Delta x^2$. An iterative scheme for Eq. (6.85) is then

$$\overset{k+1}{P_i^{n+1}} = b_i + \frac{1}{2}\beta(\overset{k}{P_{i-1}^{n+1}} - 2\overset{k}{P_i^{n+1}} + \overset{k}{P_{i+1}^{n+1}}) \tag{6.86}$$

where the superscript k indicates the iteration level in the time step $(n + 1)$. It can be proved that this equation unfortunately converges only for

$$0 < \beta \leq \tfrac{1}{2} \qquad (6.87)$$

A simple change on the right-hand side—using $P_i^{n+1\,k+1}$ instead of $P_{i+1}^{n\,k}$— produces:

$$P_i^{n+1\,k+1} = b_i + \tfrac{1}{2}\beta(P_{i-1}^{n+1\,k} - 2P_i^{n+1\,k+1} + P_{i+1}^{n\,k}) \qquad (6.88)$$

which, on collecting terms in $P_i^{n+1\,k+1}$, becomes

$$P_i^{n+1\,k+1} = \frac{b_i}{1 + \beta} + \frac{\beta}{2(1 + \beta)}(P_{i-1}^{n+1\,k} + P_{i+1}^{n+1\,k}) \qquad (6.89)$$

This equation (Eq. 6.89) converges for all values of β and converges more rapidly. This is the Jacobi method.

A refinement of this method can be effected by using the updated values of pressure as soon as they become available (see Fig. 6.10).

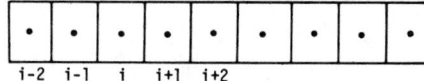

$$i\text{-}2 \quad i\text{-}1 \quad i \quad i+1 \quad i+2$$

Figure 6.10: One-dimensional grid ordering.

We are iterating on the new values of P in a systematic manner—i.e., moving from left to right; $i - 2, i - 1, i, \ldots$, etc. values of P for i less than the current value of i are all available, since they have just been calculated. They are used immediately as follows:

$$P_i^{n+1\,k+1} = b_i + \tfrac{1}{2}\beta(P_{i-1}^{n+1\,k+1} - 2P_i^{n+1\,k+1} + P_{i+1}^{n\,k}) \qquad (6.90)$$

Note the difference between Eqs. (6.90) and (6.88) in the P_{i-1}-cell term.

Therefore, collecting like terms again:

$$P_i^{n+1\,k+1} = \frac{b_i}{1 + \beta} + \frac{\beta}{2(1 + \beta)}(P_{i-1}^{n+1\,k+1} + P_{i+1}^{n+1\,k}) \qquad (6.91)$$

Equation (6.91) is the Gauss-Seidel technique, and it converges almost *twice* as fast as the Jacobi method.

The Gauss-Seidel method can be speeded up even more by giving it some added impetus—"a kick in the rear."

Consider the Gauss-Seidel form—e.g., Eq. (6.91). Adding and subtracting $P_i^{n+1\,k}$ to the right-hand side produces:

$$P_i^{k+1} = P_i^{k+1} + \left[\frac{b_i}{1+\beta} + \frac{\beta}{2(1+\beta)}(P_{i-1}^{k+1} + P_{i+1}^{k+1}) - P_i^{k+1} \right] \qquad (6.92)$$

Note that the quantity in brackets is simply the increment in each Gauss-Seidel iteration—i.e., the difference between the two sides of Eq. (6.91). This increment is overrelaxed by a quantity omega (ω) to produce:

$$P_i^{k+1} = P_i^{k+1} + \omega \left[\frac{b_i}{1+\beta} + \frac{\beta}{2(1+\beta)}(P_{i-1}^{k+1} + P_{i+1}^{k}) - P_i^{k+1} \right] \qquad (6.93)$$

where ω is a relaxation parameter. The optimum ω assumes rapid convergence at a rate faster than in the Gauss-Seidel technique.

To save some computing time, in practice, the formula shown by Eq. (6.93) is not used as such. Equation (6.91) is solved for each point iteratively, and as soon as this value is computed it is overrelaxed and updated before going to the next point. The overrelaxation is simply:

$$P_i^{k+1} = P_i^{k+1} + \omega(P_i^{k+1} - P_i^{k+1}) \qquad (6.94)$$

Convergence criteria must be set up, and a simple straightforward scheme is

$$\max \left| \frac{P_i^{k+1} - P_i^{k+1}}{P_i^{k+1}} \right| \leq \epsilon \quad \text{for all values of } i \qquad (6.95)$$

This is the relative error check during a time step. The extension of the point relaxation technique to two or three dimensions is straightforward.

Line Relaxation

The relaxation technique can be applied to a group of cells at a time, most frequently to a group along a line—i.e., a column or row. This method in effect relaxes one line at a time and is referred to as line successive overrelaxation (LSOR).

Consider a computational molecule in Fig. 6.11 obtained from Eq. (6.96):

$$\frac{\partial^2 P}{\partial x^2} + \frac{\partial^2 P}{\partial y^2} = \alpha \frac{\partial P}{\partial t} \qquad (6.96)$$

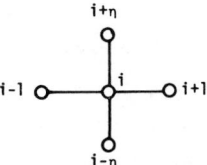

Figure 6.11: Two-dimensional grid at i.

Writing a fully implicit formulation at $(n + 1)$ time level:

$$\frac{P_{i+1}^{n+1} - 2P_i^{n+1} + P_{i-1}^{n+1}}{\Delta x^2} + \frac{P_{i-\eta}^{n+1} - 2P_i^{n+1} + P_{i+\eta}^{n+1}}{\Delta y^2} = \frac{\alpha}{\Delta t}(P_i^{n+1} - P_i^n) \quad (6.97)$$

To implement LSOR the pressure values on a line are solved simultaneously, assuming that those values on the previous lines—i.e., $i - \eta$ terms—are known, since they have just been calculated, and the values at $(i + \eta)$ are approximated by the old iteration values. Therefore,

$$\overset{k+1}{P_{i+1}^{n+1}} - (4 + C)\overset{k+1}{P_i^{n+1}} + \overset{k+1}{P_{i-1}^{n+1}} = -\overset{k+1}{P_{i-\eta}^{n+1}} - \overset{k+1}{P_{i+\eta}^{n+1}} - CP_i^n \quad (6.98)$$

where $C = \dfrac{\Delta x^2 * \alpha}{\Delta t}$

k = iteration number
n = time step number
i = cell index

This technique resolves the two-dimensional problem to a succession of temporary one-dimensional problems which can be solved rather efficiently.

Equation (6.98) is of tridiagonal form since all terms on the right-hand side are known at iteration number k. This equation when written for each point on a line produces a diagonally dominant tridiagonal matrix which is readily solvable. Equation (6.98) can be written summarily as:

$$a_i \overset{k+1}{P_{i+1}^{n+1}} - b_i \overset{k+1}{P_i^{n+1}} + c_i \overset{k+1}{P_{i+1}^{n+1}} = d_i^k \quad (6.99)$$

The mechanics of obtaining a solution can vary depending on the particular selection of the sweep direction. Consider the two-dimensional system shown in Fig. 6.12. The system is swept row by row from the bottom or top depending on the numbering system, and the P_i^{k+1}-values are solved for simultaneously on a line, as explained above. After each line computation, the next approximation to the P_i^{k+1}-values are obtained by overrelaxation as follows:

$$P_i^{k+1} = P_i^k + \omega(P_i^{k+1} - P_i^k) \quad (6.100)$$

The iterative procedure is continued until a convergence criterion similar to Eq. (6.95) is satisfied. If the relaxation parameter ω is judiciously selected, rapid convergence is assured.

Selection of Overrelaxation Parameter

For linear partial differential equations which are operative over regularly bounded figures like circles or rectangles the ω_{optimum} can be calculated by analytical means. Varga[3] shows methods for doing this. Since the number

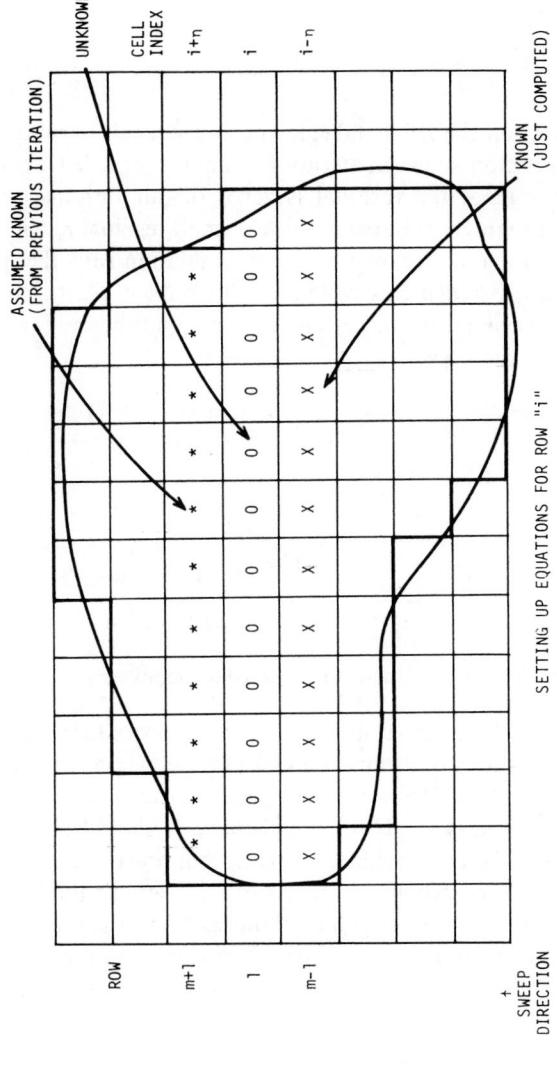

Figure 6.12: LSOR: Line successive over relaxation.

of iterations required is significantly affected by the value of ω, some care should be taken in selecting this parameter.

There is no foolproof way to obtain the correct ω for real-world problems, but a trial-and-error procedure produces a value which gives consistently good convergence times. Consider Eq. (6.101):

$$a_i P_{i+1}^{k+1} - b_i P_i^{k+1} + c_i P_{i-1}^{k+1} = d_i^k \qquad (6.101)$$

The values of P_i on the left-hand side are continuously varying with each iteration. This variation between iterations can be regarded as a residual quantity; at convergence the residual is zero. In our iterative scheme we shall not go to the limit zero, but shall pick a small residual r_m and by using varying values of ω determine by a series of simulation runs the number of iterations necessary to reach this residual. This data is plotted graphically in Fig. 6.13. The curve has a characteristic shape, and the best value of ω is. the lowest point on the curve.

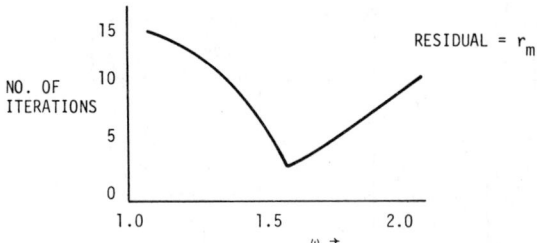

Figure 6.13: Selection of ω based on iterations.

Another method involves a similar graphical procedure except that a fixed number of iterations is set and the value of the maximum residual after these iterations is plotted versus ω. The type of plot shown in Fig. 6.14 is obtained. The optimum ω is obtained as before. It should be stressed that the ω_{opt} is an empirical factor which is hard to pin down. The experience of the engineer with the reservoir and other similar reservoir model configurations is a great help. Generally, the more homogeneous the system the closer the value of ω is to 1.0, and as the degree of anisotropy increases the more ω approaches 2.0. When ω approaches ω_{opt} from below, the curve has an al-

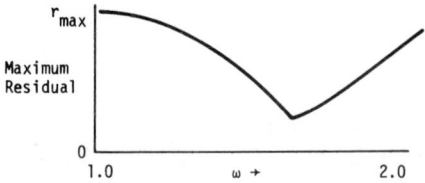

Figure 6.14: Selection of ω based on residuals.

most vertical tangent,[4] and a value of ω chosen just too small causes poor convergence. A value just above does not affect the convergence as radically, since the slope beyond ω_{opt} is close to linear. In practice, an overestimation is always better than an underestimation.

Strongly Implicit Procedure

The strongly implicit procedure (SIP) developed by Stone[10] is an iterative technique which has a significantly higher rate of convergence and which is not sensitive to the selection of iteration parameters. SIP involves the solution of the system of simultaneous linear equations by an elimination process working on a modified version of the original matrix system. The original matrix A is transformed to a matrix B which is more easily factorable into an LU product. The LU-matrices have only three nonzero elements in each row; this minimizes the work required to solve the system by elimination.

Modification of the original pentadiagonal matrix can be obtained in several different ways, and a suitable definition of the transformation must be selected which will result in rapid convergence of the iterative process. The following summarizes the application of SIP and a flow chart is given in Fig. 6.15.

Consider the two-dimensional fluid flow equations discretized over an (i, j) grid. The general equation results in the typical pentadiagonal, as shown in Eq. (6.45). The transformation of the matrix A involves the addition of two diagonals. Two new coefficients are introduced into each row:

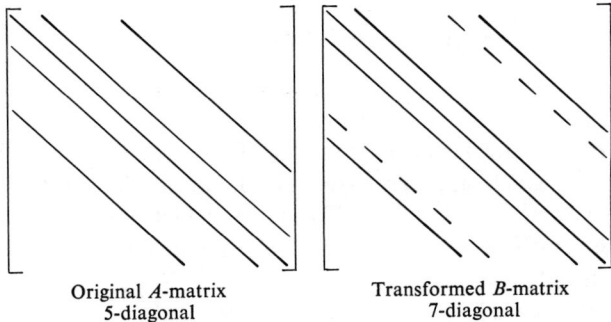

Original A-matrix
5-diagonal

Transformed B-matrix
7-diagonal

These new coefficients are associated with the cells $(i - 1, j + 1)$ and with $(i + 1, j - 1)$ in the grid system shown below:

$i - 1, j + 1$	$i, j + 1$	0
$i - 1, j$	i, j	$i + 1, j$
0	$i, j - 1$	$i + 1, j - 1$

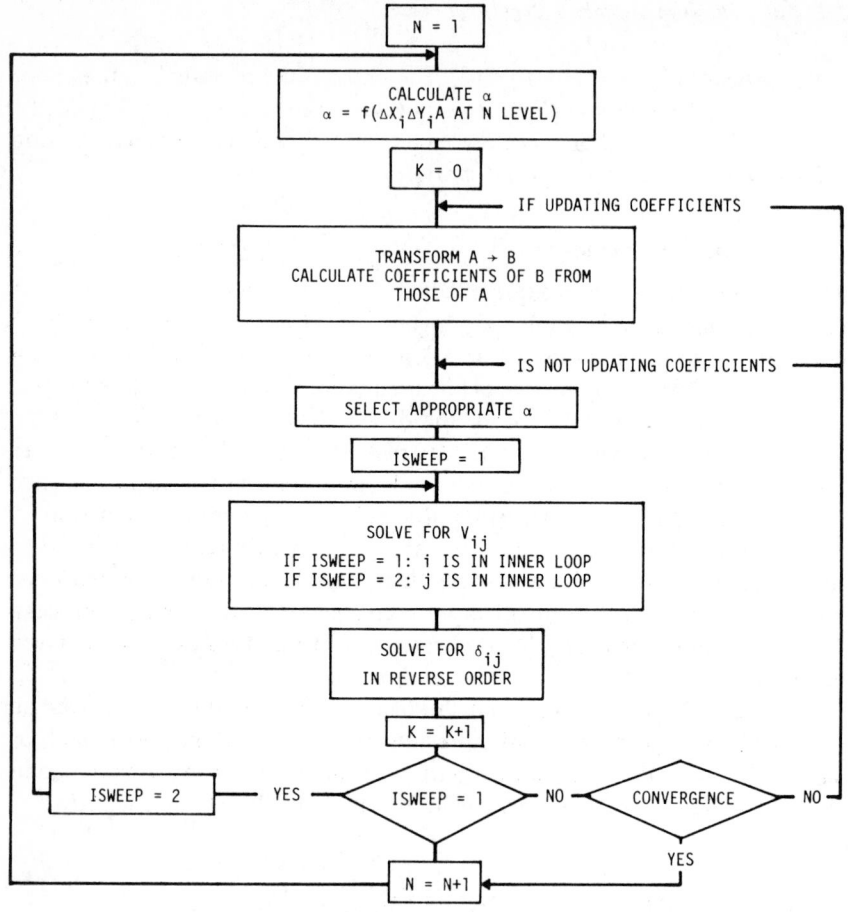

Figure 6.15: SIP flow chart.

To minimize the influence of these new coefficients they are balanced by subtracting terms which are approximately equal to the terms obtained at $(i-1, j+1)$ and $(i+1, j-1)$. The terms which are subtracted involve combinations of the five known cell pressures which are obtained by Taylor series expansions:

$$P_{i+1,j-1} = P_{i+1,j} - \Delta y_i \frac{\partial P}{\partial y} + 0(\Delta y^2) \tag{6.102}$$

$$P_{i-1,j+1} = P_{i,j+1} - \Delta x_i \frac{\partial P}{\partial x} + 0(\Delta x^2) \tag{6.103}$$

$$P_{i,j} = P_{i,j-1} + \Delta y_i \frac{\partial P}{\partial y} + 0(\Delta y^2) \tag{6.104}$$

$$P_{i,j} = P_{i-1,j} + \Delta x_i \frac{\partial P}{\partial x} + 0(\Delta x^2) \tag{6.105}$$

148

Combining Eqs. (6.102) and (6.104) we can write:

$$P_{i+1,j-1} \simeq (-P_{ij} + P_{i+1,j} + P_{i,j-1})\alpha \qquad (6.106)$$

And combining Eqs. (6.103) and (6.105) we can also write:

$$P_{i-1,j+1} \simeq (-P_{i,j} + P_{i,j+1} + P_{i-1,j})\alpha \qquad (6.107)$$

where α is a selected parameter which gives a better approximation to the pressure. The new coefficients are multiplied by ϵ_1 and ϵ_2, which minimizes their effects:

$$\epsilon_1 = P_{i+1,j-1} - \alpha(-P_{ij} + P_{i+1,j} + P_{i,j-1}) \qquad (6.107)$$
$$\epsilon_2 = P_{i-1,j+1} - \alpha(-P_{ij} + P_{i,j+1} + P_{i-1,j}) \qquad (6.108)$$

The original equation for each cell then becomes:

$$
\begin{aligned}
e_i P_{i,j-1} &+ a_i P_{i-1,j} + b_i P_{ij} + c_i P_{i+1,j} + f_i P_{i,j+1} \\
&+ g_i[P_{i+1,j-1} - \alpha(-P_{ij} + P_{i+1,j} + P_{i,j-1})] \\
&+ h_i[P_{i-1,j+1} - \alpha(-P_{ij} + P_{i,j+1} + P_{i-1,j})] \\
&= d_i
\end{aligned} \qquad (6.109)
$$

The first five terms are the original finite-difference formulation, and the last two terms reflect the additional terms which are introduced by SIP.

Iteration is carried out as follows:

$$BP^{k+1} - BP^k = AP^k - d^k \qquad (6.110)$$

or

$$\delta^{k+1} = P^{k+1} - P^k \qquad (6.111)$$
$$B\delta^{k+1} = R^k \qquad (6.112)$$

Factor B to LU:

$$hU\delta^{k+1} = R^k \qquad (6.113)$$
$$LV = R^k \qquad (6.114)$$

or

$$l_{1i}V_{i,j-1} + l_{2i}V_{i-1,j} + l_{3i}V_{ij} = R_i \qquad (6.115)$$

where l_1, l_2, and l_3 are the elements of the lower matrix obtained by factoring B.

Then from Eqs. (6.113) and (6.114), V is defined from

$$U\delta^{k+1} = V \tag{6.116}$$

or

$$\delta^{k+1} + \mu_{2i}\delta^{k+1}_{i+1,j} + \mu_{3i}\delta^{k+1}_{i,j+1} = V_{ij} \tag{6.117}$$

The new pressure increment is computed from Eq. (6.117).

Iteration Parameter α

A set of nine evenly spaced parameters is used in the iteration process. Each parameter is used twice in succession. A suggested sequence (Stone[10]) of parameters is: 9, 9; 6, 6; 3, 3; 8, 8; 5, 5; 2, 2; 7, 7; 3, 3; 1, 1. The minimum parameter used was zero, and the maximum was determined from the grid dimensions and the coefficients of the A-matrix. The maximum α_{max} is

$$\alpha_{max} = \frac{1}{NM} \sum_{i=1}^{N} \sum_{i=1}^{M} \left[1 - \min\left(\frac{(2\Delta x_i^2)\Big/\left(\sum_{i=1}^{N} \Delta x_i\right)}{1 + \dfrac{ay_{ij} + by_{ij}}{ax_{ij} + bx_{ij}}}, \frac{(2\Delta y_i^2)\Big/\left(\sum_{j=1}^{M} \Delta y_i\right)}{1 + \dfrac{ax_{ij} + bx_{ij}}{ay_{ij} + by_{ij}}} \right) \right] \tag{6.118}$$

The above equation differs slightly from that given by Stone.[10] The parameters are used recursively in order of increasing value; the other parameters are computed from

$$\alpha_i = \tfrac{1}{8}\alpha_{max} \tag{6.119}$$

The strongly implicit procedure is superior to most of the iterative methods except ADI in all but the simplest homogeneous and isotropic problems. In the problems currently encountered in practical reservoir simulation, homogeneity is more the exception than the rule, and these problems lend themselves more readily to SIP.

6.4   DIRECT METHODS IN SIMULATION PRACTICE

In recent years there have been improvements in the direct method approach to solving the pressure equations. These methods have developed as a result of breakthroughs in sparse matrix technology as used in operations research and power transmission design and other related fields. The main thrust of these new approaches is to reorder the matrices in such a way as to minimize the computational work done in the elimination process.

An algorithm is then designed to solve the system as efficiently as possible by operating only on the nonzero elements of the sparse reordered matrix. As a result of these reordering methods and computational schemes the increased speed of the execution has made these methods competitive for many reservoir problems which were formerly too large.

Optimal Reordering

There are several methods proposed for the optimal ordering of the matrix, and the work associated with each ordering scheme varies with the particular scheme. The storage allocation is also dependent on the number of nonzero elements in the original ordering and also on the nonzero elements derived from the solution process.

Coats et al.[7] have presented several schemes based on optimal ordering and the work and storage involved as functions of size (I, J).

1. *Regular row ordering:* The first scheme is the regular row ordering for Gaussian elimination where the numbering is along the shortest dimension first, as shown in Fig. 6.16. This ordering is used to minimize the band width for the application of the bandsolve routine. The work involved and

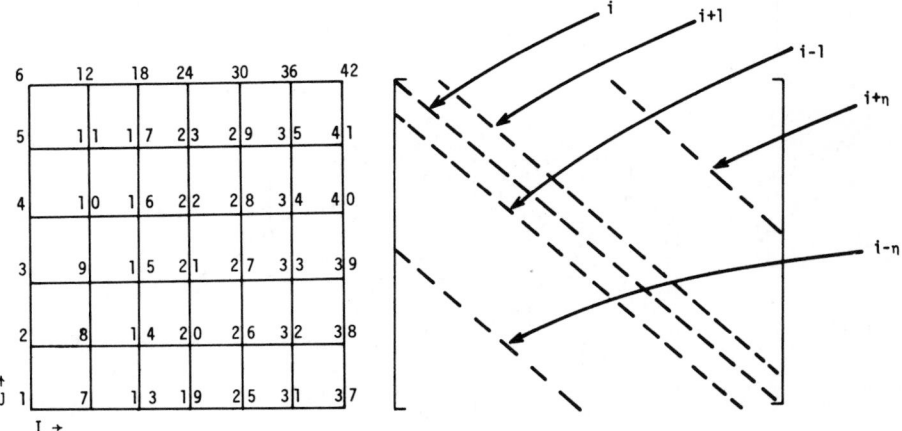

Figure 6.16: Regular ordering and matrix system.

the storage required are:

$$W = IJ^3 \qquad\qquad (6.120)$$

$$S = IJ^2 \qquad\qquad (6.121)$$

2. *Diagonal ordering* (Fig. 6.17): In this ordering the cells are numbered consecutively along the diagonals starting with the shortest direction. This method groups the cells by diagonal count, which by inspection increases as we move from the lower left through the grid to the upper right.

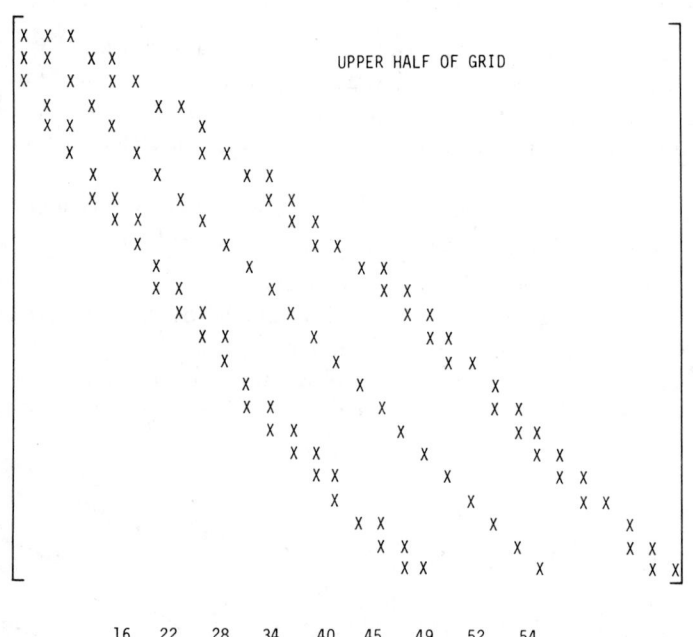

Figure 6.17: Diagonal ordering and matrix system.

The elements on each consecutive diagonal are:

Diagonal	Number of Elements	Cells
(1)	1	1
(2)	2	2, 3
(3)	3	4, 5, 6
(4)	4	7, 8, 9, 10
(5)	5	11, 12, 13, 14, 15
(6)	6	16, 17, 18, 19, 20, 21
(7)	7	22, 23, 24, 25, 26, 27
(8)	7	28, 29, 30, 31, 32, 33
(9)	7	34, 35, 36, 37, 38, 39
(10)	6	40, 41, 42, 43, 44
(11)	5	45, 46, 47, 48
.	.	.
.	.	.
.	.	.
(15)	.	54

The work and storage requirments are:

$$W = IJ^3 - \frac{J^4}{2}$$

$$S = IJ^2 - \frac{J^3}{3}$$

3. *Alternating point ordering:* This is cyclic ordering obtained by dividing the grid points into two groups, circles and squares (Fig. 6.18), then numbering the points such that no two similar cells are consecutively numbered. Each point is separated from its adjacent number by at least one other point. The work involved in the cyclic ordering scheme is:

$$W = \frac{IJ^3}{2}$$

$$S = \frac{IJ^2}{2}$$

4. *Alternating diagonal ordering scheme:* This scheme orders the grid points on alternating diagonals (Fig. 6.19), and it can be seen to be a combination of the alternating point and diagonal orderings. It has produced the greatest reduction in work of the methods examined here. The work involved

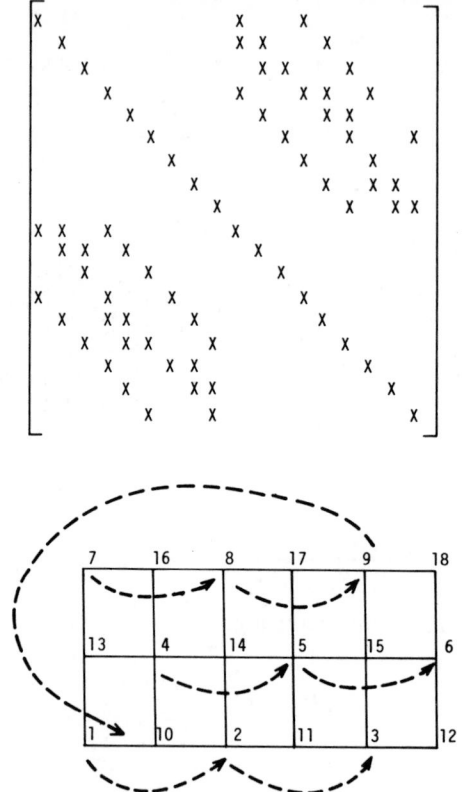

Figure 6.18: Alternative point ordering and matrix system.

is given the approximate relation:

$$W = \frac{IJ^3}{2} - \frac{J^4}{4}$$

$$S = \frac{IJ^2}{2} - \frac{J^3}{6}$$

Sparse Matrix Techniques

Another improvement on the optimal ordering technique is to use processes which are geared specifically to the direct solution of these processes. One method used in the solution of transistor-switching circuits by Gustavson et al.[8] and now being used is the *generate-and-solve* algorithms.[9] Their method makes use of the computer architecture in developing the algorithm. The basis of the method can be summarized as follows:

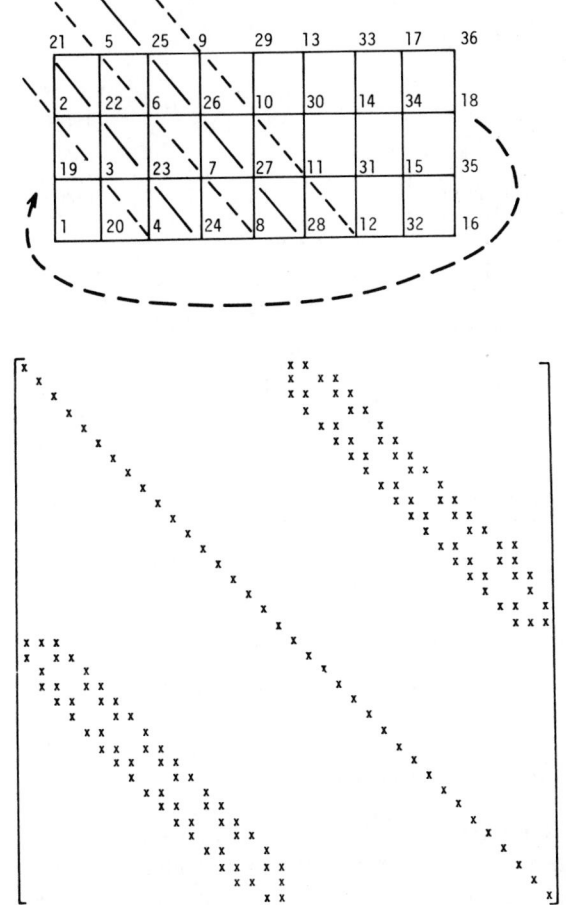

Figure 6.19: Alternate diagonal ordering and matrix system.

1. In a given matrix there is a fixed number of nontrivial operations required to solve the system by some given direct scheme.

2. The solution process produces some nonzero elements which will always be present at a given (i, j) location in the matrix.

3. If the location of all nonzero quantities is fixed a priori both in the matrix and in its solution, then a straightforward, nonbranching, nonlooping computer code can be developed to solve this system very efficiently.

The idea of generating a code and then solving the problem gives rise to the name "generate and solve." The process can be examined schematically in Fig. 6.20.

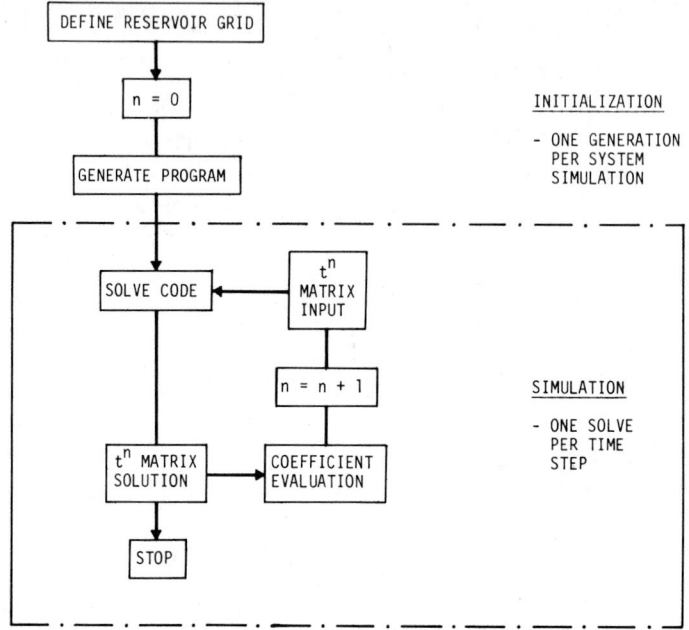

Figure 6.20: Generate and solve method.

6.5 COMPARATIVE ANALYSIS

The methods examined in this chapter present a wide range of features which may prove effective in some applications and just plain useless in others. The computational speed is a reasonable measure of the cost effectiveness of a solution algorithm, but this again is not always as easy a term to define today. In general, iterative schemes are computationally faster per time step than are direct methods. This advantage, however, can be easily lost if the required number of iterations becomes too large. Recently, however, some direct methods (see Sec. 6.4) have evolved which minimize computational labor by the efficient ordering of the matrix and also by using storage algorithms which exploit the way the machine accesses data. This technique reduces the computer overhead, which is generally billed to the user. Since the advent of virtual storage machines the efficient use of storage has become an even more critical parameter, not because of storage limitations but because of the way in which "pages" of slow memory are brought into and out of core. If the storage is inefficient, the program could expend many times the required time fetching and re-storing information required for a computation. The engineer must be familiar with the billing algorithm used to come up with the true cost of running a simulation study in order to predict study costs.

Accuracy and stability are machine-dependent to some extent due to propagation of round-off errors; they are also dependent on the time step selection, assuming the same order of error is involved in the discretization process. There is a trade-off between iterative methods and direct methods as far as stability is concerned if we can guarantee machine precision such that round-off will not be significant. The iterative techniques break down horribly if the coefficient matrix is ill conditioned (nearly singular); direct methods are still able to generate solutions in this case. The problem associated with the iterative method can be resolved by taking smaller time steps; this increases diagonal dominance. The question then arises: Is the computational time involved in n iterative runs less than that of a single direct-step run over the same number of days simulated? The engineer must address himself to this when he has to make a decision.

On the basis of programming ease the direct methods are somewhat more complicated, primarily due to the manipulation of indices and character strings in the process of storing the sparse matrices efficiently. Iterative methods are straightforward in that the coefficients are usually stored as one-dimensional vectors.

The greatest degree of difference in the methods covered here is seen in the areas of applicability. Some iterative methods, particularly the alternating-direction implicit methods, do not function at all in those reservoir problems where there are marked directional permeability trends. In these anisotropic cases the ADIP techniques become unstable and produce reversals in potentials which are physically unrealistic. Iterative ADIP sometimes overcomes this tendency to diverge, but by far the better methods are the relaxation techniques. In coning and cross-sectional studies the ADIP is not competitive with iterative ADIP or LSOR, and some of these methods have to be "helped" by the use of implicit formulations of mobility terms.

REFERENCES

1. P. D. Crout, "A Short Method for Evaluating Determinants and Solving Systems of Linear Equations with Real or Complex Coefficients," *Trans. AIEE* (1941), **60**, 1235.

2. D. W. Peaceman and H. H. Rachford, "The Numerical Solution of Parabolic and Elliptic Differential Equations," *J. Soc. Indust. Appl. Math.* (1955), **3**, 28–41.

3. R. S. Varga, *Matrix Iterative Analysis* (Englewood Cliffs, N.J.: Prentice Hall, 1962).

4. H. R. SCHWARZ, H. RUTISHAUSER, and E. STIEFEL, *Numerical Analysis of Symmetric Matrices* (Englewood Cliffs, N.J.: Prentice-Hall, 1973).

5. G. BIRKHOFF, R. S. VARGA, and D. YOUND, "Alternating Direction Implicit Methods," *Advances in Computers* (1962), **3**, 189–273.

6. E. A., BRIETENBACH, D. H. THURNAU, and H. K. VAN POOLLEN, "Solution of the Immiscible Fluid Flow Simulation Equations," *Soc. Pet. Eng. J.* (June 1969), 155–69.

7. H. S. PRICE and K. H. COATS, "Direct Methods in Reservoir Simulation," SPE 4278, paper presented at the third SPE Symposium on Numerical Simulation, Houston, Texas, Jan. 10–12, 1973.

8. F. G. GUSTAVSON, W. LINIGER, and R. WILLOUGHBY, "Symbolic Generation of an Optimal Crout Algorithm for Sparse Systems of Linear Equations," *J. ACM* (1970), **17**, 87–109.

9. P. F. WOO, S. J. ROBERTS, and F. G. GUSTAVSON, "Application of Sparse Matrix Techniques in Reservoir Simulation," SPE 4544, paper presented at the 48th Annual Meeting, Las Vegas, Nev., Oct. 1973.

10. H. L. STONE, "Iterative Solutions of Implicit Approximations of Multidimensional Partial Differential Equations," *SIAM J. on Numerical Analysis* (1968), **5**, 530.

BIBLIOGRAPHY

BJORDAMMEN, J. and K. H. COATS, "Comparison of Alternating-direction and Successive Overrelaxation Techniques in Simulation of Reservoir Fluid Flow," *Soc. Pet. Eng. J.* (March 1969), 47–58.

BRUCE, G. H., D. W. PEACEMAN, H. H. RACHFORD, JR., and J. D. RICE, "Calculations of Unsteady-state Gas Flow Through Porous Media," *Trans. AIME* (1953), **198**, 79–92.

CARTER, R. D., "Performance Predictions for Gas Reservoirs Considering Two-dimensional Unsteady-state Flow," *Soc. Pet. Eng. J.* (March 1966), 35–43; *Trans. AIME*, **237**.

CAVENDISH, J. C., H. S. PRICE, and R. S. VARGA, "Variational Methods for the Numerical Solution of Boundary Value Problems," SPE 2034, Numerical Simulation of Reservoir Performance, Dallas, Texas, April 22–23, 1968.

CAVENDISH, J. C., H. S. PRICE, and R. S. VARGA, "Galerkin Methods for the Numerical Solution of Boundary Value Problems," *Soc. Pet. Eng. J.* (June 1969), 204–20; *Trans. AIME*, **246**.

COATS, K. H. and M. H. TERHUNE, "Comparison of Alternating Direct Explicit and Implicit Procedures in Two-dimensional Flow Calculations," *Soc. Pet. Eng. J.* (Dec. 1966), 350–62; *Trans. AIME*, **237**.

CULHAM, W. E. and RICHARD S. VARGA, "Numerical Methods for Time-dependent, Nonlinear Boundary Value Problems," SPE 2806, Second Symposium on Numerical Simulation of Reservoir Performance, Dallas, Texas, Feb. 5–6, 1970.

DYKSTRA, H. and R. L. PARSONS, "Relaxation Methods Applied to Oil Field Research," *Trans. AIME* (1951), **192**, 227–32.

LAUMBACH, DALLAS D., "A High Accuracy Finite-difference Technique for Treating the Convection-Diffusion Equation", SPE 3996, 47th Annual Meeting, San Antonio, Texas, Oct. 8–11, 1972.

MATLOCK, H., "Applications of Numerical Methods to Some Structural Problems in Offshore Operations," *J. Pet. Tech.* (Sept. 1963) 1040–46; *Trans. AIME*, **228**.

PEACEMAN, D. W. and H. H. RACHFORD, "The Numerical Solution of Parabolic and Elliptic Differential Equations," *J. Soc. Indust. Appl. Math.* (1955), **3**, 28–41.

QUON, D., P. M., DRANCHUK, S. R. ALLADA, and P. K. LEUNG, "Application of the Alternating Direction Explicit Procedure to Two-dimensional Natural Gas Reservoirs," *Soc. Pet. Eng. J.* (June 1966), 137–42; *Trans. AIME*, **237**.

ROUTT, KENNETH R. and PAUL B. CRAWFORD, "A New and Fast Method for Solving Large Numbers of Reservoir Simulation Equations," SPE 4277, Third Symposium on Numerical Simulation of Reservoir Performance, Houston, Texas, Jan. 10–12, 1973.

STONE, H. L., "Iterative Solution of Implicit Approximations of Multidimensional Partial Differential Equations," *SIAM Numer. Anal.* (1968), **5**, 530.

VARGA, R. S., *Matrix Iterative Analysis* (Englewood Cliffs, N.J.: Prentice-Hall, 1962).

WATTS, J. W., "An Iterative Matrix Solution Method Suitable for Anisotropic Problems," *Soc. Pet. Eng. J.* (March 1971), 47–51; *Trans. AIME*, **251**.

7 Data Preparation

7.1 INTRODUCTION: "GETTING IT ALL TOGETHER"

There is a well-used cliché in the computer world—GIGO—an acronym that states a simple fact: "Garbage in, Garbage out." The underlying thesis is simple: the quality of the output is no better than the quality of the input. The data required to make a simulation study come from several sources and are accessible to the engineer to a greater or lesser degree. The data itself are usually in a form not directly applicable to a computer solution, and some preprocessing must be undertaken to produce the data in usable form. There are usually several sources of the same data information, and the engineer must exercise his judgment in differentiating and selecting the best data available. Sometimes there are no data available for a particular case; in a situation like this the engineer must determine some alternate means of obtaining the same information. Some of these techniques are discussed later.

The groups of data generally required in making a simulation run are as follows:

a. Fluid data
b. Rock data
c. Production data
d. Flow rate data
e. Mechanical and operational data
f. Economic data
g. Miscellaneous data

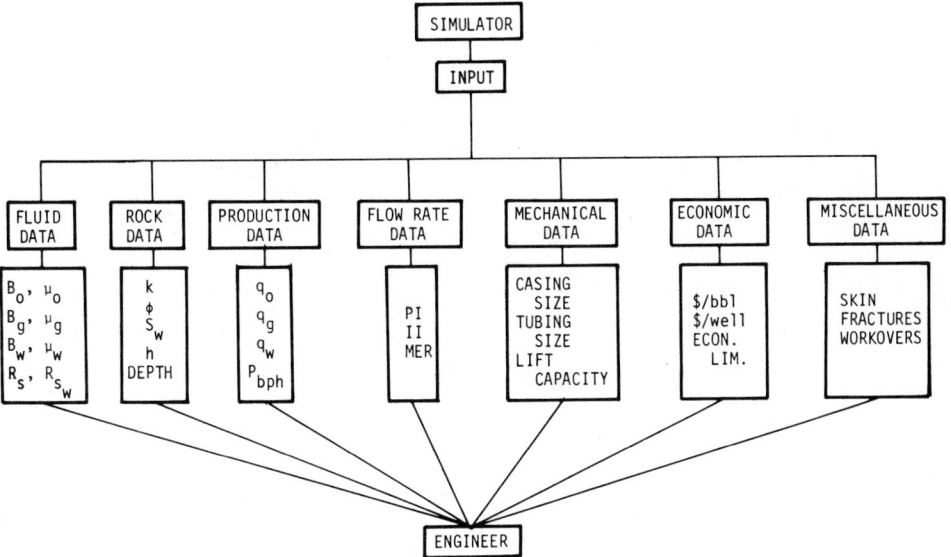

Figure 7.1: Data sources and parameters.

Figure 7.1 illustrates the sources and parameters. We shall discuss each of these areas in turn and determine how best the required data item can be obtained.

7.2 FLUID DATA

The reservoir fluids have properties which must be evaluated many times during the simulation of a reservoir under depletion or under some secondary or tertiary mechanism. An examination of the gas equation shows some of the properties which must be evaluated:

$$\frac{\partial}{\partial x}\left(\frac{k_g}{\mu_g B_g} + \frac{R_{so}k_o}{\mu_o B_o} + \frac{R_{sw}k_w}{\mu_w B_w}\right)\frac{\partial \Phi_g}{\partial x} + q_g(x, t) = \frac{\partial}{\partial t}\left[\phi\left(\frac{S_g}{B_g} + \frac{R_{so}S_o}{B_o} + \frac{R_{sw}S_w}{B_w}\right)\right]$$

(7.1)

The pressure-dependent fluid properties are:

1. Formation volume factors (Fig. 7.2)
2. Fluid viscosity (Fig. 7.3)
3. Solution gas-oil ratio (Fig. 7.4)

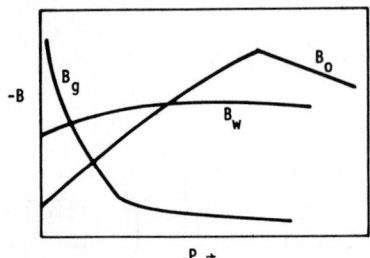

Figure 7.2: Formation volume data.

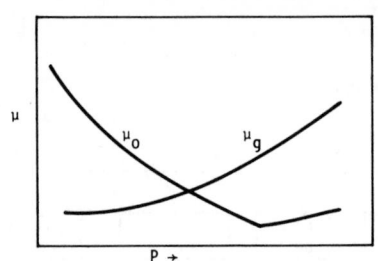

Figure 7.3: Fluid viscosity data.

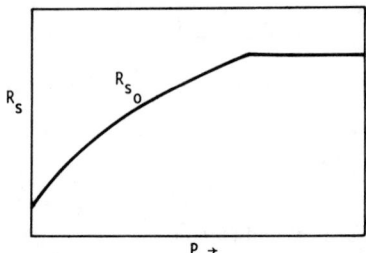

Figure 7.4: Solution gas-oil ratio.

These properties are generally obtained from laboratory studies of samples of the reservoir fluid. These studies form part of the regular pressure-volume-temperature (PVT) work done on a sample.

Input Form

Polynomial Representation: The PVT data listed above lend themselves quite readily to representations of the following type:

$$y = a_0 + a_1x + a_2x^2 + a_3x^3 \qquad (7.2)$$

where the dependent variable y can be any of the functions listed above, and x is an independent variable.

$$B_o = 1.053 + 0.000149P \qquad \text{for } P < P_b \qquad (7.3)$$

These polynomial expressions are very easy to develop with any least-squares program available on a routine basis. The engineer should be completely satisfied that the expression used generates adequate data within the range of pressures studied. The only way to be sure is to generate a table of values throughout the range. In some studies there could be significant variations in fluid properties, due to either

1. extreme areal size, or
2. high-relief reservoirs.

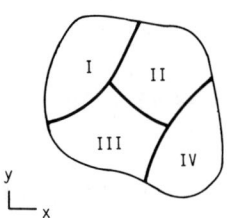

Figure 7.5: Areal variation.

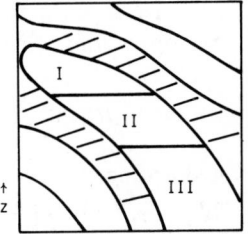

Figure 7.6: Vertical variation.

These variations are shown in Figs. 7.5 and 7.6. These changes in properties must be incorporated into the model to produce a good representation of the reservoir. The usual manner in which these are included is to define an array of *fluid data designators* (FDD) in each fluid data area (FDA), which would select the proper equation for evaluating the PVT data—e.g., as shown in Figs. 7.7 and 7.8. The PVT data could be evaluated as follows:

$$B_{o(i)} = a_{0\,(K)} + a_{1\,(K)}P_i + a_{2\,(K)}P_i^2 \qquad (7.4)$$

where K is the FDD corresponding to cell i.

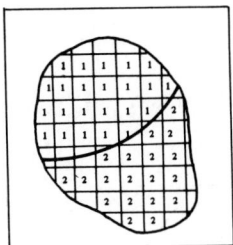

Figure 7.7: Fluid data designator (FDD).

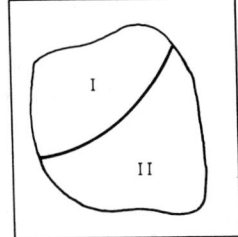

Figure 7.8: Fluid data areas (FDA).

The coefficients $a_{0,k}$, $a_{1,k}$, $a_{2,k}$ will form a matrix of values which are read into the program. In the example given, there would be two sets of *a*-coefficients used, as shown:

$$\text{Region (1):} \quad B_{o(1)} = a_{0\,(1)} + a_{1\,(1)}P + a_{2\,(1)}P^2 \qquad (7.5)$$

$$\text{Region (2):} \quad B_{o(2)} = a_{0\,(2)} + a_{1\,(2)}P + a_{2\,(2)}P^2 \qquad (7.6)$$

Within the program itself these polynomials must be efficiently evaluated, since the PVT data involve several evaluations at a given cell every time step or several times within a time step if there were iterations on pressure. An inefficient method of evaluating Eq. (7.2) is by straight exponentiation:

$$Y = A0 + A1*P + A2*P**2 + A3*P**3 + A4*P**4 \qquad (7.7)$$

This equation has too many exponentiations, and these are the most time-consuming operations. The following formulation has at most n multiplications:

$$Y = (((((A4*P + A3)*P) + A2)*P) + A1)*P + A0 \qquad (7.8)$$

This nested arrangement is as efficient as you can get without special evaluation formulas and should always be used in the evaluation of polynomials in simulators.

Table of Look-up: Some PVT data are not easily represented by a polynomial expression, either because the functions have discontinuous segments or because the order of polynomials are generally too high to be programmed economically—there are too many multiplications. The solution to this problem is to use a "look-up table." A series of pairs of ordered data points are developed to represent the data over a given range. These values have to be stored in the computer to be used on call. In order to evaluate a parameter, the array of data for the independent variable is searched until an interval is found in which the variable is bracketed; the dependent variable is calculated explicitly by interpolation (either linear or Lagrangian):

EXAMPLE [1]   Pressure = 2176; B_o required at 2176.

TABLE 7.1
Oil Formation Volume Data Used in Look-Up Table

B_o	P
1.26	2000
1.27	2050
1.28	2100
$B_o \rightarrow$ 1.30	2150
1.35	2200 ← 2176
1.40	2250
1.36	2300
1.35	2350

A simple linear interpolation produces:

$$B_{o(2176)} = B_{o(2150)} + \frac{2176 - 2150}{2200 - 2150}[B_{o(2200)} - B_{o(2150)}]$$

$$= 1.30 + \frac{26}{50}(1.35 - 1.30)$$

$$= 1.326$$

The mechanics of a table of look-up applied to a given curve are shown in Fig. 7.9. The model requires determining the value of P_c at various values of S_w. We can accomplish this by characterizing the various segments of the curve and inputting the pairs of data points as required.

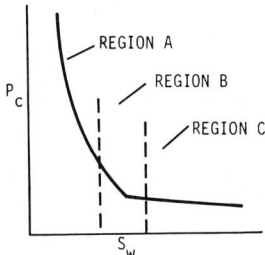

Figure 7.9: Capillary pressure versus water saturation.

Note in Fig. 7.9 that in Region A there is a slight curvature to the graph; but it shows an almost linear relationship. We shall therefore need about five data points to represent this region. In Region B the curve is very sharp showing rapidly changing S_w-values but not a great change in P_c. We should therefore use more closely spaced data points in this region. This enables the table of look-up to duplicate the curve more closely. In Region C the relation is practically a straight line, and two values, one at either end of the line segment, will suffice. The objective is to obtain accuracy consistent with the data requirements using as few points as possible.

The two most frequently used methods of interpolation are illustrated in Figs. 7.10 and 7.11, and the equations are as follows:

Linear:[1]

$$y^* = y_1 + \frac{x^* - x_1}{x_2 - x_1}(y_2 - y_1) \qquad (7.9)$$

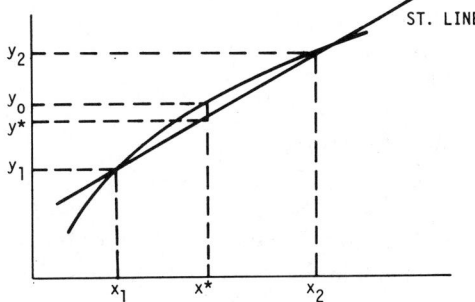

Figure 7.10: Linear interpolation.

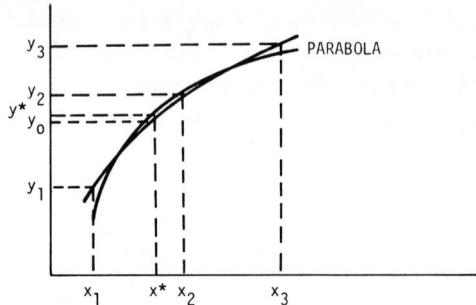

Figure 7.11: La Grangian interpolation.

Lagrangian:[1]

$$y^* = \left[\frac{x^* - x_2}{x_1 - x_2}\right]\left[\frac{x^* - x_3}{x_1 - x_3}\right]y_1 + \left[\frac{x^* - x_1}{x_2 - x_1}\right]\left[\frac{x^* - x^3}{x_2 - x_3}\right]y_2$$

$$+ \left[\frac{x^* - x_2}{x_3 - x_2}\right]\left[\frac{x^* - x_1}{x_3 - x_1}\right]y_3 \qquad (7.10)$$

Note that Lagrangian interpolation involves quite a few more multiplications, but fewer data points are required for the same range of the variables.

7.3   ROCK DATA

The various parameters needed to define the physical extent of the reservoir and to evaluate the transmissibilities during the simulation run must be input in some form. The required data are:

1. Permeability
2. Porosity
3. Formation thickness
4. Formation elevations
5. Compressibility
6. Relative permeability
7. Formation fluid saturations
8. Capillary pressure

These data form the most voluminous portion of the input data required by the simulator. Every cell in the model must be identified by a given value of each of the parameters listed earlier. The result is a matrix of values for each parameter, an example being the permeability matrix in Figs. 7.12 and 7.13.

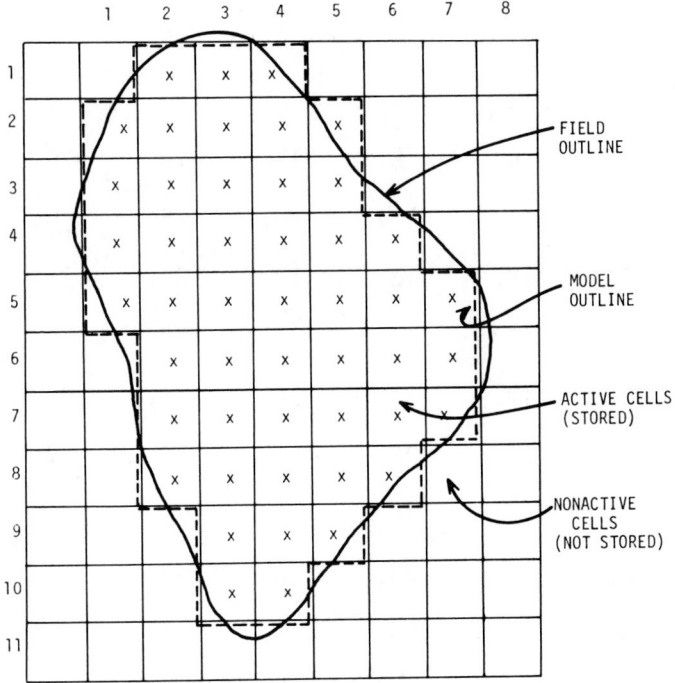

Figure 7.12: Cell identification.

Sources of Permeability Data

The absolute permeability data can be obtained from several sources:

1. Pressure build-up data (drill stem test)
2. Pressure falloff data
3. Interference tests
4. Initial potential test
5. Regression analysis (case history approach)
6. Laboratory measurements

The most important source of permeability data is pressure test analysis, and the engineer should be familiar with the current techniques available for obtaining the permeability values from these tests.

There are four commonly used methods for analysis of well test data:

1. Muskat method[2]
2. Miller-Dyes-Hutchinson method[3]
3. Horner method[4]
4. Type curve analysis[5]

	1	2	3	4	5	6	7
1	0.	4.	4.	4.	0.	0.	0.
2	4.	6.	8.	11.	13.	0.	0.
3	7.	9.	11.	15.	18.	0.	0.
4	10.	12.	13.	18.	20.	23.	0.
5	12.	14.	17.	20.	23.	25.	25.
6	0.	19.	21.	22.	26.	27.	29.
7	0.	20.	23.	24.	27.	27.	30.
8	0.	22.	25.	26.	27.	0.	0.
9	0.	0.	26.	28.	29.	0.	0.
10	0.	0.	28.	29.	0.	0.	0.

Figure 7.13: Permeability data.

The theory behind these methods is well developed throughout the literature and as such the reader is referred to the many articles[6] on this subject. We shall cover only the essential elements here in order to be able to analyze effectively well test data.

Muskat Method: Muskat proposed that a log plot of average reservoir pressure, $\bar{P}$, minus shut-in wellbore pressure, P_{ws}, versus the shut-in time of a well will produce a straight line. However, since the value of $\bar{P}$, the average reservoir pressure, is usually unknown, the procedure is a trial-and-error process. The straight line is obtained in the *later* portion of the curve. If $\bar{P}$ is too high, the curve is concave upward and for $\bar{P}$ too low it is concave downward, as shown in Fig. 7.14 The intercept at $\Delta t = 0$ is a function of kh, and the equation indicating this relationship is as follows:

For a circular drainage area:

$$kh = \frac{118.6q\mu B}{(\bar{P} - P_{ws})_{\Delta t=0}} \qquad (7.11)$$

For a square drainage area:

$$kh = \frac{94.6q\mu B}{(\bar{P} - P_{ws})_{\Delta t=0}} \qquad (7.12)$$

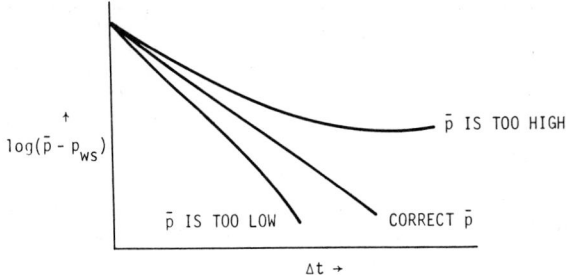

Figure 7.14: Muskat plot.

The slope m of the Muskat plot is a function of the drainage volume:
Circular:

$$\phi c \pi r_e^2 = \frac{0.00528k}{m\mu} \tag{7.13}$$

Square:

$$\phi c A = \frac{0.00471k}{m\mu} \tag{7.14}$$

The skin term can be obtained from a combination of terms:

$$S = 0.84 \frac{\bar{P} - P_{wf}}{(P - P_{ws})_{t=0}} - \ln \frac{r_e}{r_w} + 0.75 \tag{7.15}$$

EXAMPLE [2] MUSKAT METHOD: The data for the build-up test is shown in Table 7.2. From the graph in Fig. 7.15, $\bar{P} = 1860$ psig. Using Eq. (7.12):

$$kh = \frac{94.6 q \mu B}{(\bar{P} - P_{ws})_{\Delta t = 0}} \tag{7.16}$$

Then:

$$k = \frac{(94.6)(1.46)(0.85)(1.29)}{(205)(13)}$$

$$= 5.68 md \tag{7.17}$$

Using Eq. (7.15):

$$S = 0.84 \frac{\bar{P} - P_{wf}}{(P - P_{ws})_{t=0}} + \ln \frac{r_e}{r_w} + 0.75$$

$$= 0.84 \frac{1860 - 1426.9}{13} - \ln \frac{934}{0.25} + 0.75$$

$$= 8.77 \tag{7.18}$$

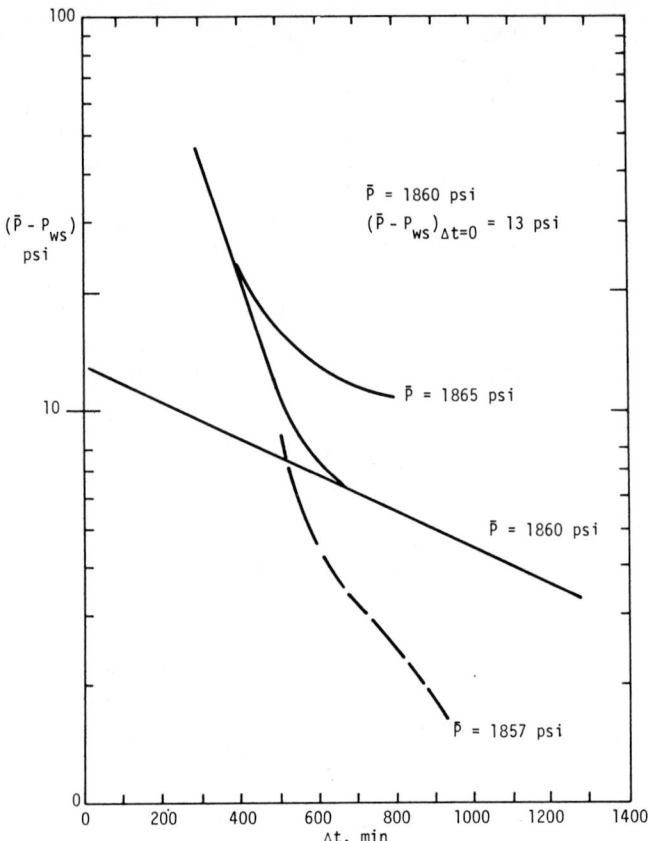

Figure 7.15: Muskat plot for Well #1.

Miller-Dyes-Hutchinson Method: This method is based on the assumption that the well has been producing long enough to reach a pseudo–steady state. With this assumption the solution to the pressure equation can be obtained in a straightforward manner. This relation is shown in Eq. (7.19):

$$\frac{kh}{162.6q\mu B}(\bar{P} - P_{ws}) = \log_{10} \Delta t + \text{Constant} \qquad (7.19)$$

Equation (7.19) suggests that a build-up pressure plot versus the logarithm of build-up time will produce a straight line whose slope is inversely proportional to the *kh*-product. From Eq. (7.19) the slope *m* is:

$$m = \frac{162.6q\mu B}{kh} \qquad (7.20)$$

TABLE 7.2
Pressure Build-up Data

	Well No. 1
Given Data	$q = 146$ bbl/day $\quad A = 20$ ac $h = 205$ ft $\qquad P_{wf} = 1426.9$ psig $\mu = 0.85$ cp* $\quad r_w = 0.25$ ft $B = 1.29$ $\qquad c_t = 12.03 \times 10^{-6}$ psi^{-1} $t = 53$ hr $\phi = 0.10$ Drainage shape: Square, no influx Well location: Center

Build-up Data:

Time		P_{ws} (psig)	$(t + \Delta t)/\Delta t$	$P_{ws} - P_{wf}$
hr	min			
0.167	10	1451.5	319.00	24.6
0.333	20	1476.0	160.00	49.1
0.500	30	1498.6	107.00	71.7
0.667	40	1520.1	80.50	93.2
0.833	50	1541.5	64.60	114.6
1.000	60	1561.3	54.00	134.4
1.167	70	1581.9	46.50	155.0
1.333	80	1599.7	40.80	172.8
1.500	90	1617.9	36.40	191.0
1.667	100	1635.3	32.80	208.4
2.000	120	1665.7	27.50	238.8
2.333	140	1691.8	23.70	264.9
2.667	160	1715.3	20.90	288.4
3.000	180	1736.3	18.70	309.4
3.333	200	1754.7	16.90	327.8
3.667	220	1770.1	15.50	343.2
4.000	240	1783.5	14.20	356.6
4.500	270	1800.7	12.80	373.8
5.000	300	1812.8	11.60	385.9
5.500	330	1822.4	10.60	395.5
6.000	360	1830.7	9.84	403.8
6.500	390	1837.2	9.15	410.3
7.000	420	1841.1	8.57	414.2
7.500	450	1844.5	8.07	417.6
8.000	480	1846.7	7.63	419.8
8.500	510	1849.6	7.23	422.7
9.000	540	1850.4	6.89	423.5
10.000	600	1852.7	6.31	425.8
11.000	660	1853.5	5.82	426.6
12.000	720	1854.0	5.42	427.1
12.667	760	1854.0	5.18	427.1
14.620	880	1855.0	4.61	428.1

*Centipoise.

and

$$kh = \frac{162.6q\mu B}{m} \qquad (7.21)$$

The skin factor can be calculated from Eq. (7.22), where the build-up pressure is read at 1-hour shut-in time, as shown in Fig. 7.16:

$$S = 1.151 \frac{P_{1hr} - P_{wf}}{m} - \log \frac{k}{\phi \mu c r_w^2} + 3.23 \qquad (7.22)$$

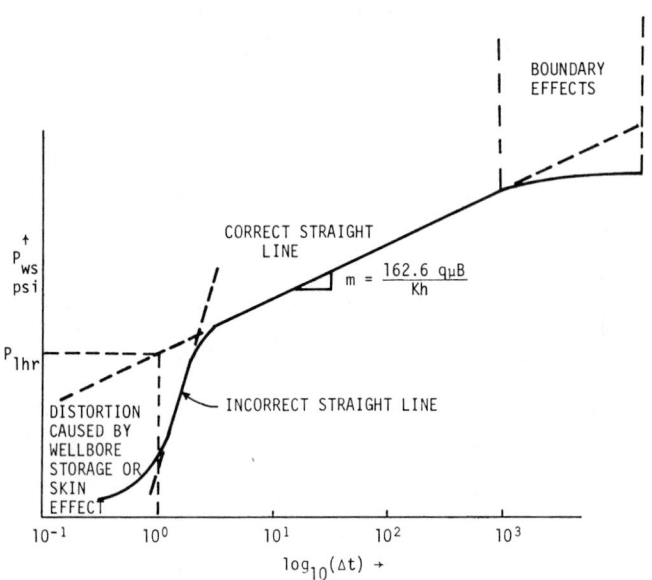

Figure 7.16: Miller-Dyes-Hutchinson plot.

EXAMPLE [3] MILLER-DYES-HUTCHINSON METHOD: The same data is used as in the Muskat example: the build-up pressure is plotted versus log build-up time; the slope is read off the build-up curve—Fig. 7.17. Permeability is:

$$m = 16 \text{ psi/cycle}$$

Using Eq. (7.20):

$$k = \frac{162.6q\mu B}{mh}$$

$$= \frac{(162.6)(146)(0.85)(1.29)}{(16)(205)}$$

$$= 7.94 \text{ md} \qquad (7.23)$$

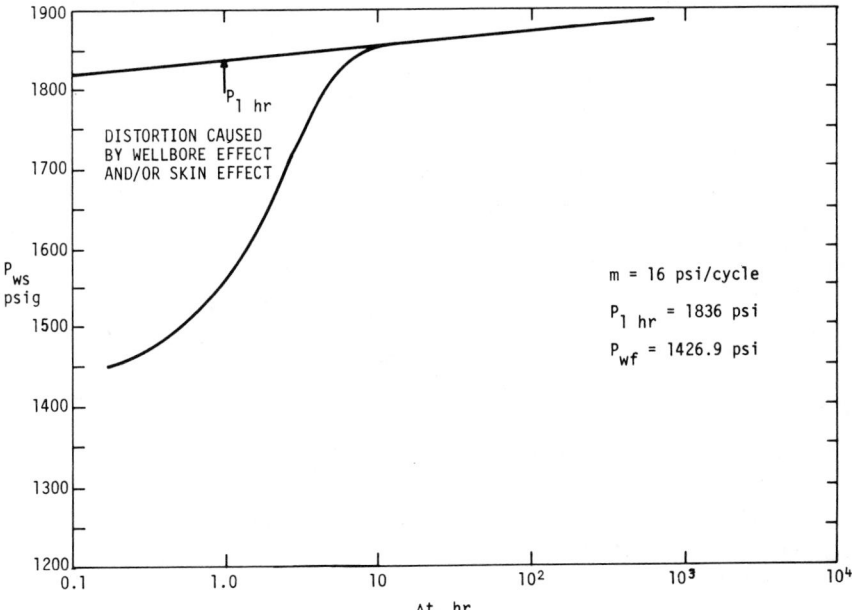

Figure 7.17: Miller-Dyes-Hutchinson plot for Well #1.

For the skin effect, using Eq. (7.22):

$$S = 1.151 \frac{P_{1hr} - P_{wf}}{m} - \log \frac{k}{\phi \mu c r_w^2} + 3.23$$

$$= 1.151 \frac{1836 - 1426.9}{16} - \log \frac{7.94}{(0.1)(0.85)(12.03 \times 10^{-6})(0.25)^2} + 3.23$$

$$= 9.43$$

Horner Method: This is without doubt the most widely used pressure analysis method. It involves a plot of shut-in pressure versus logarithm of the time ratio: $(t + \Delta t)/\Delta t$. The governing equation is:

$$\frac{kh}{141.3q\mu B}(P^* - P_{ws}) = \frac{1}{2} \ln \frac{t + \Delta t}{\Delta t} \qquad (7.24)$$

This indicates that a plot of P_{ws} versus logarithm of shut-in time will produce a graph the slope of which is inversely proportional to the permeability-thickness product:

$$kh = \frac{162.6q\mu B}{m} \qquad (7.25)$$

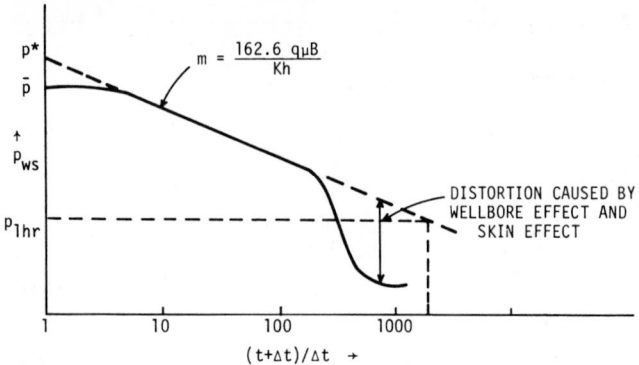

Figure 7.18: Horner plot.

The plot shown in Fig. 7.18 extrapolates to a pressure P^* at infinite shut-in time.

The skin factor can be calculated from the same equation as in the Miller-Dyes-Hutchinson method:

$$S = 1.151\frac{P_{1hr} - P_{wf}}{m} - \log\frac{k}{\phi\mu cr_w^2} + 3.23 \qquad (7.26)$$

This method requires the flowing time t prior to shut-in and is calculated below:

$$t = \frac{N_p}{q_{last}} \qquad (7.27)$$

EXAMPLE [4]   HORNER METHOD:

$$t = 53 \, \text{hr}$$

From the graph in Fig. 7.19:

$$m = 15 \, \text{psi/cycle}$$

From Eq. (7.25):

$$k = \frac{162.6q\mu B}{mh}$$

$$= \frac{(162.6)(146)(0.85)(1.29)}{(15)(205)}$$

$$= 8.47md \qquad (7.28)$$

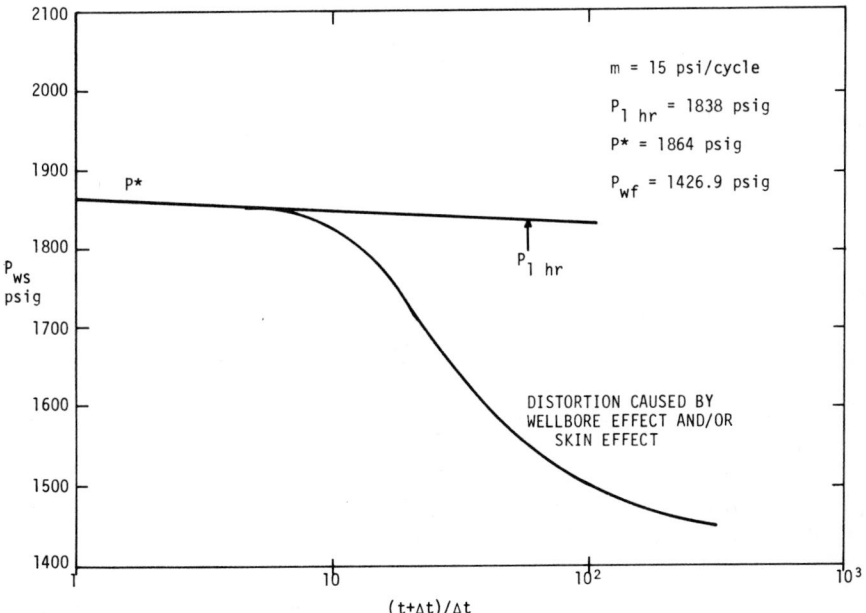

Figure 7.19: Horner plot for Well #1.

Skin effect:

$$S = 1.151\frac{P_{1hr} - P_{wf}}{m} - \log \frac{k}{\phi\mu cr_w^2} + 3.23$$

$$= 1.151\frac{1838 - 1426.9}{15} - \log \frac{8.47}{(0.1)(0.85)(12.03 \times 10^{-6})(0.25)^2} + 3.23$$

$$= 10.6 \tag{7.29}$$

Type Curve Analysis: This is a more recent method in which the log-log curve is used. A comparison is made between the actual field data and an analytically derived–type curve. The analytical curves are a family of curves generated by solving the radial flow equation to include wellbore storage effects. Build-up data is graphed as the logarithm of $(P_{ws} - P_{wf})$ versus logarithm of build-up time. The idea is to plot the field data on the same-size-coordinate tracing paper as the analytical curve and to try to match the actual curve with one of the analytical curves. Once a match has been obtained, the true kh and hydraulic diffusivity can be obtained by using the coordinates which make up the match and the defining equations for the dimensionless parameters used. These are:

Dimensionless time:

$$t_D = \frac{0.000264kt}{\phi\mu c_t r_w^2} \tag{7.30}$$

Dimensionless pressure:

$$P_D = \frac{kh(P_i - P_{wf})}{141.3q\mu B} \tag{7.31}$$

Dimensionless storage:

$$\bar{c} = \frac{5.615c}{2\pi h\phi c_t r_w^2} \tag{7.32}$$

where the unit storage constant is

$$c = \frac{qB\,\Delta t}{(P_{ws} - P_{wf})} \tag{7.33}$$

The basic reason for type curve fitting is self-evident when we look at Eqs. (7.30) and (7.31) in expanded form:

$$\log t_D = \log \frac{0.000264k}{\phi\mu c_t r_w^2} + \log t \tag{7.34}$$

and

$$\log P_D = \log \frac{kh}{141.3q\mu B} + \log (P_i - P_{wf}) \tag{7.35}$$

Then the only difference between a log-log plot of dimensionless pressures and times and the real pressures and times is a translation of both axes by the constants on the right-hand side of Eqs. (7.34) and (7.35). Type curve matching allows us to solve these two equations simultaneously to obtain the necessary parameters. The procedure is as follows:

1. Plot on the same-size-coordinate tracing paper a log-log plot of $(P_i - P_{wf})$ versus real time.
2. Check to see that the first points fall on a line of unit slope indicating that storage controlled the build-up behavior.
3. Compute the storage constant c from any point on the unit slope line using Eq. (7.33), then compute the dimensionless storage $\bar{c}$ using Eq. (7.32).
4. Translate the field data plot along the $\bar{c}$-line calculated in step 3 until the field curves and analytical curves show a reasonable match.
5. A match having been obtained, any matching point from the two curves can be used to calculate kh and $k/\phi\mu c_t$ by reading Δt, $(P_{ws} - P_{wf})$, t_D, and P_D from the curves plotted. Equations (7.30) and (7.31) are used. The skin factor is read directly off the analytical curve.

EXAMPLE [5]   TYPE CURVE METHOD: Figure 7.20 shows the type curve plot
for the same data as the earlier methods. The storage constant is calculated:

$$c = \frac{qBt}{P_{ws} - P_{wf}} \tag{7.33}$$

$$= \frac{(146)(1.29)(0.5/24)}{1498.6 - 1426.9}$$

$$= 0.0548 \text{ Reservoir Barrels/psi}$$

$$\bar{c} = \frac{c}{2h\pi c_t r_w^2} \tag{7.32}$$

$$= \frac{(5.615)(0.0548)}{2(3.14)(205)(0.1)(12.03 \times 10^{-6})(0.25)^2}$$

$$= 3.18 \times 10^3$$

Figure 7.20 indicates a match point at:

$$t = 100 \text{ min} \qquad P_{ws} - P_{wf} = 1000 \text{ psi}$$
$$t_D = 2 \times 10^4 \qquad P_D = 42$$

Thus, from Eq. (7.30),

$$k = \frac{141.3 q\pi B P_D}{h(P_{ws} - P_{wf})} \tag{7.36}$$

$$= \frac{(141.3)(146)(0.85)(1.29)(42)}{(205)(1000)}$$

$$= 4.64 \text{ md}$$

From type curve, by inspection:

$$S = 11 \tag{7.37}$$

Summary: Table 7.3 summarizes the four basic methods for analyzing
well tests and indicates the data required by each. In Table 7.4 the parameters
obtained by the evaluation of each method are classified. It should be noted
that some of the data—e.g., $\bar{P}$, the average pressure—are really required
during the history-matching phase, and others—e.g., S, skin factor—are
required for defining the wellbore hydraulics, which allow a better definition
of the model.

The flow diagram of the processes involved in each method is shown
in Fig. 7.21

Some of the other methods available to determine the value of the rock
permeability are indicated below. These methods are not as definitive as the
pressure transient analysis methods just investigated, but they do produce
figures which can be used as a good starting point.

TABLE 7.3

Data Required by Method for Analyzing Well Tests

		Muskat	M-D-H	Horner	Type curve
Pressure Build-up Data $P_{ws} - \Delta t$		yes	yes	yes	yes
Field data plot	Semilog Log-log	$\log(\bar{P} - P_{ws})$ vs. Δt no	P_{ws} vs. $\log \Delta t$ no	P_{ws} vs. $\log \dfrac{t+\Delta t}{\Delta t}$ no	no $\log(P_{ws} - P_{wf})$ vs. $\log \Delta t$
Producing time before shut-in, t		no	no	yes	no
Trial-and-error method used		no	no	yes	no
PVT data	Formation volume factor B	yes	yes	yes	yes
	Viscosity μ, cp	yes	yes	yes	yes
	Compressibility c (psi^{-1})	yes	yes	yes	yes
	P_{BP}	yes	yes	yes	yes
	R_S^*	no	no	no	no
	GOR*	no	no	no	no

		ΔP_{DMDH} vs. Δt_{DA} chart	ΔP_{DMBH} vs. t_{DA} chart	P_D vs. t_D analytical type curve
Reservoir data	h	yes	yes	yes	yes
	ϕ	no	yes	yes	yes
	$A(r_e)$	yes	yes	yes	no
	Saturation*	no	no	no	no
	C_f*	no	no	no	no
	T_R*	no	no	no	no
	Drainage shape and well location	yes	yes	yes	no
Production data	q	yes	yes	yes	yes
	N_p	no	no	yes	no
	r_w	yes	yes	yes	yes
	P_{wf}	yes	yes	yes	yes
Dimensionless chart used		no	ΔP_{DMDH} vs. Δt_{DA} chart	ΔP_{DMBH} vs. t_{DA} chart	P_D vs. t_D analytical type curve
Pseudo–steady state needed before shut-in		yes	yes	yes	no

*Data required only for multiphase well or gas well. R_s = solution gas. GOR = gas-oil ratio.

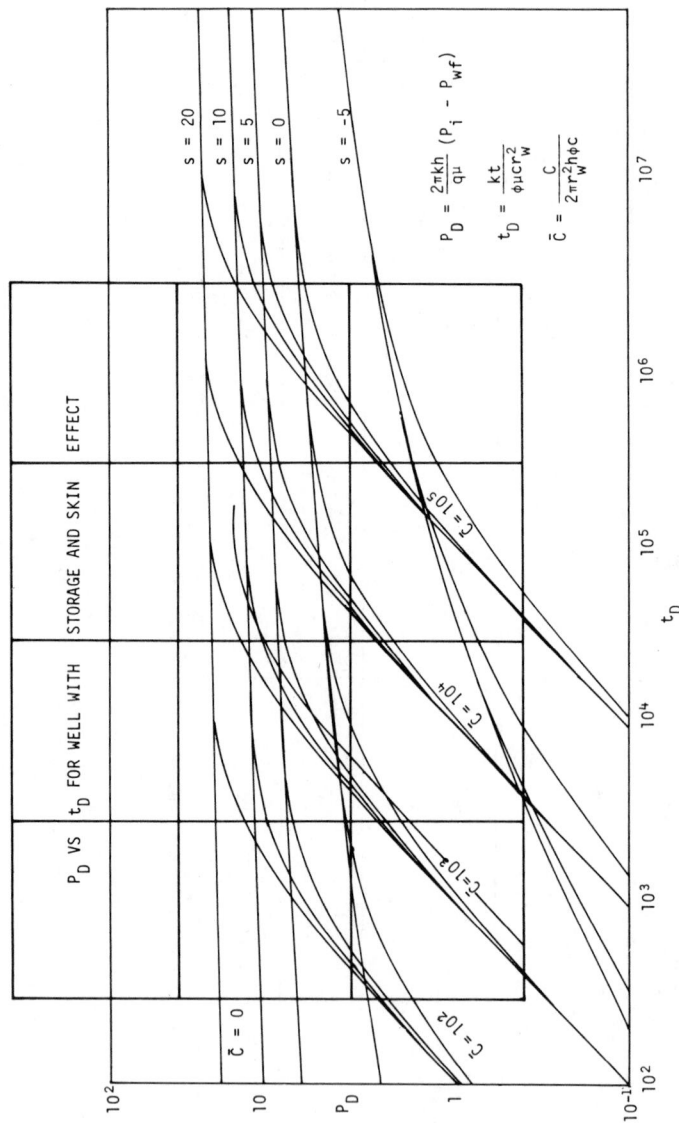

Figure 7.20: Type curve analysis for Well #1.

TABLE 7.4
Parameters Obtained by Methods for Analyzing Well Tests

	Muskat	M-D-H	Horner	Type Curve
Slope of straight line	$m = \dfrac{0.000264ka_1}{2.303\phi\mu c r_e^2}$	$m = \dfrac{162.6q\mu B}{kh}$	$m = \dfrac{162.6q\mu B}{kh}$	Unit slope for early build-up points
Intercept	$(\bar{P} - P_{ws})_{\Delta t=0} = \dfrac{141.2q\mu B}{khb_1}$	—	$P^* = P_{ws}$ at $\log\dfrac{t + \Delta t}{\Delta t} = 1$	—
P_{1hr}	—	Extend from the straight line on field data plot	Extend from the straight line on field data plot	—
k	$k = \dfrac{94.6q\mu B}{(\bar{P} - P_{ws})_{\Delta t=0}}$ (For square)	$k = \dfrac{162.6q\mu B}{mh}$	$k = \dfrac{162.6q\mu B}{mh}$	$k = \dfrac{141.4q\mu BP_D}{h(P_{ws} - P_{wf})}$
$t_{DA} = \dfrac{0.000264kt}{\phi\mu cA}$	—	Calculate t_{DA}	Calculate $\forall\, t_{DA}$	—
$\Delta t_{DA} = \dfrac{0.000264k\,\Delta t}{\phi\mu cA}$	—	Calculate Δt_{DA}	—	—
$\Delta t_{De} = \Delta t_{DA} \times \pi$	—	Calculate Δt_{De}	—	—
ΔP_D	—	ΔP_{DMDH}	ΔP_{DMBH}	—
$\bar{P}$	By trial and error	$\bar{P} = \dfrac{m}{1.151}\Delta P_{DMDH} + P_{ws}$	$\bar{P} = P^* - \dfrac{m}{2.303}\Delta P_{DMBH}$	—
S	$S = 0.84\dfrac{\bar{P} - P_{wf}}{(\bar{P} - P_{ws})_{\Delta t=0}} - \ln\dfrac{r_e}{r_w} + 0.75$	$S = 1.151\left(\dfrac{P_{1hr} - P_{wf}}{m} - \log_{10}\dfrac{k}{\phi\mu r_w^2 c} + 3.23\right)$	$S = 1.151\left(\dfrac{P_{1hr} - P_{wf}}{m} - \log_{10}\dfrac{k}{\phi c r_w^2 \mu} + 3.23\right)$	Read directly from the analytical type curve

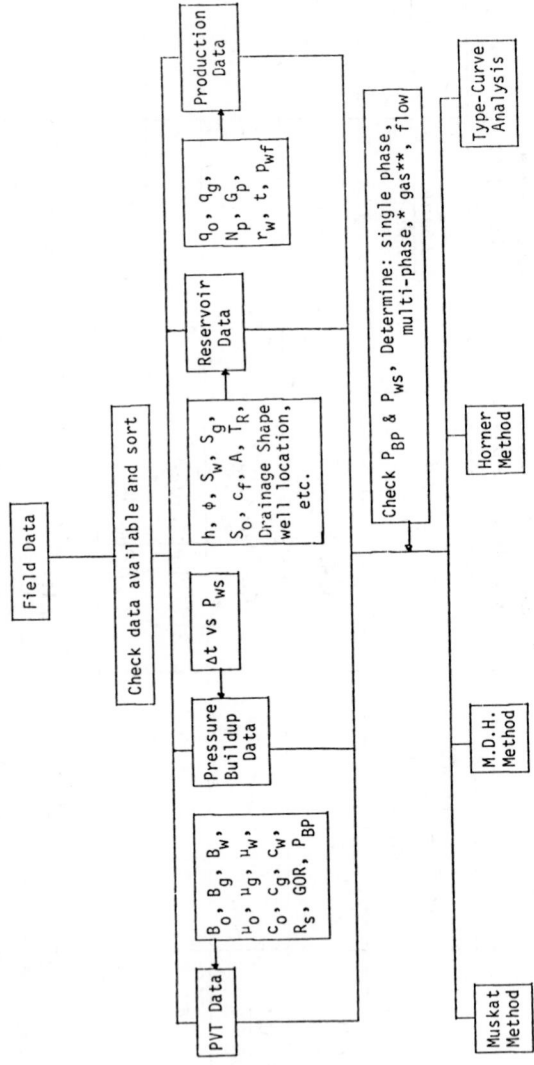

Figure 7.21: Outline of comparative analysis of each method available. *For multi-phase well, we need more data, such as s_o, s_w, s_g, c_o, c_g, c_f, B_o, B_g, B_w, q_o, q_g, q_w, to calculate total compressibility C_t and total mobility $(K/\mu)_t$. **For gas well, we need to know gas gravity, T_R, T_{sc}, P_{sc}, T_c, P_c, to calculate Z, B_g, and to estimate μ_g, c_g.

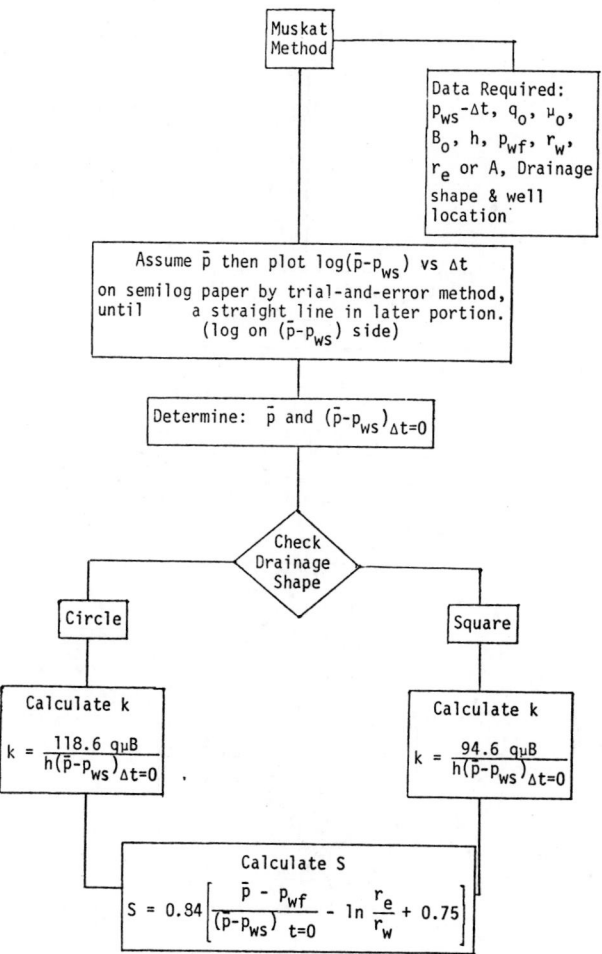

Figure 7.21 (Cont'd.)

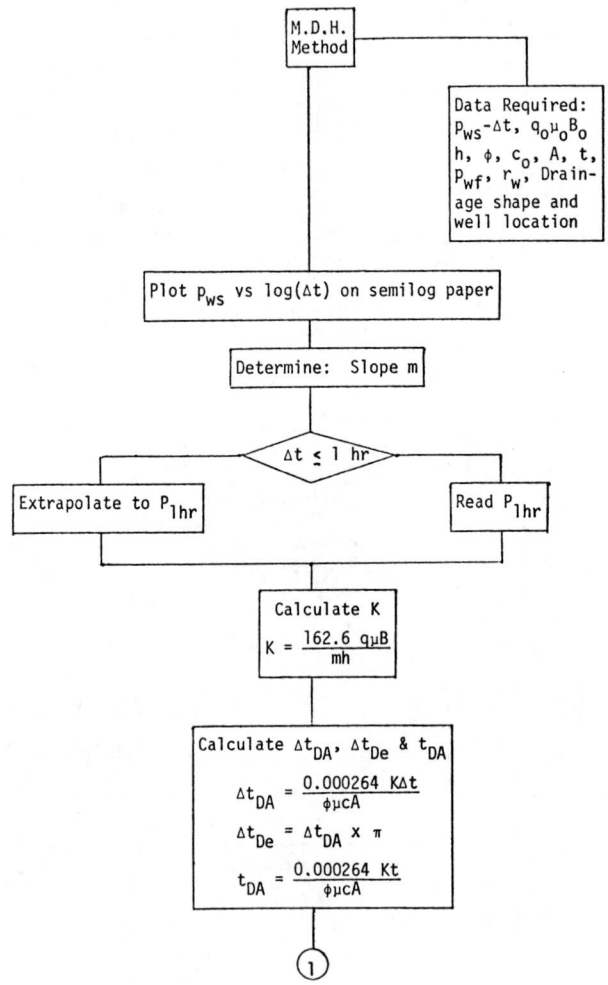

Figure 7.21 (Cont'd.)

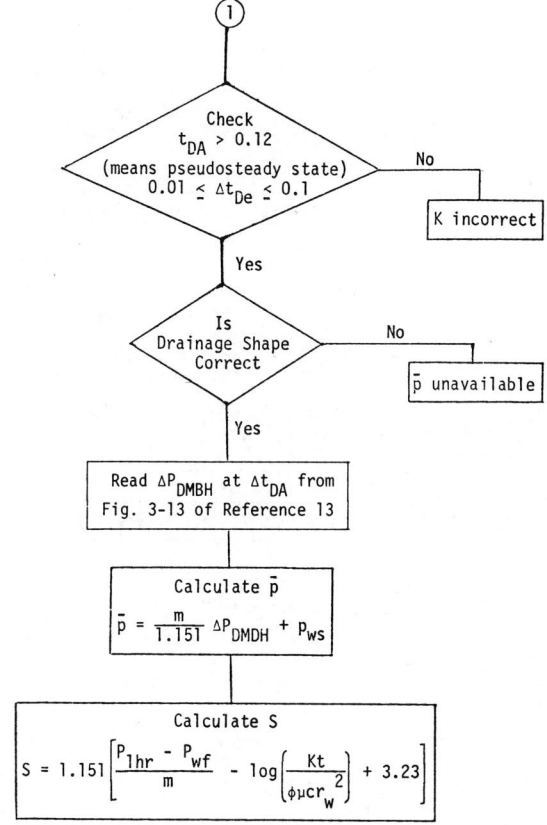

Figure 7.21 (Cont'd.)

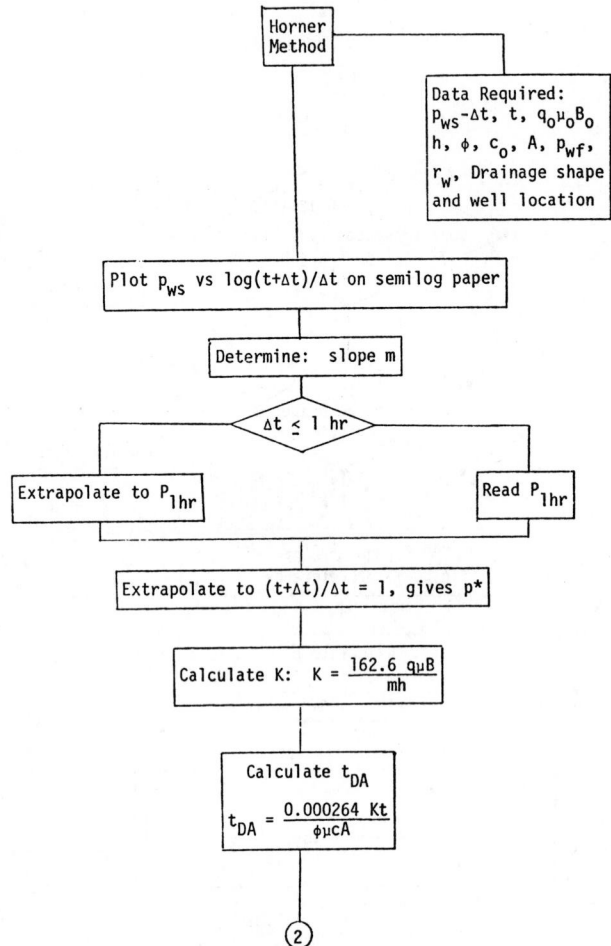

Figure 7.21 (Cont'd.)

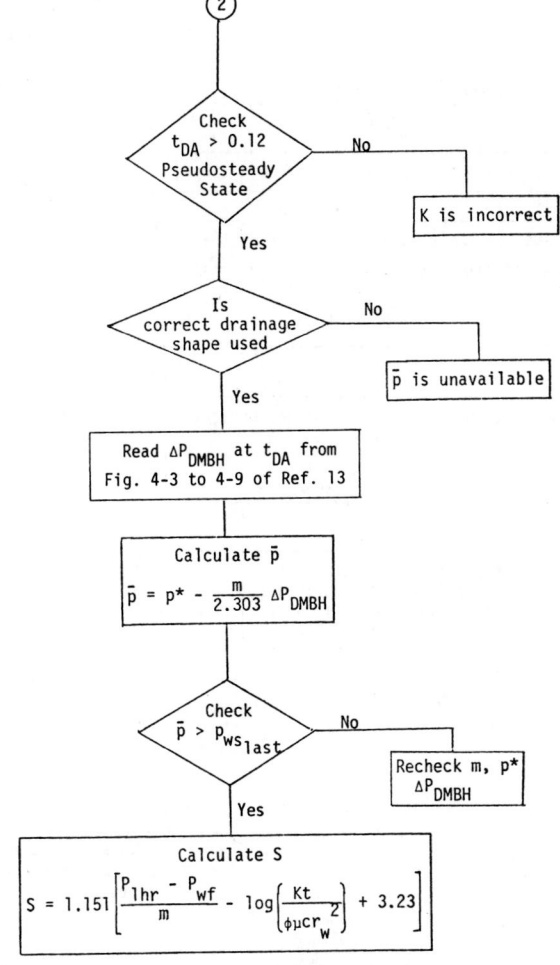

Figure 7.21 (Cont'd.)

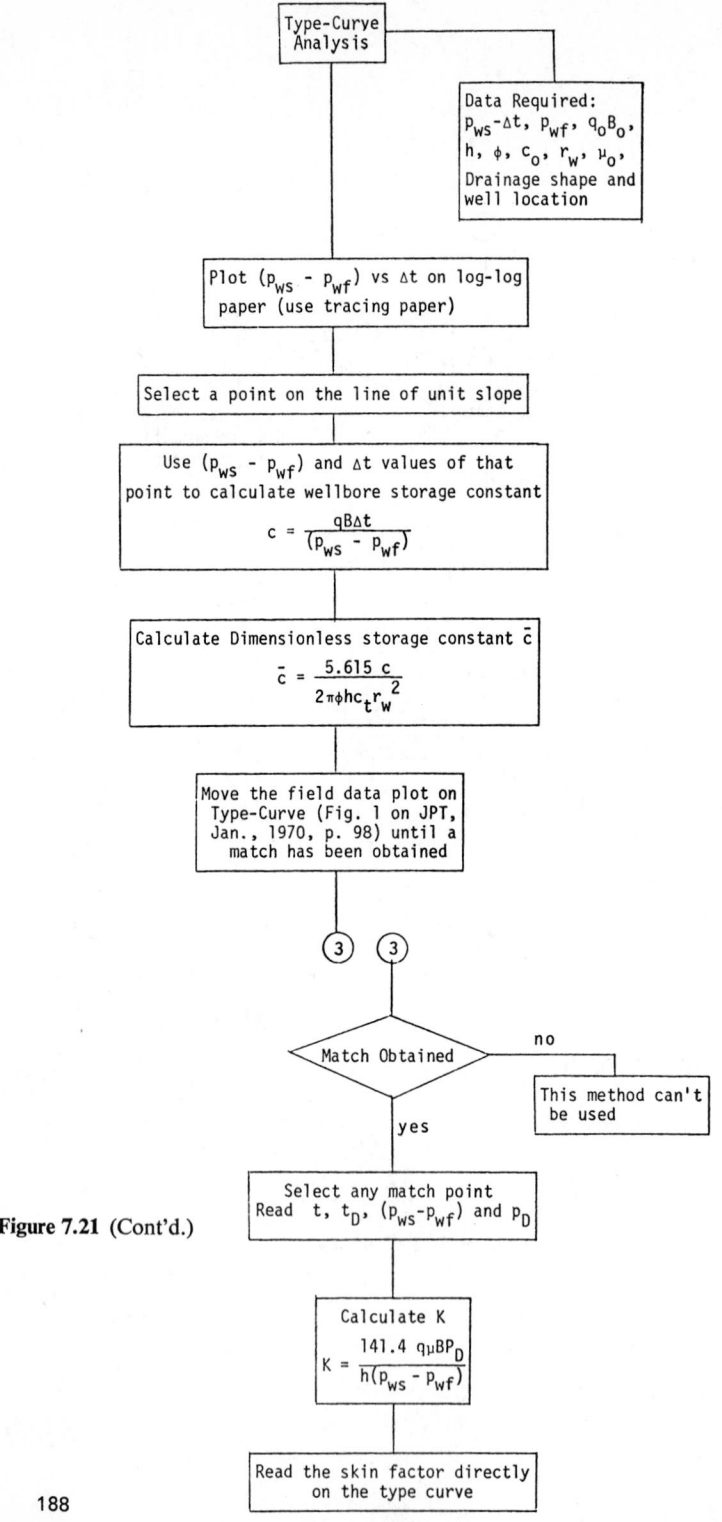

Figure 7.21 (Cont'd.)

The flowchart contains the following elements:

Type-Curve Analysis

Data Required:
$p_{ws}-\Delta t$, p_{wf}, $q_o B_o$, h, ϕ, c_o, r_w, μ_o,
Drainage shape and well location

Plot $(p_{ws} - p_{wf})$ vs Δt on log-log paper (use tracing paper)

Select a point on the line of unit slope

Use $(p_{ws} - p_{wf})$ and Δt values of that point to calculate wellbore storage constant
$$c = \frac{qB\Delta t}{(p_{ws} - p_{wf})}$$

Calculate Dimensionless storage constant $\bar{c}$
$$\bar{c} = \frac{5.615\ c}{2\pi\phi h c_t r_w^2}$$

Move the field data plot on Type-Curve (Fig. 1 on JPT, Jan., 1970, p. 98) until a match has been obtained

③ ③

Match Obtained — no → This method can't be used

yes

Select any match point
Read t, t_D, $(p_{ws}-p_{wf})$ and p_D

Calculate K
$$K = \frac{141.4\ q\mu B p_D}{h(p_{ws} - p_{wf})}$$

Read the skin factor directly on the type curve

The Initial Potential Test: This test could produce an estimate of the permeability of a region. The flow during an initial test is strictly radial, and the Darcy flow equation could be rearranged to solve for k:

$$Q = 7.082 \frac{kh\,\Delta P}{\mu B \ln (r_e/r_w)} \tag{7.38}$$

$$k = \frac{Q\mu B \ln (r_e/r_w)}{7.082 h\,\Delta P} \tag{7.39}$$

Regression Analysis:[7] The use of regression analysis techniques is recommended in those situations where no data are available from other sources. Consider the situation illustrated in Fig. 7.22. Areas A and B are two

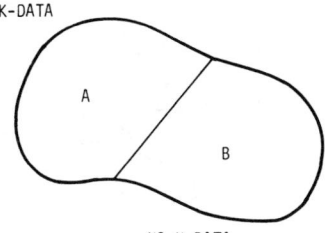

K-DATA

A

B

Figure 7.22: Determination of K data by regression.

NO K-DATA

units of a common reservoir owned by different operators. A simulation study is being set up for A and B, but permeability data is available in A only, where pressure build-up studies had been made since company A was a diligent operator. Since the reservoir is the same geologic unit in both areas, it is feasible to postulate that the factors affecting permeability in A are similar to those in B. A regression analysis is run on the permeability data in A, trying to define a relation between permeability and other known reservoir parameters—e.g., permeability as a function of porosity and water saturation:

$$k = a\phi + b\phi^2 + cS_w + dS_w^2 \tag{7.40}$$

The values of coefficients a, b, c, and d which give the best fit to the observed data are determined with a routine regression analysis program. This equation is used to calculate permeability values for reservoir B, and these values constitute our best approximation at this time of the permeabilities in area B. An example of a regression analysis is Eq. (7.41):

$$k = 250 \frac{\phi^3}{S_w} \tag{7.41}$$

Sources of Porosity Data

The porosity parameter is usually obtained from one of the following sources:

1. Logging data
2. Laboratory measurements
3. Published correlations

Logging Data: Logging data[8] in the form of sonic or acoustic logs as shown in Fig. 7.23 are obtained by measuring the travel time of sound

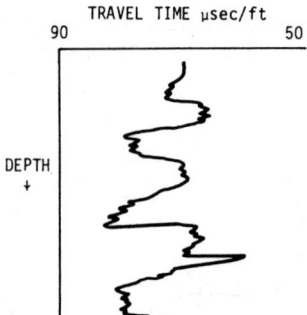

TRAVEL TIME μsec/ft

Figure 7.23: Sonic logs.

through the formation. The travel time is directly affected by the fluids which inhabit the pore space of the rock as indicated by

$$t = \frac{1}{V} = \frac{\phi}{V_f} + \frac{1 - \phi}{V_m} \qquad (7.42)$$

where t is the interval travel time for sound, V is the measured velocity, and V_f and V_m are the velocity of sound in the formation fluid and matrix rock, respectively. Solving Eq. (7.42) for ϕ, the porosity, produces:

$$\phi = \frac{\frac{1}{V} - \frac{1}{V_m}}{\frac{1}{V_f} - \frac{1}{V_m}} \qquad (7.43)$$

The typical acoustic log records the interval transit time, and Eq. (7.42) becomes:

$$\phi = \frac{\Delta t^* - \Delta t_m}{\Delta t_f - \Delta t_m} \qquad (7.44)$$

where Δt^* is the recorded interval transit time, usually in microseconds/foot.

Laboratory Measurements: Porosity measurements in the laboratory are based on the determination of any two of the following three parameters: bulk volume, grain volume, and pore volume. Usual methods determine pore volume by either the introduction of fluid into a rock or the removal of fluid from a rock. There are several types of devices, generally called porosimeters, which are routinely used to determine porosities. The determination of porosity is based on the following. If

$$\text{Dry weight of core sample} = W_d$$

$$\text{Saturated weight} = W_s$$

Then

$$\text{Fluid weight} = W_s - W_d$$

$$\text{Fluid volume} = \frac{W_s - W_d}{\rho_f}$$

$$= \text{Effective pore space}$$

$$\text{Bulk volume core} = \frac{W_d}{\rho_s}$$

And then

$$\text{Porosity} = \frac{\text{Pore volume}}{\text{Bulk volume}}$$

$$= \frac{(W_s - W_d)/\rho_f}{W_d/\rho_s}$$

$$\phi = \frac{\rho_s(W_s - W_d)}{\rho_f W_d} \tag{7.45}$$

where ρ_f = fluid density

ρ_s = sand grain density

Published Correlations: Several correlations have been published relating porosity to depth of burial for different types of rocks. Krumbein and Sloss[9] have a porosity depth curve based on natural compaction.

Sources of Formation Thickness Data

Formation thickness data are obtained from gross isopach maps or net isopach maps. Most simulators usually use gross isopach maps in computing the flow characteristics of the model. The gross isopach map gives the correct vertical dimension necessary for evaluating the correct potential head; however, for calculating oil-in-place figures which are based on net oil sand, it is customary either to include a net/gross factor which allows computation of net oil in place or to use a separate program to calculate the oil in place based on net thicknesses. Formation thickness can also be obtained from the structural data by subtracting the structural contours on the formation bottom from those at the top of the formation.

Sources of Formation Elevations Data

Formation elevations data are obtained from the subsurface structural maps of the reservoir. These data are initially compiled from:

    a. Log data
    b. Drilling records

Plotting of Reservoir Rock Data: The data sets previously enumerated —namely,

    1. permeability,
    2. porosity,
    3. formation thickness, and
    4. formation elevation

—are obtained at discrete locations in the reservoir. For these data to be used in the simulator, they must be available at every point in the reservoir. The data are thus plotted and contoured to obtain an overall distribution within the reservoir limits. In this contouring process the engineer uses all the known geological data in evaluating and contouring these rock parameters. The use of computer contouring programs is recommended as a good starting point for development of the contour maps; however, care should be exercised when interpreting the results of some "canned" contour programs, especially where the data are widely separated and in those locations where the data peters out. (Close to the boundaries, some programs use inaccurate approximations.)

Contouring Methods: There are two basic contouring techniques used in obtaining a two-dimensional representation of data.

Interpretative contouring is such that the contours honor the known geological information and the contour lines parallel the known geologic trends. This technique is almost always used in contouring formation thicknesses and elevations. It is an intuitive method.

Mechanical contouring is nonintuitive and is based on some computing process or algorithm in which the data is fitted, extrapolated, or interpolated between known points to obtain values at unknown locations. Mechanical contouring is generally used for those parameters which do not depend on the geological configuration of the rock—e.g., porosity. Most multiple-parameter contours are mechanically contoured.

Usually the data to be contoured is available as a single unit—e.g., h, ϕ; in some cases, however, only combinations of data are available—e.g., kh, ϕh. In determination of original oil in place under the leases or the total reservoir, the following equation must be evaluated:

$$N = 7758\frac{1}{B_o} \sum_{i=1}^{M} (A_i h_i \phi_i S_{o_i}) \qquad (7.46)$$

It is sometimes advisable to plot multiple-parameter maps to determine exactly where the oil is and consequently where the reservoir energy is located.

Single-Parameter Contouring: Several reservoir parameters are usually contoured singly:

1. The isopach map is interpretively contoured.
2. The isoporosity map is mechanically contoured.
3. The isopermeability map is mechanically contoured.
4. The oil saturation map is mechanically contoured. This segment can be omitted if the oil saturation is reasonably uniform.

Multiple-Parameter Contouring: This process combines several parameters into a more complex function which is then contoured:

1. Porosity-thickness (ϕh)
   a. At each location the porosity-thickness product is evaluated.
   b. A mechanically contoured map is made of this parameter.
   c. A rectangular grid is laid over the map and the ϕh-product evaluated at each cell:

$$N = 7758\frac{S_{oavg}}{B_o} \sum_{i=1}^{M} A_i(\phi h)_i \qquad (7.47)$$

2. Porosity-saturation-thickness ($\phi S_o h$)
   a. At each location the porosity-saturation-thickness product is evaluated.
   b. A mechanically contoured map is made.
   c. ($\phi S_o h)_i$ is evaluated at each point in a rectangular net:

$$N = \frac{7758}{B_0} \sum_{i=1}^{M} A_i(\phi S_o h)_i \qquad (7.48)$$

Digitizing of Contoured Data: As indicated in Chapter 4, the reservoir is divided into several cells by superimposing a rectangular grid over the region. The discretization process is required because the finite-differencing scheme requires the definition of all the reservoir properties at every given cell in the model. Within any given cell the parameters which describe the reservoir in the model are generally assumed to be uniform throughout that particular cell. For example, if a thickness value of 20 ft is used in a 10-acre cell in the model, this implies that the average thickness over that 10-acre region is 20 ft. The determination of these discretized values is best done manually by the engineer and can be obtained with reasonable facility with a little effort on the engineer's part.

The determination of the average value to put into a given cell can be obtained as follows. The "parameter volume" of a cell may be defined as the volume of that prism delineated between the surfaces of the contoured parameter and the vertical sides of the cell as shown in Figs. 7.24 and 7.25. If we could define a rectangular prism (parallelipiped) on the same base

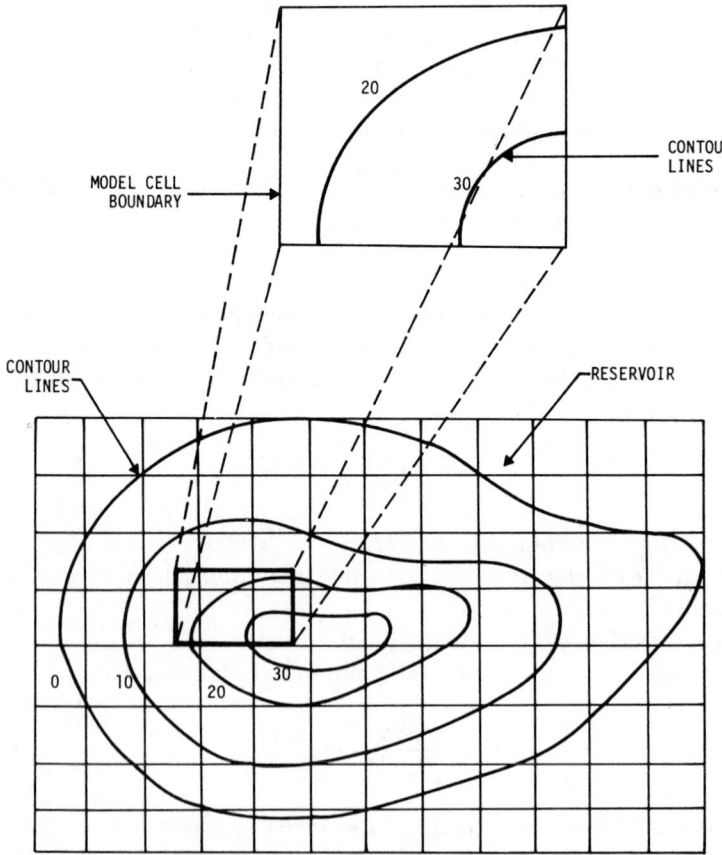

Figure 7.24: Contour map and enlarged cell view.

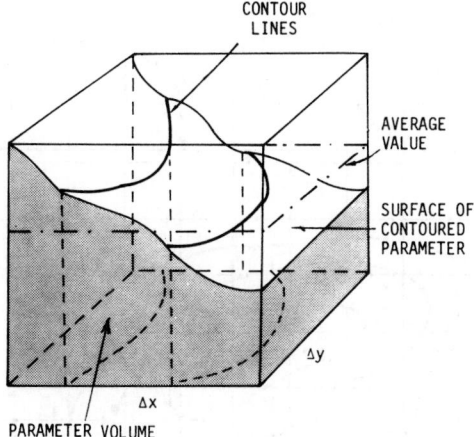

Figure 7.25: Determination of average parameter value.

and having the "parameter volume," then the height of this rectangular prism will be the average height of the cell. This is the value which would be encoded in the matrix of input data.

In those areas where the zero contour line passes through a cell, if that cell is part of the model, the parameter is averaged over the area of nonzero thickness, and this value is then averaged over the whole cell. This averaging must be done because the simulator treats the whole cell as a part of the model and the information is distributed throughout the cell.

The digitizing process can be expedited by using a Mylar or other transparent material overlay (see Fig. 7.26) on which an identical grid has been marked. The overlay allows the engineer to digitize the contour map without actually defacing the original map by drawing or writing on it. Since ink on the Mylar overlay is erasable, corrections can be easily made. A photographic copy of the completely digitized map can then be made for a permanent record.

A record of all digitized maps should be kept, since the engineer will refer back to several of these maps in doing a history match.

Sources of Rock Compressibility Data

Rock compressibility data are obtained from laboratory analyses of the reservoir rock or from published correlations.

Sources of Relative Permeability Data

Relative permeability is often one of the more difficult pieces of data to evaluate or obtain. The relations usually required by simulators are the following:

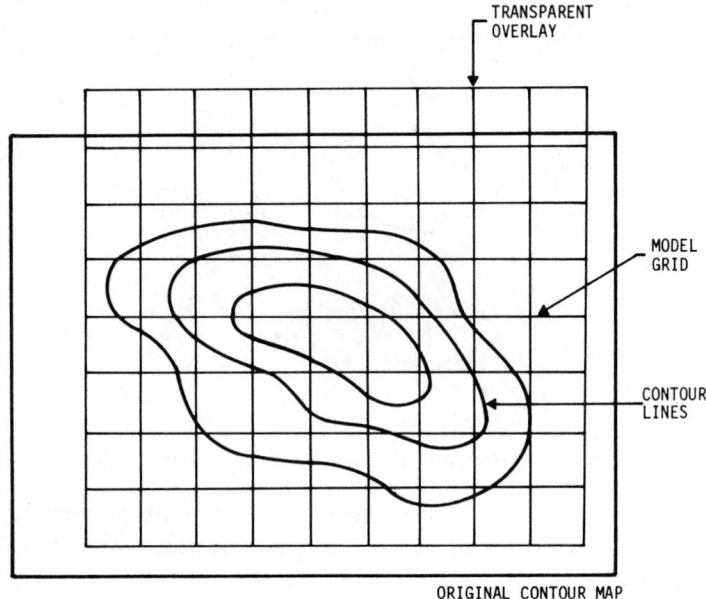

Figure 7.26: Use of transparent overlay.

a. Gas-oil relative permeability
b. Oil-water relative permeability
c. Gas-water relative permeability

The relative permeability data can be obtained from one of five means:

1. Laboratory measurements using steady-state displacement processes
2. Laboratory measurements using unsteady-state displacement processes.
3. Calculations from capillary pressure data.
4. Calculations from field data.
5. Calculations from published correlations.

Review of Methods for Determining Relative Permeability: Steady-state techniques involve the desaturation of a given core which was originally 100% saturated by the wetting phase. Amyx, Bass, and Whiting[10] survey several methods for the steady-state determination. Basically, it consists of the following for an oil/water system:

1. Saturate the core 100% with wetting phase.
2. Select a given injection ratio of nonwetting to wetting phase.
3. Inject at this rate until the outflow ratio is the same as the inflow ratio.

4. Determine and record saturations in the core.
5. Calculate relative permeability of each phase from:

$$f_o = \frac{q_o}{q_T} = \frac{q_o}{q_o + q_w}$$

$$= \frac{\dfrac{k_o}{\mu_o}}{\dfrac{k_o}{\mu_o} + \dfrac{k_w}{\mu_w}}$$

$$= \frac{1}{1 + \dfrac{k_w}{k_o}\dfrac{\mu_o}{\mu_w}} \tag{7.49}$$

6. The process is repeated by going back to step 1 and using a different injection ratio.

By this iterative process a complete set of relative permeability curves can be developed.

Unsteady-state techniques have been developed by Welge and also by Johnson, Bossler, and Naumann[11] which enable a rapid determination of relative permeability. These methods are usually referred to as dynamic displacement relative permeability measurements and are based on the measurement of rates and/or injection pressures to determine the fractional flow. The saturation is obtained by a material balance computation. The relationships involved are basically the following.

At any time in the displacement process:

$$\frac{dS_{w_{avg}}}{dW_i} = \frac{dN_p}{dW_i} = \frac{q_o}{q_T} = f_o \tag{7.50}$$

i.e., the change of average water saturation with water injected must equal the production ratio, since only two fluids are present and a Buckley-Leverett–type displacement is occurring. Also:

$$\frac{k_{rw}}{k_{ro}} = \frac{f_w}{f_o}\frac{\mu_w}{\mu_o} \tag{7.51}$$

Figure 7.27 shows the typical response of a displacement-type process. Note the initial slope of unity which always occurs until breakthrough; after breakthrough the curve decreases in slope and finally reaches a limit where $S_{w_{avg}} = 1 - S_{or}$. The other relation needed is that for the change in average water saturation. This relation is derived as follows.

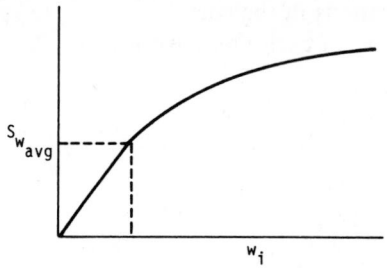

Figure 7.27: Dynamic displacement response.

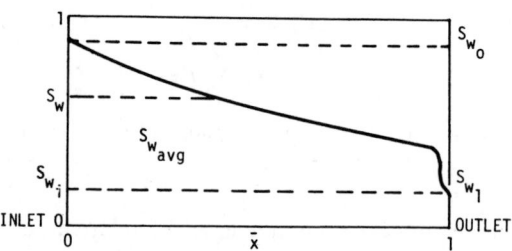

Figure 7.28: Water saturation distribution at breakthrough in a linear core.

Consider Fig. 7.28. The oil recovery is:

$$N_p = \phi AL(\Delta S_{avg}) \tag{7.52}$$

$$N_p = \phi AL \int_{S_{wi}}^{S_{wo}} \bar{x} \, dS_w \tag{7.53}$$

In two steps this is:

$$N_p = \phi AL\left(\Delta S_1 + \int_{S_{L1}}^{S_{wo}} \bar{x} \, dS_w\right) \tag{7.54}$$

Using the frontal advance equation:

$$N_p = \phi AL\left(\Delta S_1 + \int_{S_{wi}}^{S_{wo}} w_i \frac{df_w}{dS_w} \, dS_w\right)$$

$$= \phi AL\left(\Delta S_1 + w_i \int_{f_{w1}}^{1} df_w\right)$$

$$N_p = \phi AL[\Delta S_1 + w_i(1 - f_{w1})] \tag{7.55}$$

Then

$$\Delta S_{avg} = \Delta S_1 + w_i(1 - f_{w1}) \tag{7.56}$$

Equations (7.56) and (7.51) form the basis for these methods. The determination of the relative permeability is done very rapidly using a graphical technique on a plot similar to Fig. 7.27. The process is as follows:

1. Construct a plot of water injected and average water saturation. Note:

$$S_{w_{avg}} = S_{w_i} + \frac{N_p}{PV}$$

where N_p is measured as a cumulative volume. This plot is shown in Fig. 7.29.

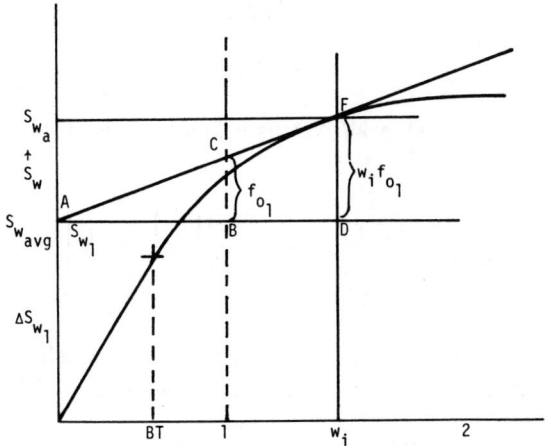

Figure 7.29: Graphical determination of f_o, f_w using $S_{w_{avg}}$.

2. Pick any value of $S_{w_{avg}}$ as shown in Fig. 7.29. Draw vertical and horizontal lines to S_{w_a} and W_i, respectively. Draw a tangent to the curve letting this tangent intersect the ordinate (S_w-axis).
3. Read S_{w_1}, S_{w_a} on the ordinate. Then, since

$$\begin{aligned}
\Delta S_{avg} &= S_{w_a} - S_{w_1} \\
&= \Delta S_1 + w_i(1 - f_{w_1}) \quad \text{from Eq. (7.55)} \\
&= \Delta S_1 + w_i f_{o_1}
\end{aligned}$$

4. Therefore, read $w_i f_{o_1}$ as shown in Fig. 7.29. Then using the principle of similar triangles, f_{o_1} is read on the vertical axis at $w_i \equiv 1$. (Note that the triangles ABC and ADE are similar.)
5. A table of f_o-, f_w-, and $S_{w_{avg}}$-values is then obtained, and the relative permeability values can be computed.

The other dynamic displacement approach uses a measure of the injection pressure ratio as a function of cumulative water injected. The needed relation between these terms is:

$$\frac{f_o}{k_{r_o}} = \frac{d\left(\frac{1}{w_i I_r}\right)}{d\left(\frac{1}{w_i}\right)} \tag{7.57}$$

where I_r, the injection pressure ratio, is:

$$I_r = \frac{\left(\frac{q}{\Delta P}\right)_{t=0}}{\left(\frac{q}{\Delta P}\right)_{t>0}} \tag{7.58}$$

The curve actually plotted is $1/I_r$ versus w_i, and this is graphically resolved. Note that:

$$\frac{f_o}{k_{r_o}} = \frac{d\left(\frac{1}{w_i I_r}\right)}{d\left(\frac{1}{w_i}\right)} = \left[\frac{-w_i \, d\left(\frac{1}{I_r}\right)}{dW_i} + \frac{1}{I_r}\right] \tag{7.59}$$

Using a similar technique to the $S_{w_{avg}}$ method we can develop the fractional flow data and relative permeability data as shown in Fig. 7.30.

Capillary Pressure: Several theoretical treatments are available for determining relative permeability from capillary pressure data. Purcell presented the following equations:

$$k_{r_{wt}} = \frac{\int_{S=0}^{S=S_{wt}} dS/(P_c)^n}{\int_{S=0}^{S=1.0} dS/(P_c)^n} \tag{7.60}*$$

$$k_{r_{nwt}} = \frac{\int_{S=S_{wt}}^{S=1} dS/(P_c)^n}{\int_{S=0}^{S=1} dS/(P_c)^n} \tag{7.61}†$$

For Purcell's equation $n = 2$, while Fatt and Dykstra produced a similar relation with $n = 3$.

Field Data: These can be used to complete relative permeability ratios. This method is based on the concept of Darcy flow being applicable to each

---

*The subscript wt means wetting.

†The subscript nwt means nonwetting.

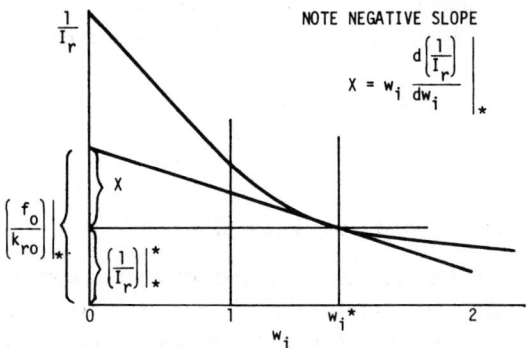

Figure 7.30: Graphical relative permeability data using injection.

flowing phase independently. Consider two-phase flow of oil and gas in a radial system:

$$Q_g = 7.08 \frac{k_g h}{\mu_g B_g} \frac{P_e - P_w}{\ln(r_e/r_w)} \quad \text{Free gas} \tag{7.62}$$

$$Q_o = 7.08 \frac{k_o h}{\mu_o B_o} \frac{P_e - P_w}{\ln(r_e/r_w)} \quad \text{Oil} \tag{7.63}$$

The producing gas-oil ratio is

$$R_p = \frac{Q_g}{Q_o} \tag{7.64}$$

$$= \frac{k_g}{k_o} \frac{\mu_o}{\mu_g} \frac{B_o}{B_g} 5.615 + R_s \tag{7.65}$$

$$= \text{Free gas} + \text{Solution gas}$$

Solving for relative permeability ratio:

$$\frac{k_g}{k_o} = \frac{(R_p - R_s)}{5.615} \frac{B_g}{B_o} \frac{\mu_g}{\mu_o} \tag{7.66}$$

The corresponding oil saturation is determined from a material balance at the cumulative produced oil:

$$S_o = \left(1 - \frac{N_p}{N}\right) \frac{B_o}{B_{o_i}} (1 - S_w) \tag{7.67}$$

where N_p = cumulative oil production
N = original oil in place

This method can be used to determine several values of relative permeability data.

Relative Permeability from Published Correlations: These have already been discussed in Chapter 2. The general equation types are:

$$k_o = (1 - S)^n$$
$$k_D = S^k(2 - S)$$

Sources of Formation Fluid Saturations Data

In a reservoir there are two possible planes of interest which can be used to evaluate saturations of reservoir fluids: the gas/oil contact and water/oil contact. The saturations are generally computed from the locations of the contacts within a cell (see Fig. 7.31).

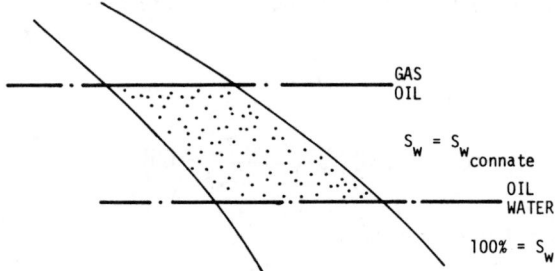

GAS
OIL

$S_w = S_{w_{connate}}$

OIL
WATER

$100\% = S_w$

Figure 7.31: Saturations and fluid contacts.

Above the oil/water contact the water saturation will be essentially constant and equal to the connate saturation. The average saturation in those blocks which are traversed by the oil/water contact has to be computed using a weighted thickness of 100% water-saturated sand. Connate water can be evaluated from:

1. Core data
2. Electric logs
3. Capillary pressure data

Sources of Capillary Pressure Data

Capillary data are needed to evaluate pressures in the various phases during the IMPES calculation and also to set up the equations in the simultaneous solution. Capillary pressure is usually determined from laboratory data. (See Fig. 7.32.)

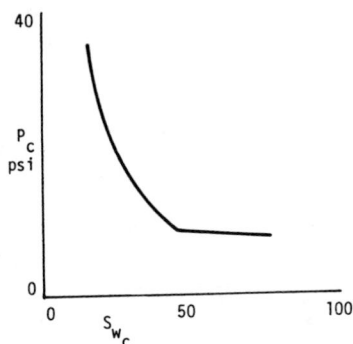

Figure 7.32: Capillary pressure.

7.4 PRODUCTION DATA

Production data are required to operate the simulator in history mode. The information required is obtained from well production records. The following is required for each well:

1. Oil production vs. time
2. Water production vs. time
3. Gas production vs. time
4. Any measured pressures vs. time

Most well records have complete *oil* production data; however, some of the data on water and gas production are usually missing. The missing data must be determined to be used in the history mode. The simplest procedure used entails plotting the production data and *smoothing* the curve through the missing points (Fig. 7.33).

The input format requires production data at every time step, and a table of values is generally the best means of encoding these data—e.g., as in

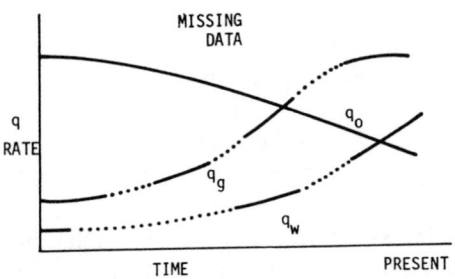

Figure 7.33: Smoothing missing data.

TABLE 7.5
Production Data

Well	I	J	Time	Oil	Gas	Water	Pressure
A2	10	17	1960.25	155	600	35	2180
A2	10	17	1960.50	180	650	40	2160
A2	10	17	1960.75	200	700	45	2140
A2	10	17	1961.00	190	715	50	—
A2	10	17	1961.50	180	800	60	—

Table 7.5. This well has had the production of oil, gas, and water during the times shown. In those cases where the measured pressures are absent, no values are encoded. Some simulators do not require gas or water production data but compute those values of gas and oil which are produced based on relative permeability relations and existing reservoir conditions.

7.5 FLOW RATE DATA

Flow rate data are required by the simulator to compute producing capacity of a well within the system. These data are generally based on the following:

1. Productivity index
2. Injectivity index
3. Optimum flow rates
4. Maximum allowable drawdowns

Flowing wells and gas lift wells generally show some rate sensitivity to gas-oil ratio (GOR), bottom hole pressure (BHP), and flow rate. A correlation of the types developed by Ros, Poettman-Carpenter, Orkiszewski, and others[12] is essential to obtain a true representation of the fluid behavior at the well-bore/reservoir interface. This procedure involves a simultaneous solution of the reservoir flow equation and two-phase flow in vertical pipe shown in Fig. 7.34. It is possible that the fluid volume capable of being produced by the formation exceeds the capacity of the flow lines. The typical curves are shown in Figs. 7.35 and 7.36. A surface fit of flowing bottom hole pressure versus flow rate and GOR is necessary to determine the flow parameters in the wellbore during simulation.

It is also possible to develop the flow rate correlation from a set of computations within the simulator. This is most efficiently done by using a subroutine which accepts reservoir and flow string data and calculates the flow

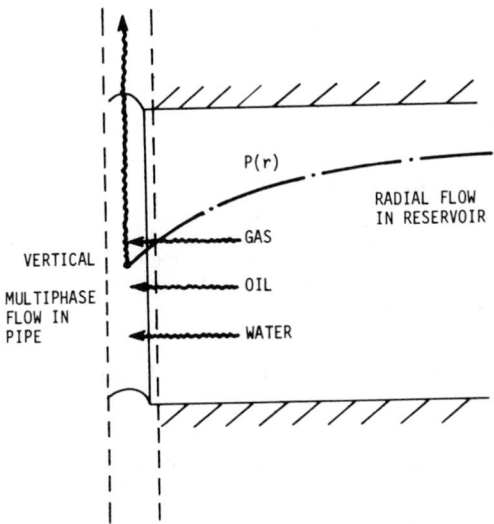

Figure 7.34: Tubing flow equation and reservoir flow equation satisfied simultaneously.

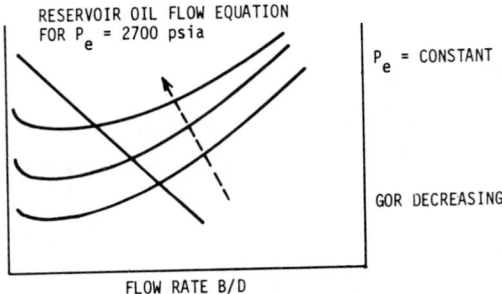

Figure 7.35: Typical tubing flow equations and reservoir flow equation.

rates. This approach is usually better, and it allows for incorporation of an optimization process into the model.

An example of a surface fit for flowing bottom hole pressure is:

$$\text{FBHP} = a_0 + a_1 x + a_2 x^2 + a_3 y + a_4 y^2 + a_5 xy$$

where $x = $ oil production rate
$y = $ gas-oil ratio
FBHP $= $ flowing bottom hole pressure

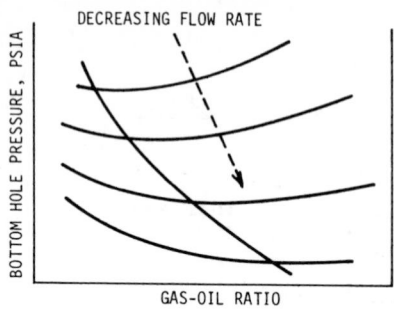

Figure 7.36: Near optimum gas-oil ratio.

7.6 CASE STUDY: SIMULATION OF A GAS RESERVOIR

The following study is designed to show the procedures involved in putting together a simulation of a small dry gas reservoir. The engineer is provided with production and pressure data for a given period of history and the necessary rock and fluid data. The structure is believed to be a gentle anticline. The flow data during history are shown in Fig. 7.37 and 7.38; and the

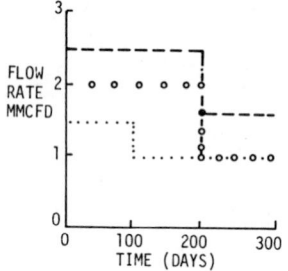

Figure 7.37: Flow rates for wells during history.

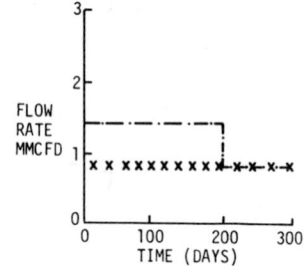

Figure 7.38: Flow rates during history.

measured pressures are shown in Fig. 7.39. The basic reservoir data are indicated at each well location in Fig. 7.40. Figure 7.41 is a plot of the gas deviation factor and viscosity with pressure variation.

The basic data are contoured to determine an areal distribution throughout the region. The formation thickness data are contoured interpretatively as shown in Fig. 7.42. The porosity and permeability data are contoured mechanically; these plots are shown in Figs. 7.43 and 7.44. A rectangular grid is superimposed on the reservoir with an effort being made to get the wells in the center of cell blocks. The selected grid is then used to digitize all the reservoir rock data for the model. Fig. 7.48 shows a sample of the input data and Fig. 7.49 a sample of the output results.

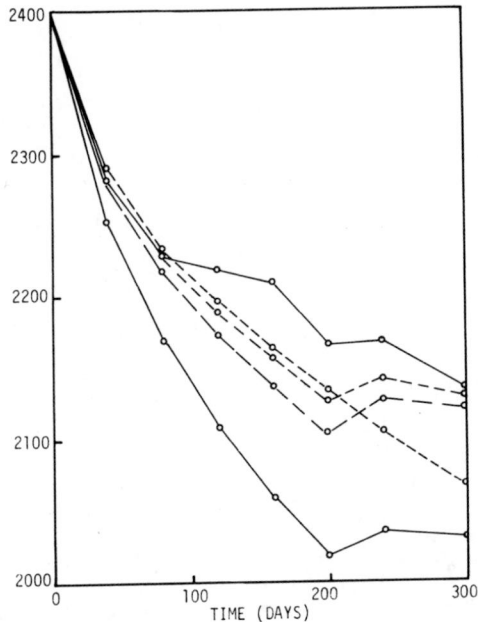

Figure 7.39: Measured well pressures during history.

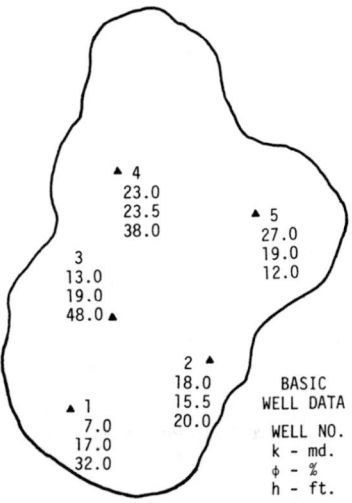

Figure 7.40: Basic reservoir data.

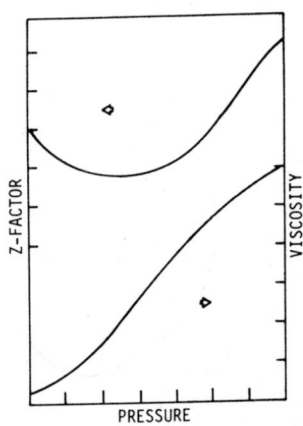

Figure 7.41: Gas deviation factor and viscosity.

207

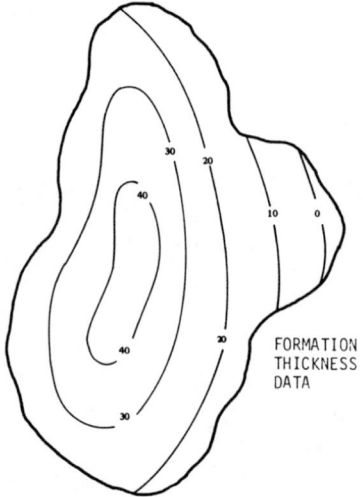

Figure 7.42: Isopach map.

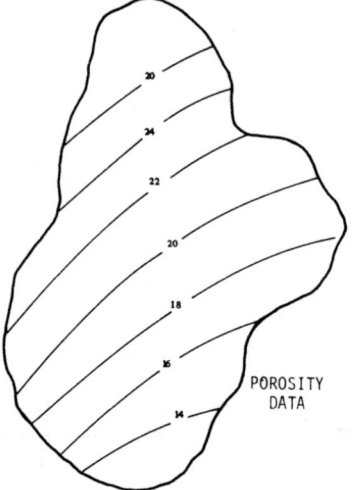

Figure 7.43: Isoporosity map.

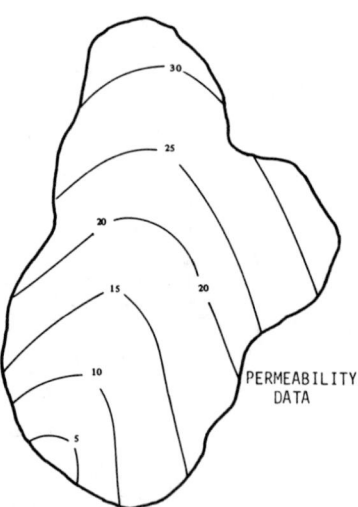

Figure 7.44: Permeability data.

THICKNESS   DISTRIBUTION MATRIX

	1	2	3	4	5	6	7
1	0.	0.	25.	15.	0.	0.	0.
2	0.	0.	27.	23.	13.	0.	0.
3	0.	25.	34.	27.	15.	0.	0.
4	0.	25.	38.	33.	21.	10.	2.
5	0.	27.	42.	35.	21.	12.	4.
6	23.	36.	46.	37.	24.	13.	5.
7	27.	37.	48.	34.	24.	14.	0.
8	25.	36.	38.	30.	20.	0.	0.
9	22.	32.	32.	25.	10.	0.	0.
10	0.	22.	25.	18.	0.	0.	0.

Figure 7.45: Thickness distribution matrix.

POROSITY   DISTRIBUTION MATRIX

	1	2	3	4	5	6	7
1	0.	0.	27.	27.	0.	0.	0.
2	0.	0.	27.	26.	25.	0.	0.
3	0.	26.	25.	24.	23.	0.	0.
4	0.	25.	24.	23.	22.	21.	21.
5	0.	23.	23.	21.	20.	19.	19.
6	23.	22.	21.	20.	19.	18.	7.
7	22.	22.	19.	18.	17.	17.	0.
8	21.	19.	18.	17.	16.	0.	0.
9	19.	17.	16.	15.	14.	0.	0.
10	0.	15.	14.	13.	0.	0.	0.

Figure 7.46: Porosity matrix.

PERMEABILITY   DISTRIBUTION MATRIX

	1	2	3	4	5	6	7
1	0.	0.	32.	32.	0.	0.	0.
2	0.	0.	28.	28.	30.	0.	0.
3	0.	27.	26.	26.	28.	0.	0.
4	0.	25.	23.	22.	26.	29.	31.
5	0.	22.	21.	20.	24.	27.	32.
6	20.	17.	16.	18.	22.	26.	28.
7	16.	14.	14.	17.	19.	25.	0.
8	13.	10.	13.	15.	18.	0.	0.
9	7.	7.	12.	14.	17.	0.	0.
1C	0.	7.	11.	13.	0.	0.	0.

Figure 7.47: Permeability matrix.

```
2400.
  5
 10      7
2640.   2640.

        THICKNESS MATRIX GAS STUDY              1.0        THICK 1
              25.   15.                                    THICK 2
              27.   23.   13.                              THICK 3
        25.   34.   27.   15.                              THICK 4
        25.   38.   33.   21.   10.   7.                   THICK 5
        27.   42.   35.   21.   12.   4.                   THICK 6
  23.   36.   46.   37.   24.   13.   5.                   THICK 7
  27.   37.   48.   34.   24.   14.                        THICK 8
  25.   36.   38.   30.   20.                              THICK
  22.   32.   32.   25.   10.                              THICK 10
        22.   25.   18.

        PERMEABILITY MATRIX GAS STUDY           1.0        PERM  1
              32.   32.                                    PERM  2
              28.   28.   30.                              PERM  3
        27.   26.   26.   28.                              PERM  4
        25.   23.   22.   26.   20.   31.                  PERM  5
        22.   21.   20.   24.   27.   32.                  PERM  6
  20.   17.   16.   18.   22.   26.   28.                  PERM  7
  16.   14.   14.   17.   19.   25.                        PERM  8
  13.   10.   13.   15.   18.                              PERM
  7.    7.    12.   14.   17.                              PERM 10
        7.    11.   13.

        POROSITY MATRIX GAS STUDY               1.0        POR   1
              26.8  26.8                                   POR   2
              26.5  26.   25.                              POR   3
        26.   25.   24.   23.                              POR   4
        25.   23.5  22.5  21.5  21.   20.5                 POR   5
        23.   22.5  21.   20.   19.   18.6                 POR   6
  23.   22.   20.8  19.5  18.6  18.1  7.                   POR   7
  22.   21.5  19.   19.   17.   16.5                       POR   8
  20.5  19.   17.5  16.8  15.5                             POR
  19.   17.   15.8  14.5  14.                              POR  10
        15.   14.   13.                                    A
.01153E.00  0.          .00137E.00  -.00810E.00           B
.98337E.00  -.05959E.0 -.07457E.00  0.02892E.00
  25   1000.
   6    4
   1    20.              2    0
 902  2000000.
 805   800000.
 702  3200000.
 505   750000.
 303  2500000.
   2    20.              2    0
   3    20.              2    0
   4    20.              2    0
   5    20.              2    0
   6    20.              2    0
   7    20.              2    0
   8    20.              2    0
   9    20.              2    0
  10    20.              2    0
```

Figure 7.48: Gas reservoir simulation study.

PRESSURE

	1	2	3	4	5	6	7
1	0.	0.	2396.	2398.	0.	0.	0.
2	0.	0.	2385.	2395.	2398.	0.	0.
3	0.	2383.	2325.	2386.	2395.	0.	0.
4	0.	2394.	2387.	2394.	2391.	2396.	2397.
5	0.	2396.	2396.	2393.	2359.	2391.	2390.
6	2394.	2385.	2396.	2397.	2391.	2396.	2395.
7	2381.	2285.	2386.	2395.	2388.	2396.	0.
8	2393.	2376.	2393.	2389.	2330.	0.	0.
9	2386.	2280.	2386.	2394.	2381.	0.	0.

WELL PRODUCTION TABLES

CUMULATIVE TIME = 20. DAYS

WELL	LOCATION		FLOW RATE	AVG. PRES.	BHP	CUMULATIVE
1	9	2	2,000,000.	2280.	2070.	40,000,000.
2	8	5	800,000.	2330.	2279.	16,000,000.
3	7	2	3,200,000.	2285.	2140.	64,000,000.
4	5	5	750,000.	2359.	2324.	15,000,000.
5	3	3	2,500,000.	2325.	2260.	50,000,000.

TOTAL GAS = 185,000,000.

Figure 7.49

PROBLEMS

Using an appropriate curve fitting program determine an equation to efficiently calculate each of the following parameters:

1. *Oil formation volume factor*

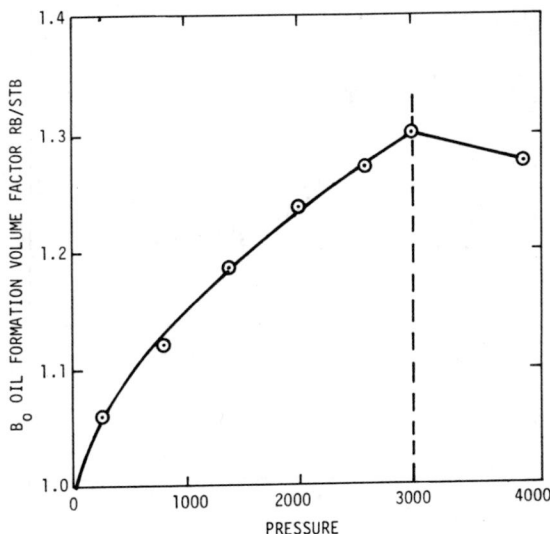

2. *Solution Gas-Oil Ratio*

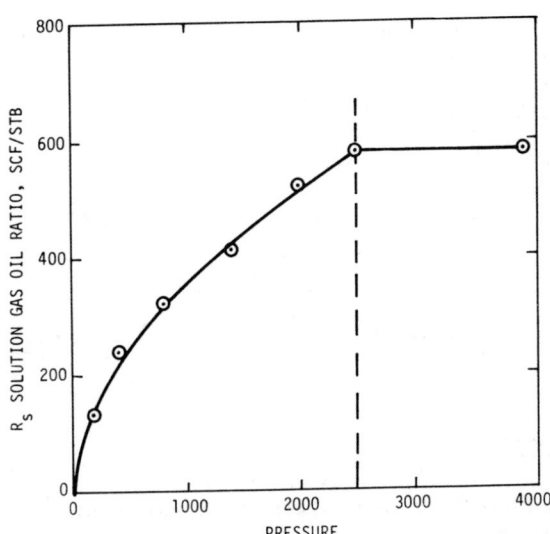

3. *Capillary Pressure*

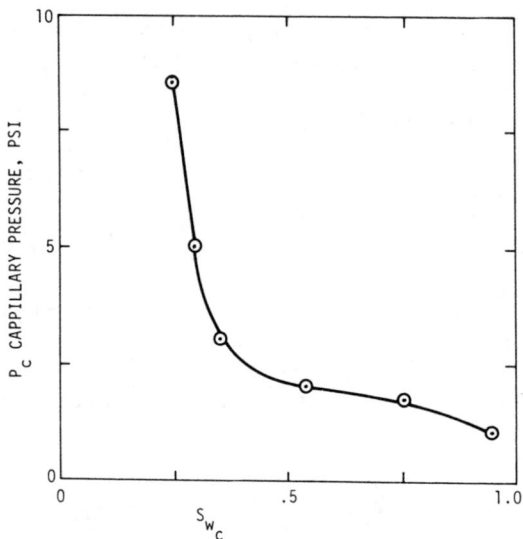

4. *Relative Permeability*

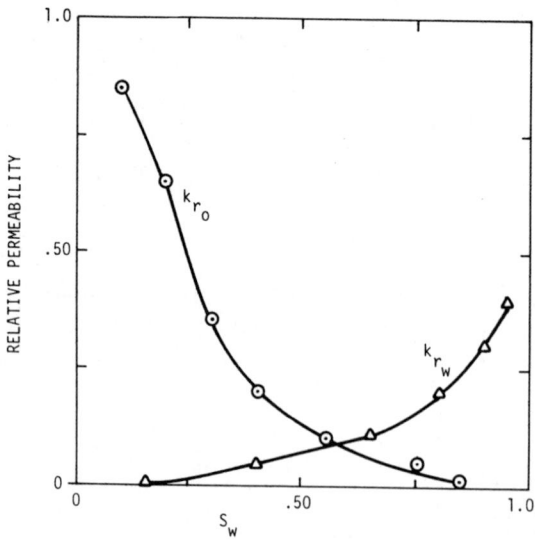

REFERENCES

1. W. S. DORN and D. D. MCCRACKEN, *Numerical Methods with Fortran Case Studies* (New York: Wiley, 1972).

2. M. Muskat, "Use of Data on Buildup of Bottom Hole Pressures," *Trans. AIME* (1937), **123**, 44.

3. C. C. Miller, A. B. Dyes and C. A. J. Hutchinson, "The Estimation of Permeability and Reservoir Pressure from Bottom Hole Pressure Buildup Characteristics," *Trans. AIME* (1950), **189**, 91.

4. D. R. Horner, "Pressure Buildup in Wells," in *Proc. Third World Pet. Congress* (Leiden: Brill, 1951), **2**, 503.

5. R. G. Agareval, R. Al-Hussainy, and H. J. Ramey, Jr., "An Investigation of Wellbore Storage and Skin Effect in Unsteady Liquid Flow, I: Analytical Treatment," *Soc. Pet. Eng. J.* (Sept. 1970), 279–90.

6. AIME, *Pressure Analysis Methods*, Petroleum Transactions Reprint Series No. 9, (Dallas: 1967).

7. B. Carnahan, H. A. Luther, and J. O. Wilkes, *Applied Numerical Methods* (New York: Wiley, 1969).

8. S. J. Pirson, *Handbook of Well Log Analysis* (Englewood Cliffs, N.J.: Prentice-Hall, 1963).

9. W. C. Krumbein and L. L. Sloss, *Stratigraphy and Sedimentation* (San Francisco: Freeman, 1951).

10. J. W. Amyx, D. M. Bass, Jr., and R. C. Whiting, *Petroleum Reservoir Engineering Physical Properties* (New York: McGraw-Hill, 1960).

11. E. F. Johnson, D. P. Bossler, and V. O. Naumann, "Calculation of Relative Permeability from Displacement Experiments," *Trans. AIME* (1959), 370.

12. G. W. Govier and K. Aziz, *The Flow of Complex Mixture in Pipes* (New York: Van Nostrand Reinhold, 1972).

13. C. S. Matthews and D. G. Russel, *Pressure Buildup and Flow Tests in Wells*, SPE of AIME, (Dallas: 1967).

BIBLIOGRAPHY

Asafari, Ardeshir and Paul A. Witherspoon, "Numerical Simulation of Naturally Fractured Reservoirs," SPE 4290, Third Symposium on Numerical Simulation of Reservoir Performance, Houston, Texas, Jan. 10–12, 1973.

Aufricht, W. R. and E. H. Koepf, "The Interpretation of Capillary Pressure Data from Carbonate Reservoirs," SPE of AIME (1957), 402.

Bailey, H. R. and W. B. Gogarty, "Diffusion Coefficients from Capillary Flow," SPE of AIME (1963), II-256.

Baker, P. E., "Discussion of Effect of Viscosity Ratio on Relative Permeability," SPE of AIME (1960), 404.

Bixel, H. C. and H. K. Van Poollen, "Pressure Drawdown and Buildup in the Presence of Radial Discontinuities," *Soc. Pet. Eng. J.* (Sept. 1967), 301–9; *Trans. AIME*, **240**.

BRILL, J. P., A. T. BOURGOYNE, and T. N. DIXON, "Numerical Simulation of Drill-stem Tests as an Interpretation Technique," *J. Pet. Tech.* (Nov. 1969), 1413–20.

BURDINE, N. T. "Relative Permeability Calculations from Pore Size Distribution Data," SPE of AIME (1953), 71.

BURNS, W. A., JR., "New Single-well Test for Determining Vertical Permeability," SPE of AIME (1969), I-743.

CALHOUN, JOHN C. and FRANK T. BETHEL, "Capillary Desaturation in Uncon-solidated Beads," SPE of AIME (1953), 197.

COLONNA, J., F. BRISSAND, and J. L. MILLET, "Evolution of Capillary and Relative Permeability Hysteresis," SPE of AIME (1972), II-28.

COREY, A. T., C. H. RATHJENS, J. H. HENDERSON, and M. R. J. WYLLIE, "Three-Phase Relative Permeability," SPE of AIME (1956), 349.

EARLOUGHER, R. C., JR., and K. M. KERSCH, "Field Examples of Automatic Tran-sient Test Analysis," *J. Pet. Tech.* (Oct. 1972), 1271–77.

FATT, I., "The Effect of Overburden Pressure on Relative Permeability," SPE of AIME (1953), 325.

FATT, I. and H. DYKSTRA, "Relative Permeability Studies," *Trans. AIME,* (1951), **192.**

HANDY, L. L., "Determination of Effective Capillary Pressure for Porous Media From Imbibition Data," SPE of AIME (1960), 75.

HARSTOCK, J. H. and J. E. WARREN, "The Effect of Horizontal Hydraulic Fracturing on Well Performance," *J. Pet. Tech.* (Oct. 1961), 1050–56; *Trans. AIME,* **222.**

HARTING, WILLIAM H., "Theoretically Indicated Methods of Wetting Liquid Rela-tive Permeability Measurement," SPE of AIME (1953), 316.

HAYNES, R. D. and W. K. SAWYER, "Analysis of Pressure Interference Tests Using Computer Graphics," SPE 4570, 48th Annual Meeting, Las Vegas, Nev., Sept. 30–Oct. 3, 1973.

HEARN, C. L., "Simulation of Stratified Waterflooding by Pseudo Relative Perme-ability Curves," *J. Pet. Tech.* (July 1971), 805–13.

HERNANDEZ, VICTOR M. and GEORGE W. SWIFT, "A Method for Determining Reservoir Parameters From Early Drawdown Data," SPE 3982, 47th Annual Meeting, San Antonio, Texas, Oct. 8–11, 1973.

HOFFMAN, R. N., "A Technique for the Determination of Capillary Pressure Curves Using a Constantly Accelerated Centrifuge," SPE of AIME (1963), II-227.

HOVANESSIAN, S. A. and F. J. FAYERS, "Linear Water Flood with Gravity and Capillary Effects," SPE of AIME (1961), II-32.

HUPPLER, JOHN D., "Numerical Investigation of the Effects of Core Heterogenities on Waterflood Relative Permeabilities," *Soc. Pet. Eng. J.* (Dec. 1970), 381–92; *Trans. AIME,* **249.**

JOHNSON, E. F., D. P. BOSSLER, and V. O. NAUMANN, "Calculation of Relative Permeability from Displacement Experiments," *Trans.* AIME (1959), 370.

KAZEMI, H., "Pressure Transient Analysis of Naturally Fractured Reservoirs with Uniform Fracture Distribution," *Soc. Pet. Eng. J.* (Dec. 1969), 451–62.

KAZEMI, H., L. S. MERRILL, and J. R. JARGON, "Problems in Interpretation of Pressure Fall-off Tests in Reservoirs With and Without Fluid Banks," *J. Pet. Tech.* (Sept. 1972), 1147–56.

KAZEMI, H. and M. S. SETH, "Effect of Anisotropy and Stratification on Pressure Transient Analysis of Wells with Restricted Flow Entry," *J. Pet. Tech.* (May 1969), 639–47; *Trans. AIME,* **246.**

KAZEMI, H., M. S. SETH, and G. W. THOMAS, "The Interpretation of Interference Tests in Naturally Fractured Reservoirs With Uniform Fracture Distribution," *Soc. Pet. Eng. J.* (Dec. 1969), 463–72; *Trans. AIME,* **246.**

KNOPP, C. R., "Gas-Oil Relative Permeability Ratio Correlation From Laboratory Data," SPE of AIME (1965), I-111.

LAND, C. S., "Calculation of Imbibition Relative Permeability for Two Three-phase Flow From Rock Properties," SPE of AIME (1968), II-149.

LAND, C. S., "Comparison of Calculated with Experimental Imbibition Relative Permeability," SPE of AIME (1971), II-419.

LOREN, J. D. and J. D. ROBINSON, "Relations Between Pore Size, Fluid, and Matrix Properties and NML Measurements," SPE of AIME (1970), II-268.

LUFFEL, D. L., "A Technique for the Determination of Capillary Pressure Curves Using a Constantly Accelerated Centrifuge; Discussion," SPE of AIME (1964), II-191.

MAY, R. A. and C. A. CHASE, "Parametric Pulsing of Oil and Gas Reservoirs—An Evaluation of the Merits," *J. Pet. Tech.* (Sept. 1963), 1040–46; *Trans. AIME,* **228.**

MELROSE, J. C., "Wettability as Related to Capillary Action in Porous Media," SPE of AIME (1965), II-259.

MORROW, N. R. and C. C. HARRIS, "Capillary Equilibrium in Porous Materials," SPE of AIME (1965), II-15.

MORSE, R. A. and W. D. VON GONTEN, "Productivity of Vertically Fractured Wells Prior to Stabilized Flow," *J. Pet. Tech.* (July 1972), 807–11.

MUELLER, T. D., J. E. WARREN, and W. J. WEST, "Analysis of Reservoir Performance K_g/K_o Curves and a Laboratory K_g/K_o Curve Measured on a Core Sample," *Trans. AIME* (1955), **204,** 128–31.

MUNGAN, N., "Relative Permeability Measurements Using Reservoir Fluids," SPE of AIME (1972), II-28.

NAAR, J. and R. J. WYGAL, "Three-phase Imbibition Relative Permeability," SPE of AIME (1961), II-254.

ODEH, A. S., "Effect of Viscosity Ratio on Relative Permeability," SPE of AIME (1959), 364.

ODEH, A. S. and J. M. McMILLEN, "Pulse Testing, Mathematical Analysis, and

Experimental Verification," SPE 3536, 46th Annual Meeting, New Orleans, Oct. 3–6, 1971.

OWENS, W. W., D. R. PARRISH, and W. E. LAMOREAUX, "An Evaluation of a Gas Drive Method for Determining Relative Permeability Relationships," SPE of AIME (1956), 275.

PARSONS, R. W., "Directional Permeability Effects in Developed and Unconfined Five Spots," *J. Pet. Tech.* (April 1972), 487–94; *Trans. AIME,* **253**.

PERKINS, F. M., JR., "An Investigation of the Role of Capillary Forces in Laboratory Water Floods," SPE of AIME (1957), 409.

PICKELL, J. J., B. F. SWANSON, and W. B. HICKMAN, "Application of Air-Mercury and Oil-Air Capillary Pressure Data in the Study of Pore Structure and Fluid Distribution," SPE of AIME (1966), II-55.

PIRSON, S. J., E. M. BOATMAN, and R. L. NETTLE, "Prediction of Relative Permeability Characteristics of Intergranular Reservoir Rocks from Electrical Resistivity Measurements," SPE of AIME (1964), I-564.

PRATS, M., "A Method for Determining the Net Vertical Permeability Near a Well from In-situ Measurements," SPE of AIME (1970), I-637.

PRATS, M. and J. S. LEVINE, "Effect of Vertical Fractures on Reservoir Behavior— Results on Oil and Gas Flow," *J. Pet. Tech.* (Oct. 1963), 1119–26; *Trans. AIME,* **228**.

PURCELL, W. R., "Capillary Pressures—Their Measurement Using Mercury and the Calculation of Permeability Therefrom," *Trans. Aime* (1949), **186**.

RAGHAVEN, R., J. D. T. SCORER, and F. S. MILLER, "An Investigation by Numerical Methods of the Effect of Pressure-dependent Rock and Fluid Properties on Well Flow Tests," *Soc. Pet. Eng. J.* (June 1972), 267–75; *Trans. AIME,* **253**.

RAGHAVEN, R., H. N. TOPALOGLU, WILLIAM M. COBB, and HENRY J. RAMEY, JR., "Well Test Analysis for Wells Producing from Two Commingled Zones of Unequal Thickness," SPE 4559, 48th Annual Meeting, Las Vegas, Nev., Sept. 30–Oct. 3, 1973.

ROCKWOOD, S. H., G. H. LAIR, and B. J. LANGFORD, "Reservoir Volumetric Parameters Defined by Capillary Pressure Studies," SPE of AIME (1957), 252.

RUSSELL, D. G. and N. E. TRUITT, "Transient Pressure Behavior in Vertically Fractured Reservoirs," *J. Pet. Tech.* (Oct. 1964), 1159–70.

SANDBERG, C. R., L. S. GOURNEY, and R. F. SIPPEL, "The Effect of Fluid-flow Rate and Viscosity on Laboratory Determinations of Oil-Water Relative Permeabilities," SPE of AIME (1958), 36.

SARAF, D. N. and I. FATT, "Three-phase Relative Permeability Measurement Using a Nuclear Magnetic Resonance Technique for Estimating Fluid Saturation," SPE of AIME (1967), II-235.

SAREM, A. M., "Three-phase Relative Permeability Measurements by Unsteady-state Method," SPE of AIME (1966), II-199.

SAWYER, W. K., C. D. LOCKE, and W. K. OVERBEY, JR., "Simulation of a Finite-capacity Vertical Fracture in a Gas Reservoir," SPE 4593, 48th Annual Meeting, Las Vegas, Nev., Sept. 30–Oct. 3, 1973.

SIBLEY, W. P., "A Method for Handling Spatially Varying Fluid Properties in a Simulation Model for a Fissured Reservoir," *Soc. Pet. Eng. J.* (March 1970), 25–32.

STEFFENSON, R. J. and M. SHEFFIELD, "Reservoir Simulation of a Collapsing Gas Saturation Requiring Areal Variation in Bubble-point Pressure," SPE 4275, Third Symposium on Numerical Simulation of Reservoir Performance, Houston, Texas, Jan. 10–12, 1973.

THOMEER, J. H. M., "Introduction of a Pore Geometrical Factor Defined by the Capillary Pressure Curve," SPE of AIME (1960), 354.

WATTENBARGER, R. A. and H. J. RAMEY, JR., "An Investigation of Wellbore Storage and Skin Effect in Unsteady Liquid Flow, II: Finite Difference Treatment," *Soc. Pet. Eng. J.* (Sept. 1970), 291–97; *Trans. AIME*, **249**.

WEIGE, H. J., "A Simplified Method for Computing Oil Recovery by Gas or Water Drive," *J. Pet. Tech.* (April 1952), 91.

8 Making a Simulation Study

8.1 MECHANICS OF SIMULATION RUN

The engineer today is coming ever closer to the computer as an everyday working tool. The "systems approach" has entered his way of life, regardless of his acceptance or not. Familiarity with the simulator of necessity entails a certain degree of acquaintance with the operations of the programs. It is for this reason that the following detailed account of the simulator flow-chart will be undertaken.

The objective of the flowchart is to give a somewhat coherent account of the computational operations required and to show any areas where errors or omissions may affect results.

The simulator is generally divided up into three main areas (see Fig. 8.1):

1. Input
2. Simulation calculations
3. Output

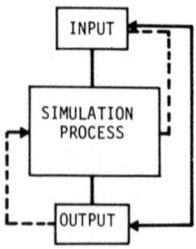

Figure 8.1: System overview.

Each area will be discussed, and the interfacing with the concepts already studied will be indicated.

A simulation program is by its very nature a complex collection of computer subroutines tied together by a well-defined logical set of operations which allow the simulator to perform and to perform adequately.

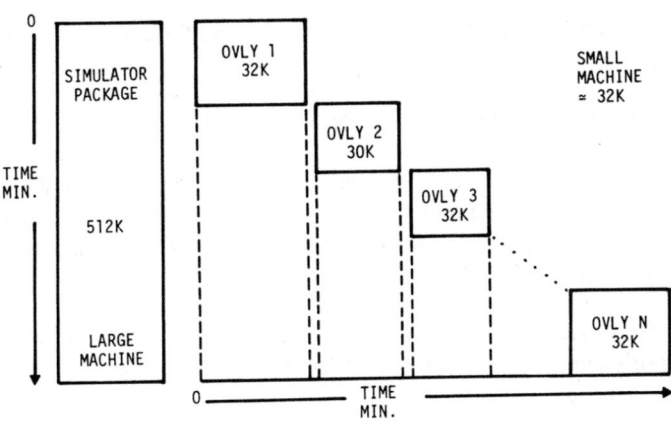

Figure 8.2: Overlay process.

Simulator programs are very large and generally occupy several hundred thousand words of computer storage. The vast bulk of these systems limits their effectiveness to large machines. There are ways in which smaller machines—i.e., 32K words—can be used by using a series of overlays or similar linking processes. The overlay process is shown schematically in Fig. 8.2.

Automatic Monitoring of Input Data

The input data required by the simulator usually run into thousands of cards, and with a data set this large there is ample room for error both by the engineer and by the keypunch operator. There are several checkpoints which are usually built into a good simulator package which check the input for errors and indicate these errors to the user. This process is essential, since the input phase can be executed in a matter of minutes while the simulation phase may run for an hour or more. A bad run on incorrect data is worse than no run at all.

Listed below are some features which make the error checking of the input data easier.

Range of Validity: Every bit of information should have a range within which it should lie; e.g., on the permeability matrix, there should be a location in the input where the engineer can input his k_{max} and k_{min}. The input program will then check every k-value that is read in to ensure that

$$k_{min} \leq k \leq k_{max}$$

In the event that data are mispunched or missing, the location is identified and printed—e.g.:

***** PERM LESS THAN MIN AT I $= 3$, J $= 18$.

Polynomial Evaluation and Printout: All polynomial data used in the simulation should be evaluated over the range of parameters expected in the run and this data output. The look-up tables should be likewise printed out for visual inspection.

Input Grid Data: All matrix data must be printed out before the run and the grids checked for empty grid locations. The outlines of every grid should be identical.

Modifications of Basic Data: It should be possible within the input program to perform any transformation of the basic grid data by simple operations—e.g.:

1. Addition of a constant ($+$ or $-$)
2. Multiplication by a constant ($*$ or $\div$)
3. Obtaining one data set from another by correlation—e.g.:

$$k_i = \phi_i^a * S_{w_i}^b$$

where the k-matrix is obtained from the porosity and water saturation matrix.

Ordering of Independent Variables: In a table of look-up the independent variables should be consistently ordered in increasing order of magnitude. As the data are read in, the computer should check that these values are in ascending order. This minor check ensures that correct interpolations will be made later when the program is executed.

Maintaining the Input File

During the course of a history match several runs will be made on modified versions of the input data. Each version should be progressively closer to the correct description of the reservoir. The engineer must endeavor to keep clearly identifiable records of each run; by comparing the results from these runs, he can make new changes in the data. The data are generally kept on magnetic tape as a series of sequential files.

The tape can be visualized as follows as shown in Fig. 8.3: Each file is a separate entity sequentially located on the tape; the engineer has access to any of the files by using the "skip" or "backspace" feature of tape handling.

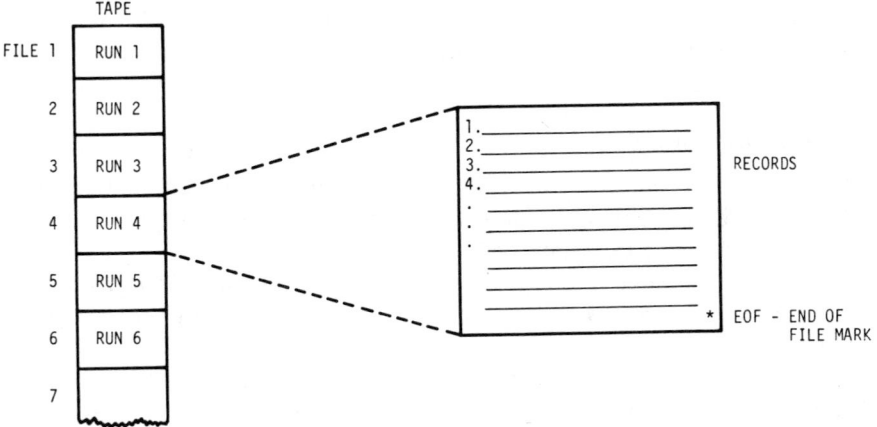

Figure 8.3: Maintaining tape file of input.

The required file can be copied to a working area and modified for another run. If each record in the file is an image of an input card sequentially numbered, it is a simple operation to replace or modify a given record or series of records by the use of a utility program. The utility program will read the corrected data cards as input and insert, write over, or delete the appropriate cards in the original tape. The modified tape will then be the basis for the next run.

Restart Procedures

The simulator program performs the simulation in a sequential manner starting at the beginning and proceeding to the end of the run as indicated by a given cumulative time. After the program has passed through a time step, the parameters calculated or used in the time step are literally "lost" unless the data is saved on some temporary storage device like a tape, disc, or drum. At the end of the run when the program is removed from core everything will be lost. The engineer will be faced with rerunning everything from scratch if he has to make some operational changes. The use of "restart" or "recovery" procedures alleviate most of these problems.

The restart procedure quite simply allows the simulator to leave "footprints" behind. In essence, the engineer indicates at selected time steps that he needs to have a restart record written. At that particular time the simulator dumps out onto a storage file all the information needed to continue the run from that point. This information is identical in form to the input data as required by the simulator except that it is compressed onto the tape, usually in binary notation. Several restart records may be written during a run, and each is sequentially identifiable by an end-of-file record.

Reasons for Restart

The most obvious reason for using restart records is machine failure. Although such occurrences are few and far between, systems sometimes die and jobs may be inadvertently aborted.

In "history" mode a job may blow up at a given point due to incorrect time step size; in such cases, instead of losing the complete run, only a portion of it is lost.

By far the best use of restart records is in prediction. By selecting a set of points in time throughout prediction the engineer can make scheduled changes in his operating plan and so optimize his project; e.g., he can

1. convert producers to injectors,
2. drill new wells,
3. fracture wells,
4. work over, or
5. perform any change that does not modify the basic reservoir description.

An example set of runs using restart procedure is shown in Fig. 8.4.

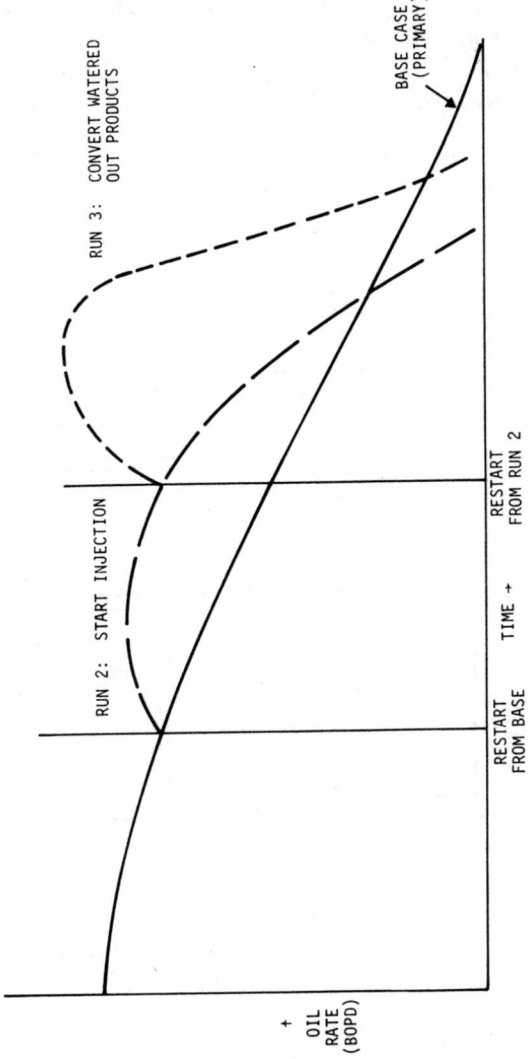

Figure 8.4: Use of restart procedures.

8.2 SIMULATOR FLOWCHART

The simulator program is a very complex set of routines. The following flow-chart attempts to show the sequence of operations that are involved in putting together a simulator. Some of the statements may seem cryptic, but the engineer should be able to follow the logic by reading between the lines. The IMPES procedure as developed in Chapter 5 is used to formulate the flowchart.

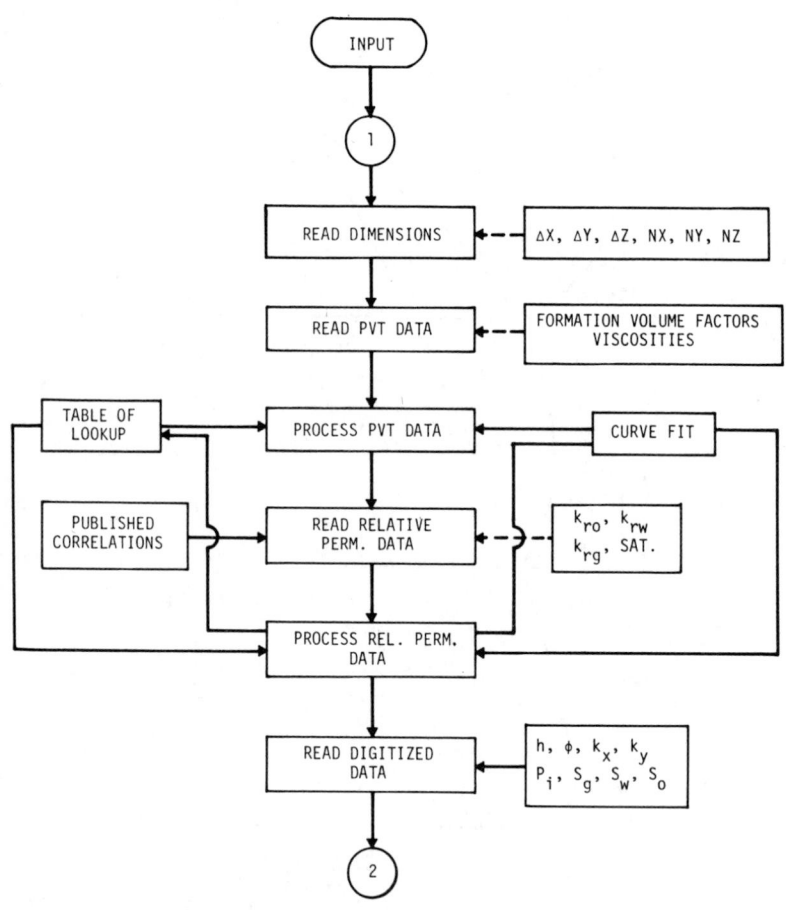

SIMULATOR FLOW CHART

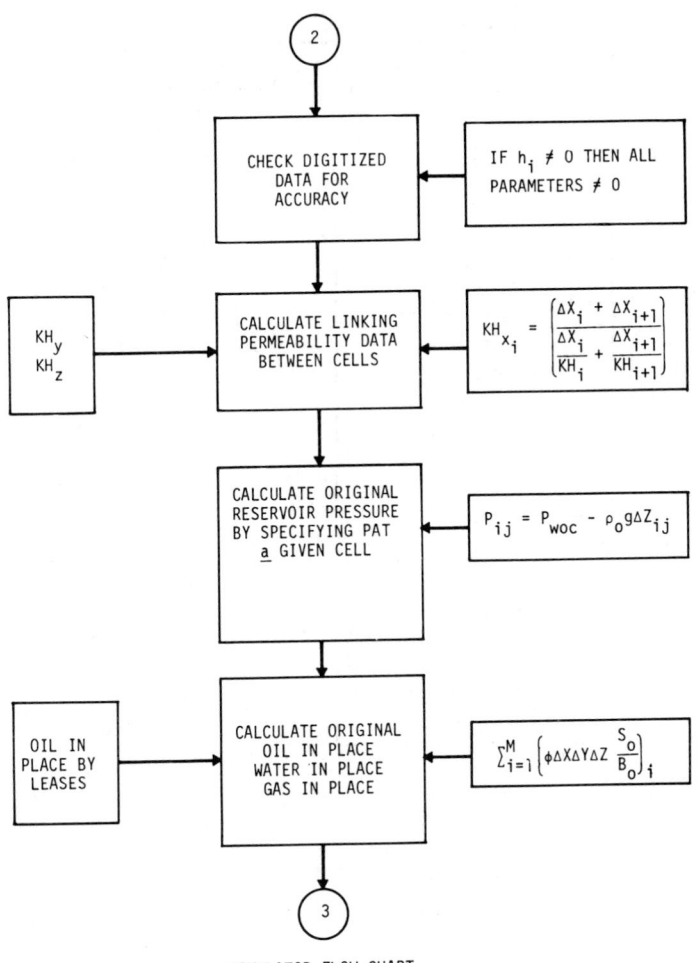

SIMULATOR FLOW CHART

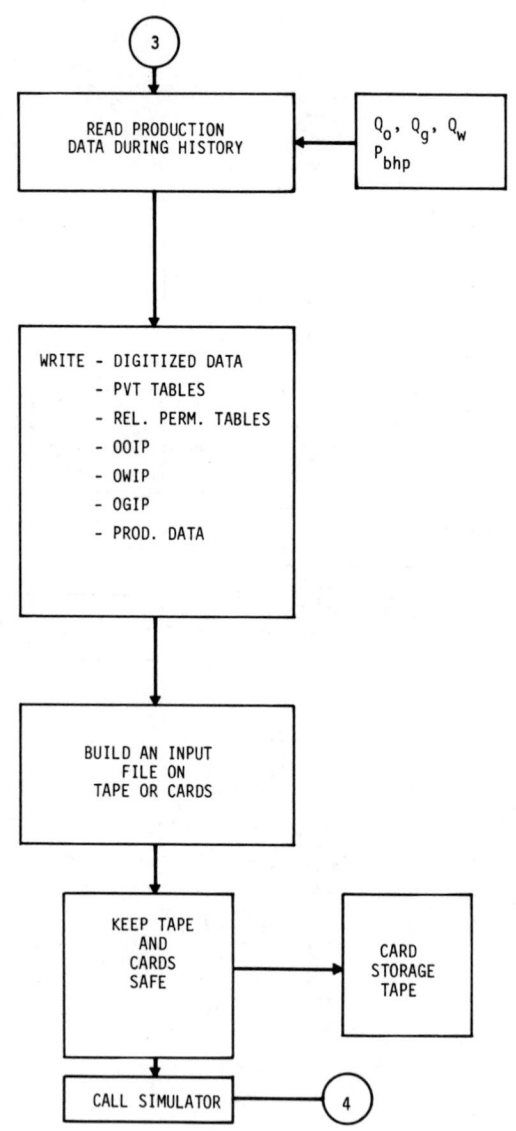

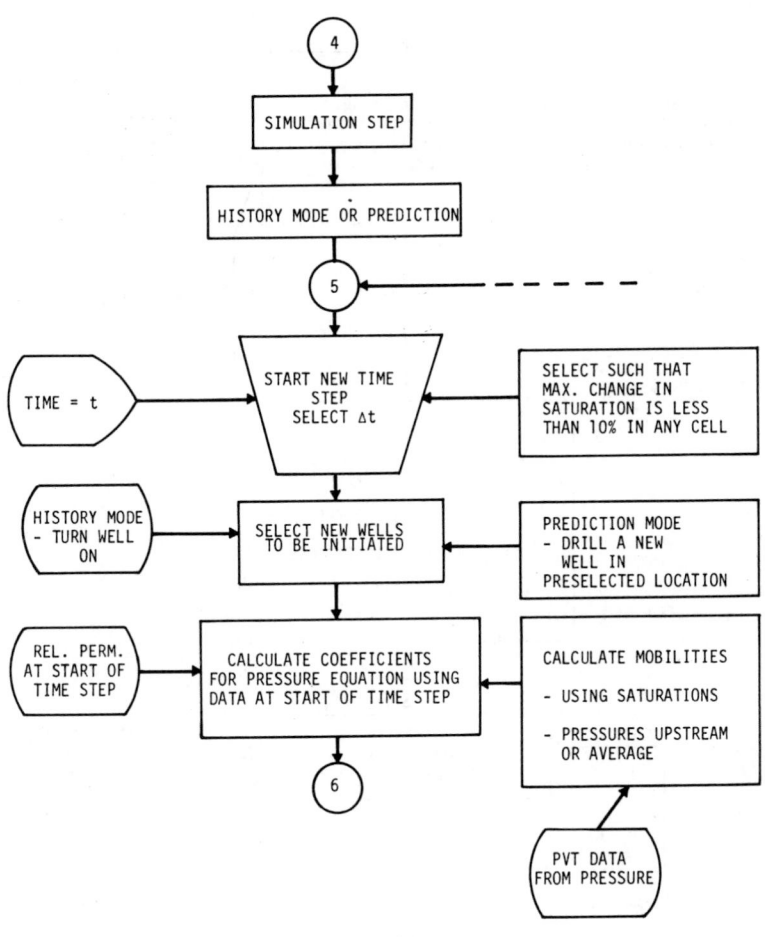

SIMULATOR FLOW CHART

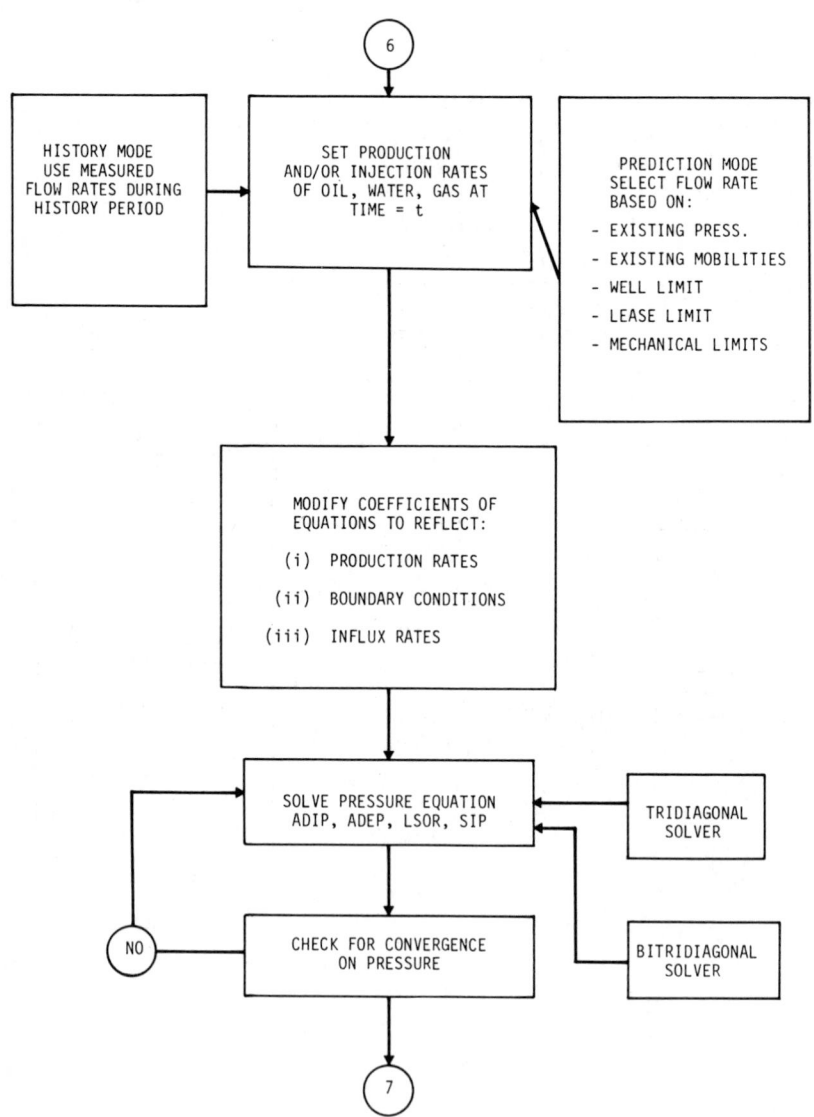

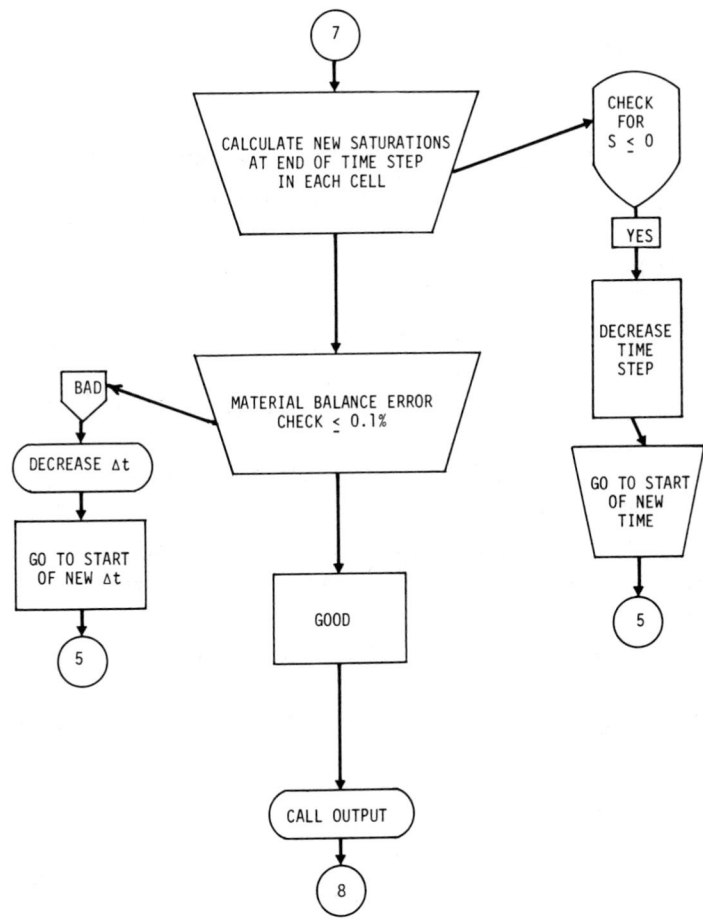

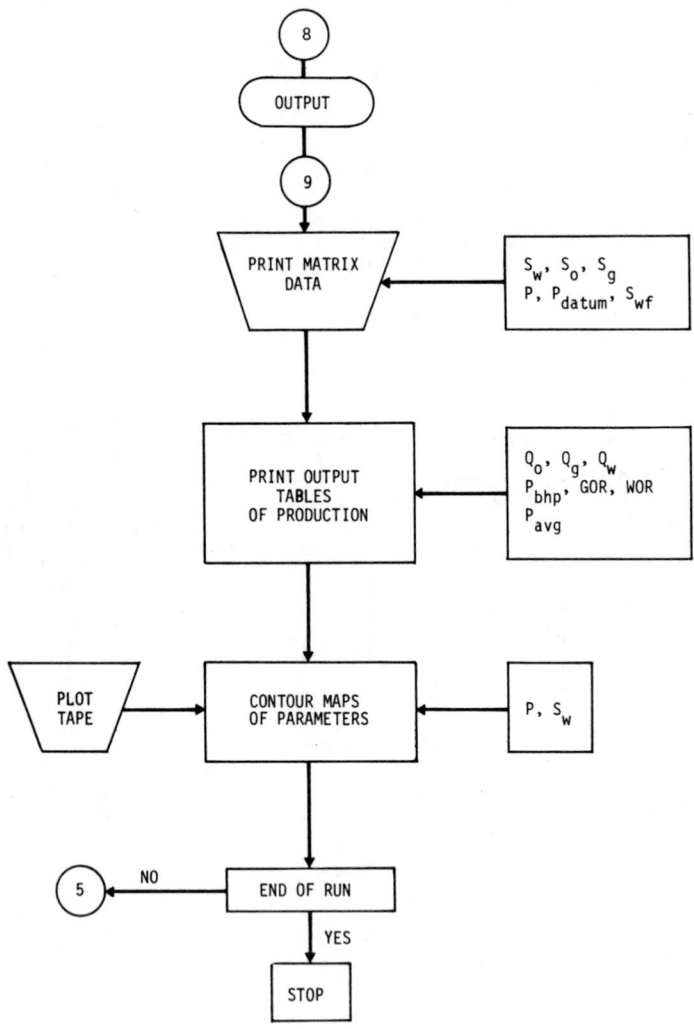

SIMULATOR FLOW CHART

232

	1	2	3	4	5	6	7	8	9	10	11
1	0.	0.	2872.	2952.	0.	0.	0.	0.	0.	0.	0.
2	0.	2927.	2760.	2914.	2956.	2956.	0.	0.	0.	0.	0.
3	0.	2773.	2504.	2838.	2905.	2851.	2977.	0.	0.	0.	0.
4	0.	2935.	2828.	2914.	2901.	2360.	2880.	2981.	0.	0.	0.
5	0.	2983.	2946.	2972.	2966.	2859.	2954.	2959.	2844.	2987.	0.
6	0.	0.	2989.	2986.	2928.	2595.	2864.	2866.	2339.	2838.	0.
7	0.	0.	0.	2999.	2989.	2897.	2847.	2931.	2866.	2978.	3000.
8	0.	0.	0.	0.	2971.	2819.	2331.	2855.	2926.	2988.	0.
9	0.	0.	0.	0.	0.	2976.	2914.	2980.	2990.	0.	0.

Figure 8.5: Pressure matrix.

	1	2	3	4	5	6	7	8	9	10	11
1	0.	0.	18.	18.	0.	0.	0.	0.	0.	0.	0.
2	0.	18.	24.	36.	30.	24.	0.	0.	0.	0.	0.
3	0.	24.	42.	60.	48.	18.	12.	0.	0.	0.	0.
4	0.	30.	48.	60.	48.	18.	12.	6.	0.	0.	0.
5	0.	30.	30.	30.	30.	24.	30.	30.	24.	6.	0.
6	0.	0.	12.	12.	12.	30.	42.	60.	42.	24.	0.
7	0.	0.	0.	6.	12.	30.	60.	84.	48.	30.	18.
8	0.	0.	0.	0.	12.	30.	42.	42.	42.	18.	0.
9	0.	0.	0.	0.	0.	12.	12.	12.	12.	0.	0.

Figure 8.6: Permeability matrix.

	1	2	3	4	5	6	7	8	9	10	11
1	0.	0.	5.	5.	0.	0.	0.	0.	0.	0.	0.
2	0.	6.	6.	8.	11.	14.	0.	0.	0.	0.	0.
3	0.	10.	10.	13.	14.	15.	16.	0.	0.	0.	0.
4	0.	13.	14.	14.	14.	15.	17.	17.	0.	0.	0.
5	0.	17.	17.	17.	17.	18.	19.	21.	23.	25.	0.
6	0.	0.	21.	21.	20.	20.	21.	22.	24.	26.	0.
7	0.	0.	0.	22.	22.	23.	23.	24.	26.	27.	28.
8	0.	0.	0.	0.	25.	25.	25.	26.	27.	28.	0.
9	0.	0.	0.	0.	0.	28.	28.	28.	28.	0.	0.

Figure 8.7: Porosity matrix.

	1	2	3	4	5	6	7	8	9	10	11
1	0.	0.	3.	4.	0.	0.	0.	0.	0.	0.	0.
2	0.	8.	10.	10.	8.	4.	0.	0.	0.	0.	0.
3	0.	4.	14.	15.	16.	8.	4.	0.	0.	0.	0.
4	0.	3.	13.	17.	23.	15.	6.	3.	0.	0.	0.
5	0.	3.	10.	15.	25.	28.	16.	10.	4.	2.	0.
6	0.	0.	3.	6.	17.	26.	24.	17.	10.	5.	0.
7	0.	0.	0.	2.	6.	17.	17.	17.	12.	7.	3.
8	0.	0.	0.	0.	3.	7.	12.	13.	7.	3.	0.
9	0.	0.	0.	0.	0.	2.	3.	3.	2.	0.	0.

Figure 8.8: Formation thickness matrix.

236

8.3 SELECTION OF THE MODEL

The engineer has to make a decision in selecting the optimum model to simulate the reservoir under study. His selection must be made systematically and with an analysis of all the parameters involved. The parameters which are significant in a model selection are the following:

a. Reservoir type
b. Reservoir geometry and dimensionality
c. Data availability
d. Type of secondary or tertiary process being modeled
e. Manpower requirements
f. Computer availability
g. Cost effectiveness of model

The overriding concern of the engineer in making a model study is to adequately simulate the reservoir with the minimum of effort. This does not necessitate "cutting corners," but simply means that all due care should be exercised not to "overkill" the model study by using a simulator which is obviously too sophisticated and expensive.

The Selection Process

The simulator selector shown in flow diagram form in Fig. 8.9 spells out the modus operandi of optimum selection.

 a. System Definition: Reservoirs exist usually in three generic groups:

1. Gas
2. Oil (black)
3. Condensate

Gas reservoirs may be present with or without active (mobile) water systems. If there is no active water, a single-phase model is adequate. Oil reservoirs which have minimal mass transfer between oil and its associated gas can be handled by the black oil simulators. The presence of mobile water may necessitate the need for the inclusion of the water phase and hence a two-phase model. When the mass transfer between the hydrocarbon phases is significant, it is imperative that a compositional model be used to account for the physical processes of mass transfer.

1. *Gas*—Single-phase gas with no active water drive
2. *Black oil*—Saturated or undersaturated system

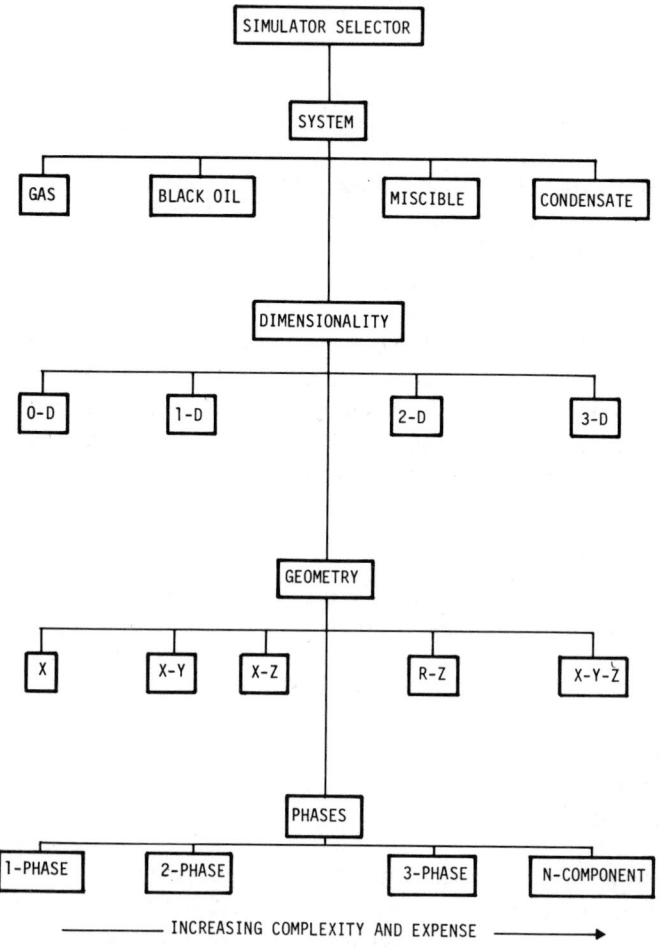

Figure 8.9: Simulator selector.

    a. No mass transfer
    b. Presence or absence of water drive
    c. Above or below the bubble point
3. *Condensate*—Compositional behavior
    a. Significant mass transfer between phases
    b. Very volatile hydrocarbon systems
    c. Gas-cycling operations

    b. Reservoir Geometry and Dimensionality: Perhaps the easiest phase of the selection process is the determination of the model dimensionality. There are only four possible combinations. First, a *zero-dimensional model* indicates that reservoir properties do not change with location in the reservoir; the reservoir is essentially homogeneous, isotropic, and uniform in every sense. In this case we use a zero-dimensional simulator, which is a material balance equation.

    If the engineer is trying to simulate a pilot project or a simple linear segment of the reservoir, a *one-dimensional model* is adequate. This one-dimensional model can be rotated in either the vertical, horizontal, or curvilinear direction depending on the need to simulate particular effects.

    Figure 8.10 illustrates the typical one-dimensional application of a simulator. Note the presence of the two fluid contacts. Figure 8.11 shows a modification of the linear model to account for a dipping reservoir. This

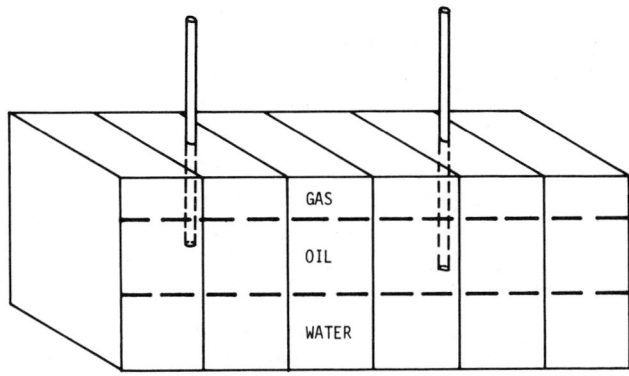

ONE DIMENSIONAL HORIZONTAL
- SIMULATES RESERVOIR SECTIONS
- SPECIALIZED STUDIES
    - LINE DRIVE BEHAVIOR
    - MISCIBLE FLOODING
    - PILOT FLOOD SIMULATIONS

Figure 8.10: One dimensional horizontal.

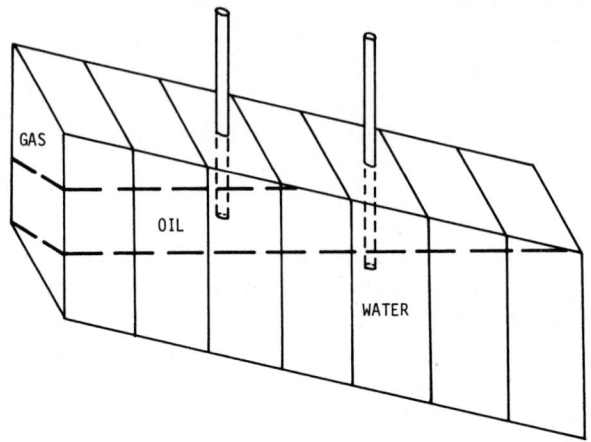

ONE DIMENSIONAL - DIPPING

- SIGNIFICANT GRAVITY OVER-RIDE
- SAME IN GENERAL AS HORIZONTAL
    ONE DIMENSIONAL MODEL
- UPDIP GAS INJECTION
- FLANK INJECTION OF WATER

Figure 8.11: One dimensional—dipping.

application is suited to those areas where significant gravity override exists or when the engineer wants to examine the updip injection of gas or the flank injection of water into a section of his reservoir.

A *two-dimensional model* is best suited for large studies where effects of areal changes are important. Versions of the two-dimensional model are available for special studies. Figure 8.12 shows the most common reservoir model. It is the general-purpose simulator which is used on more studies than any other. It allows the simulation of large multiwell structures where the engineer is interested in the total behavior of the system. This model handles wide variations in rock and fluid properties areally, but assumes that there is not a great variation in these properties vertically. Because of the very size of the area modeled, the engineer can look at fluid migration across lease lines and the effects of aquifer interference and other outside influences on the reservoir behavior. Recently, methods have been proposed whereby these two-dimensional models can be used to simulate three-dimensional flow by the selection of a set of relative permeability curves which would account for the vertical effects of the flow dynamics. These pseudo–relative permeability data are being used to economically predict three-dimensional behavior without the prohibitive incurred cost of a three-dimensional model. Figure 8.13 is another extension of the two-dimensional model.

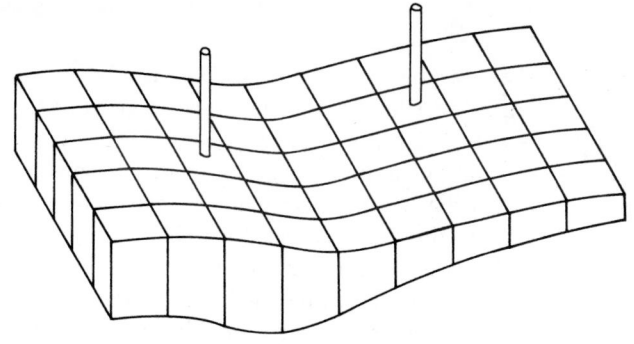

TWO DIMENSIONAL HORIZONTAL

MOST GENERALIZED AND ALL PURPOSE MODEL
- SIMULATION OF LARGE MULTI-WELL STRUCTURE
- LARGE RESERVOIR SIMULATIONS OF MULTI-UNIT SYSTEMS
- HETEROGENEOUS ROCK PROPERTIES
- SLIGHT VERTICAL VARIATION IN FLUID PROPERTIES
- ANALYSIS OF MIGRATION ACROSS LEASE LINES
- SELECTION OF OPTIMUM OPERATIONAL SCHEMES IN
  SECONDARY RECOVERY AND PRESSURE MAINTENANCE

Figure 8.12: Two dimensional horizontal.

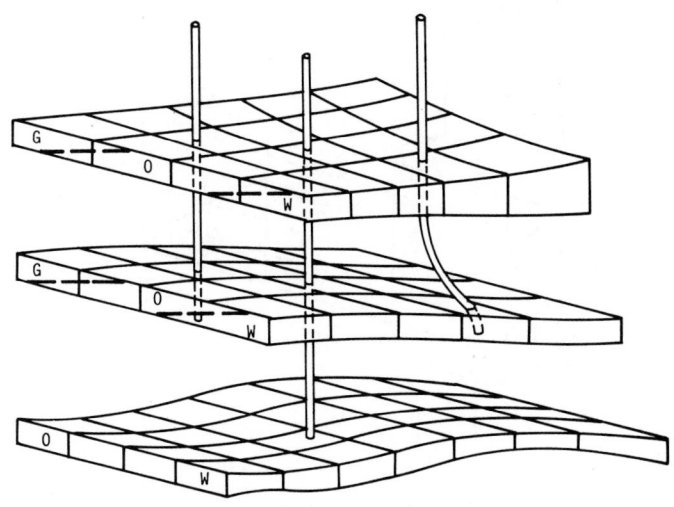

THREE DIMENSIONAL LAYERED

- SIMULATION OF LARGE RESERVOIRS CONSISTING OF
  SEVERAL PRODUCING HORIZONS
- COMMINGLED OR NON-COMMINGLED PRODUCTION
- MULTIPLE COMPLETIONS
- THIS MODEL IS IN EFFECT SEVERAL 2-D MODELS STACKED
  TOGETHER WITH SPECIAL WELL BORE HYDRAULICS ROUTINES

Figure 8.13: Three dimensional layered.

In this application the reservoir to be modeled consists of a sequence of producing horizons each of which is individually two-dimensional. The areal simulator can be used to solve this system as long as the locations which contain the wells are suitably handled. Since the wells present the only common region between these layers, it is clear that a special section of the model must be used to compute the wellbore behavior. The computed conditions in each layer then form the inner boundary condition for each cell containing a well. This application can be used with or without commingled production.

Figure 8.14 is a very special two-dimensional model. It is the r-z model used in simulating single-well behavior with radial symmetry and vertical heterogeniety. The coning model looks at single-well parameters such as completion effects, critical production rates to prevent gas or water coning, and deliverability response of gas wells. A rather novel use of the coning study is to back-calculate a pressure build-up study. This exercise is a history-

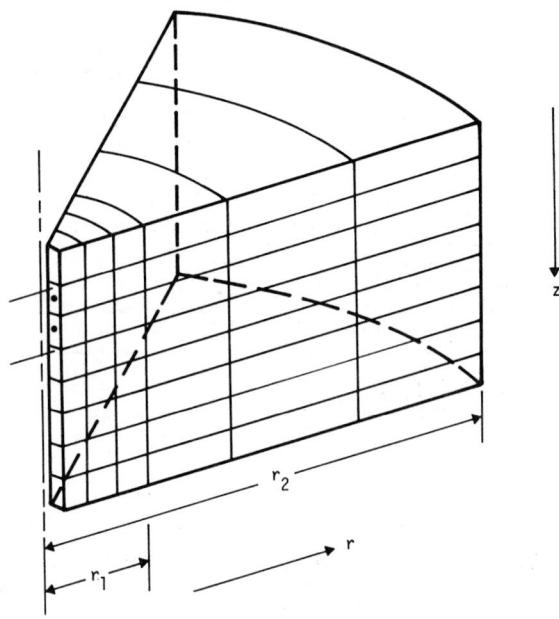

TWO DIMENSIONAL CONING MODEL

- SINGLE WELL OPTIMIZATION STUDIES
- LOCATION OF COMPLETION INTERVALS
- MAXIMUM EFFICIENT RATES
- DELIVERABILITY STUDIES
- WELL TEST ANALYSIS

Figure 8.14: Coning model.

matching procedure in order to determine in situ permeability characteristics.

Figure 8.15 indicates the *x-z*–type model where the study of a reservoir cross section is made. This approach can analyze single- or multiple-well completions and examine gravity segregation and the effects of crossflow and anisotropy on frontal displacement processes. For example, a *coning* study by its very nature requires an *r-z* model to allow a proper representation of the converging flow patterns.

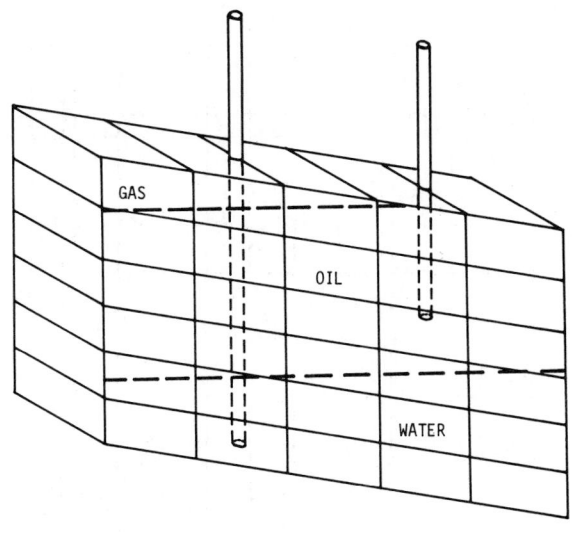

TWO DIMENSIONAL VERTICAL

- SINGLE OR MULTIPLE WELL SIMULATION
- CROSS-SECTION ANALYSIS OF RESERVOIR
 FOR (1) GRAVITY SECREGATION
      (2) EFFECT OF ANISOTROPY ON
          FRONTAL PLACEMENT

Figure 8.15: Two dimensional vertical.

Three-dimensional models are required in situations where the reservoir relief is great, fluid properties vary vertically, and flow patterns are complicated by shale breaks and other impediments. Figure 8.16 indicates a typical three-dimensional study. The reservoir consists of a large area with reasonably thick sand sections. There is significant variation of both rock and fluid properties vertically, indicating the need for the incorporation of the third dimension. Figure 8.17 shows a portion of a three-dimensional model where the oil/water contacts are shifted, where considerable faulting has modified the normal reservoir configuration.

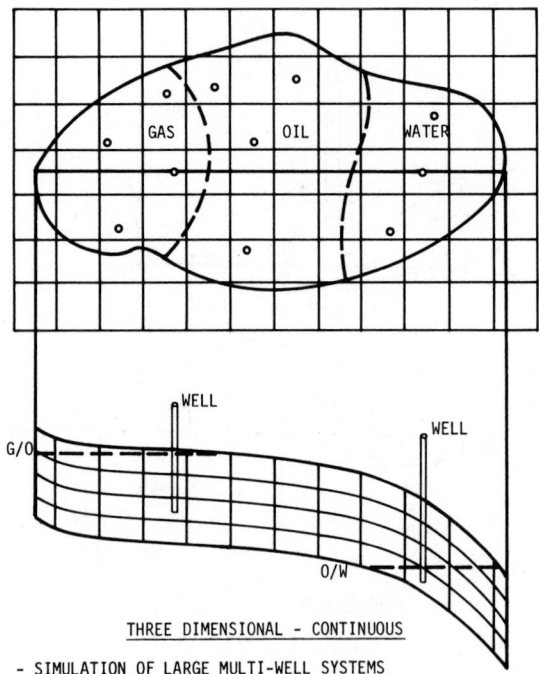

THREE DIMENSIONAL - CONTINUOUS

- SIMULATION OF LARGE MULTI-WELL SYSTEMS
- THICK RESERVOIR PAY SECTIONS
- SIGNIFICANT VARIATION OF ROCK PROPERTIES VERTICALLY
- SIGNIFICANT VARIATION OF FLUID PROPERTIES VERTICALLY
- LAYERED RESERVOIR SYSTEMS WITH COMMON AQUIFER
  OR SIGNIFICANT VERTICAL CROSS FLOW

Figure 8.16: Three dimensional—continuous.

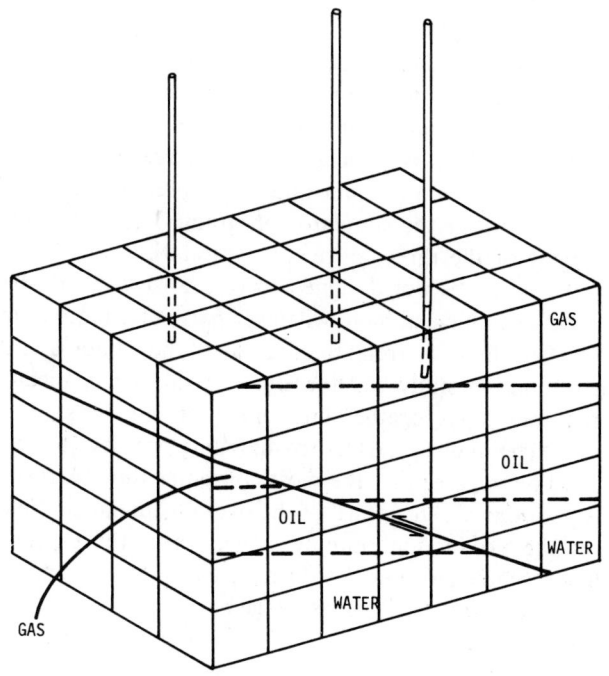

THREE DIMENSIONAL - CONTINUOUS (SECTION)

- ESSENTIALLY SAME AS 3-D CONTINUOUS
- FAULTED SYSTEMS

Figure 8.17: Three dimensional—continuous faulted.

c. Data Availability: A sufficient quantity of data is required before any simulation should be attempted. In the absence of rock parameters and production history data, no attempt should even be made at simulation. The more complex the model becomes, the more detailed is the data requirement. The minimum data required should be enough to define the reservoir adequately in the area of interest.

d. Type of Secondary or Tertiary Process: The most common processes are water flood or gas injection. In these cases a three-phase model is necessary to account for the mobility of all fluid phases. In some instances a miscible flood may be initiated and the model must be modified to reflect the miscibility of the injected slug in the hydrocarbon phases. In polymer flood studies the permeability saturation data must be set up to reflect the fluid behavior of polymers in water. This usually entails modification of the relative permeability data.

   e. *Manpower Requirements:*   Complex studies involve a lot of data gathering and compilation. Without an adequate staff it may be impossible to do the job in a reasonable amount of time.

   The simulation process is basically a three-stage process. The primary stage involves the gathering of data and the subsequent reduction of this data to a form usable by the simulator program. The gathering process is by far the longest of the three, since the exploitation and development of reservoirs in the past was not designed to centralize the data bases but to allow each engineer to have at hand whatever particular piece of information he required. Today, there is a trend to consolidate these data, and the start-up process in a simulation study will therefore be shorter. Typically, six man-months are required to put together data for a moderate size study.

   The second stage involves the history-matching process; this stage, though shorter in man-hours, consumes the most computer time. A reasonable estimate for this period is about a third of the overall simulation time. The final stage involves the production running stage, where the engineer uses the completed history-matched model to generate a series of runs to obtain an optimum operational scheme. This last stage normally involves about a quarter of the total simulation time.

   f. *Computer Availability:*   A large machine is essential for rapid turnaround and efficient completion of a study. The minimum requirement is a 32K machine. On any smaller device it becomes physically impossible to run even a small study. Tape changing, disc access, and computing time increase exponentially as the machine size decreases.

   The resurgence of virtual memory in the computer marketplace has extended the use of larger models to smaller machines. Since the virtual-memory machine maintains only that fraction of the program in real core that is needed and spools in continuously from the "virtual" area those "pages" of memory as the program executes, it is possible to run large models on smaller machines. The overhead involved in swapping these pages of memory into and out of real core can become significant if the computer program is not coded efficiently for the use of virtual memory.

   g. *Cost Effectiveness of Model:*   The models used for reservoir simulation increase in complexity and expense as we go from one-phase to three-phase, as we go from one-dimensional to three-dimensional, and as the applications become more and more specialized. A general all-purpose model is more cost-effective for all-round reservoir engineer work, and it is far better than any method used to date for reservoir engineering. In areas where several hundred thousand dollars will be invested on a project, it makes good engineering sense to conduct a simulator study which allows the optimum development scheme. The saving accrued from not drilling a *single* well could pay for the entire study.

9 History Matching

9.1 INTRODUCTION

The reservoir simulator, as such, cannot be used to predict the performance of a reservoir under any operating scheme unless the data built into it adequately describe the geometrical configuration, the rock and fluid properties, and the flow mechanics of the reservoir system.

The original data built into the simulator are the engineer's best estimate of all the parameters which describe the reservoir. Unless he is very lucky, this data will not be exactly representative of the reservoir as a whole. These data must be modified until the simulator reproduces the behavior of the reservoir to an acceptable degree. The process of modifying the existing model data until a reasonable comparison is made with the observed data is called *history matching*. The process is a necessary prelude to making any sensible predictions with the simulator, because the same mechanisms which were operative in the history period of the reservoir should still be operative in the prediction period.

The process of history matching is one of the more time-consuming aspects of a simulation study. As much time is spent on making a match as is spent on compiling and preparing the data for the simulator. A good match is as important as good data.

The process of history matching is characterized by a feedback loop in which the engineer reformulates his basic conception of the reservoir as a result of the responses of those parameters which he is using as a measure of the system behavior. By analysis of the effect of changes made after a particular run n, the engineer then decides on the form of the input data for run $(n + 1)$. This process has all the earmarks of a classic control problem, and recently this observation has led to several algorithms of varying degrees of effectiveness which attempt automatically to perform the history match. The feedback loop has been tightened considerably, with the net result that the engineer is now squeezed out of the loop completely. Some engineers seem to prefer it this way, but there is still some need for the subjective approach of the trained mind. We shall discuss some of these automatic techniques later.

History-Matching Parameters

Several parameters are available for determining a good history match:

1. Pressures
2. Flow rates
3. Gas-oil ratios
4. Water-oil ratios

The objective of the engineer is to determine the reservoir description which will minimize the difference between the observed parameter as indicated above and that predicted by the simulator program. It can be argued very effectively that there is really no unique set of descriptive parameters which fit a reservoir. While this is true, it is also noteworthy that the practicing engineer is not completely in the dark; he is not solving an unconstrained problem. Whether or not he quantifies it, he has instinctively set in the back of his mind a list of lower cutoff values and upper limits of all his variables. This set of constraints, whether written down or not, are part of the engineering know-how essential in making a good simulation study. A good history match is, in essence, that set of rock, fluid, and relative permeability data which acting together produce the most reasonable results at a given point in time. As more data are amassed with the passage of time, the description can be refined even more. This updating process is essential in a simulation study not only for continual monitoring but also for being able to make predictions based on any unforeseen operational changes in the future.

Mechanics of History Matching

There are several parameters which can be varied either singly or collectively to minimize the differences between the observed data and those calculated by the simulator. Modifications are usually made on the following areas:

1. Rock data modifications
   a. Permeability
   b. Porosity
   c. Thicknesses
   d. Saturations

2. Fluid data modifications
   a. Compressibilities
   b. PVT data
   c. Viscosity
3. Relative permeability data
   a. Shift in relative permeability curve
   b. Shift in critical saturation data
4. Individual well completion data
   a. Skin effect
   b. Bottom hole flowing pressure

The two fundamental processes which are controllable in history matching are as follows:

1. *The quantity of fluid* in the system at any time and its *distribution* within the reservoir, and
2. The *movement of fluid* within the system under existing potential gradients.

The manipulation of these two processes enables the engineer to modify any of the earlier-mentioned parameters which are criteria for history matching. It is mandatory that these modifications of the data reflect good engineering judgment and be within reasonable limits of conditions existing in that area. The expertise of the engineer and his familiarity with the particular reservoir can markedly reduce the total time spent on history matching.

9.2 MODIFICATIONS USING ROCK DATA

Rock data are generally input as a matrix of values over a two- or three-dimensional grid. The data were derived by digitizing a contour map of some kind. Grid data modifications are never made on a single-cell basis, but over an area wherein the necessary changes are required. The parameter being changed—e.g., permeability—is recontoured to produce the desired change and this new map again redigitized. This method ensures a certain smoothness and continuity of rock properties and prevents the building up of discontinuities, which may not be present. There may be instances, however, where faults or pinchouts may cause discontinuities and, as such, must be incorporated in the model.

The following is an example of grid data modification to remove a localized pressure abnormality. The mechanics of grid data modification are similar for all the various parameters which constitute the basic rock data set, and this example should familiarize the engineer with the "modus operandi." Figure 9.1 illustrates the permeability data as input to the simulator for a given region of model. The computed pressure distribution indicated in Fig. 9.2 shows a localized high area and a localized low area in the region selected.

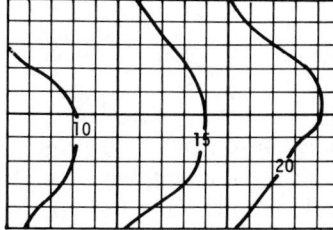

Figure 9.1: Isopermeability map.

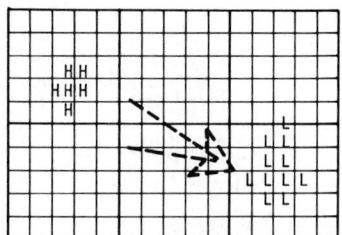

Figure 9.2: Pressure distribution from model at given time.

The remedial action required here is to lower the high-pressure area and raise the low-pressure area. This could be achieved by one or more of the following procedures:

1. Move fluid from the high-pressure to the low-pressure zone by a change in rock permeability.
2. Decrease the oil in place in the high-pressure area by either
    a. decreasing porosity,
    b. decreasing thicknesses,
    c. decreasing oil saturation, or
    d. all of the above.
3. Increase the oil in place in the low-pressure area by either
    a. increasing porosity,
    b. increasing thicknesses,
    c. increasing oil saturation, or
    d. all of the above.

The engineer at this point must decide which procedure is most likely to create the desired action and not disturb the model at other locations nor to create additional problems in these locations at later times.

The most likely procedure is to move fluid across the reservoir from the high- to the low-pressure area. This is accomplished by recontouring the permeability map. This maintains the same original oil-in-place figure, since we have not varied ϕ or h.

The permeability map is recontoured in such a manner as to move the fluid in the general direction of the low-pressure region. The "noses" of the contour lines are toward the low-pressure zone. This gives some directional permeability into the low-pressure region as shown in Fig. 9.3.

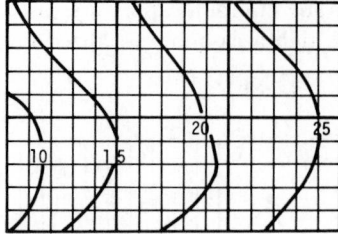

Figure 9.3: New permeability map.

This map is redigitized and the data repunched to make a new run. The new pressure map shown in Fig. 9.4 may need some modifications in data again. These high-pressure zones can be easily modified by slight porosity changes in that general area, as shown in Fig. 9.5.

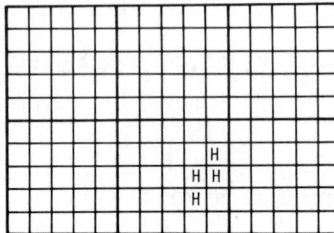

Figure 9.4: New pressure map at given time.

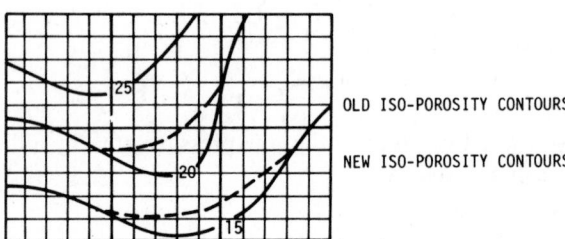

Figure 9.5: Modification of isoporosity lines in localized areas.

One of the more definitive matching criteria is pressure, since it is also possible to obtain pressure data with reasonable accuracy. The computed pressures can be in error in several ways—namely:

1. The *pressure levels throughout the reservoir* may be either too high or too low.

2. The *pressure distribution* may be too discontinuous—i.e., "jagged" when looking at a cross-sectional profile.

3. *Localized well pressure data* may be too high or too low, indicating specific localized imbalances.

These abnormalities can be corrected as follows.

Pressure levels too high throughout usually indicate that the oil-in-place figure is too high in the reservoir; a blanket change in porosity could reduce the overall oil in place, thereby reducing the total expansive energy in the system. This is illustrated in Fig. 9.6.

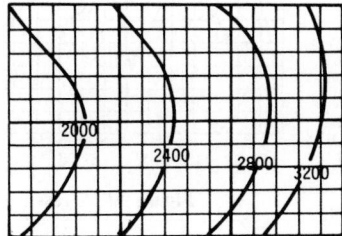

Figure 9.6: Pressures at time *t*.

The pressures are too high overall by a factor of, say, 10%. We therefore perform a reduction of every porosity value by a constant factor—say, 0.96:

$$Por_{ij} = Por_{ij} * Factor$$

Note that the contours in Fig. 9.7 are exactly in the same place. The numbers associated with them are different. Note an overall decrease in the pressures, as shown in Fig. 9.8.

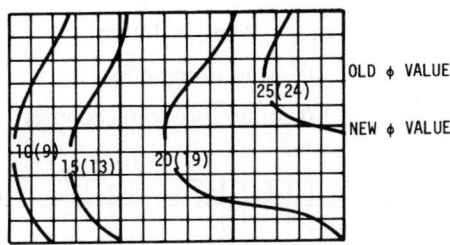

Figure 9.7: Porosity contour map used in determination of Fig. 9.6 pressures (new values).

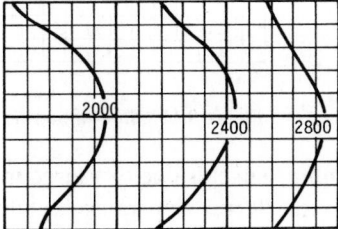

Figure 9.8: New pressure map.

A pressure distribution that is too discontinuous usually indicates incorrect permeability values within the system. A cross section of the reservoir showing the pressure profile is shown in Figs. 9.9 and 9.10.

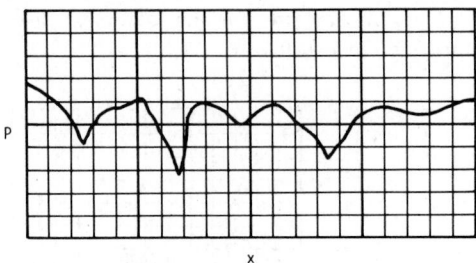

Figure 9.9: Pressure profile.

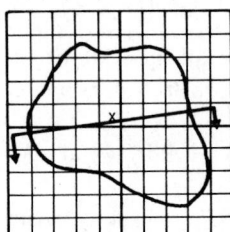

Figure 9.10: Traverse.

By making a blanket change in permeability by using a constant factor, the pressure "highs" and "lows" are smoothed out as indicated in Fig. 9.11:

$$\text{Perm}_{ij} = \text{Perm}_{ij} * \text{Factor}$$

Factor > 1.0—smoothing effect

Factor < 1.0—more pronounced effect

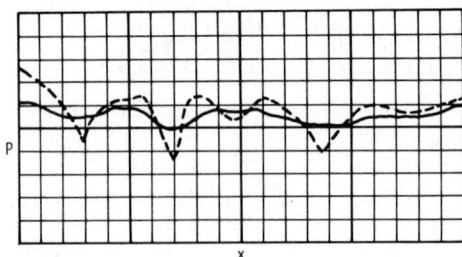

Figure 9.11: Smoothing the pressure profiles.

This "smoothing" effect is due to a large transmissibility product throughout the reservoir; the larger the transmissibility, the closer the model behaves as a steady-state model and less as a transient or unsteady-state model. The model tends to react more like a "tank" of uniform properties the larger the permeability.

9.3 MODIFICATIONS USING FLUID SATURATIONS

The fluid saturations within the reservoir are usually reasonably well known from various sources:

1. Core samples
2. Electric logs
3. Correlations.

The quantity of expansive energy within the porous rock is directly related to the saturations of the three fluid phases, and the effects of reservoir energy are shown in several ways:

1. Reservoir pressure level
2. Production rates
3. Gas-oil ratios
4. Water-oil ratios.

During the course of a history match several events may be noticeable which can be attributed to errors in fluid saturations within the pore space. The most obvious is the inability to maintain a given production rate from an area and consequent drop in reservoir pressure and fluid saturation.

This is generally due to insufficient oil in place and insufficient influx in the given locality. As mentioned earlier, this can be corrected by changes in pore volume; an increase in oil saturation is recommended to correct this discrepancy. It is difficult to say precisely which change is more realistic at this time, and only an overall analysis of the model behavior will indicate the more reasonable change.

Another event that can occur in a history match when there is insufficient gas in place in the reservoir is that the model runs out of gas. The wells have been making the observed gas production, but there is not enough gas originally in place in the system. An increase in S_{g_i} by some factor would offset this tendency to deplete the gas in place and maintain reasonable production rates throughout the prediction mode. Before this increase is made, the engineer should reexamine the location of the fluid contacts within the reservoir, particularly the gas/oil contact. A shift of a few feet in this contact can make a significant difference in the quantity of gas that can be produced by depletion.

9.4 MODIFICATIONS USING FLUID DATA

Fluid data are generally well known in a simulation study. These data include the following:

1. Formation volume factors
2. Viscosity
3. Compressibility
4. Solution gas data.

The usual errors involving these data are caused by *faulty input*. A misplaced decimal point or an incorrect exponent can cause an order-of-magnitude error in the input quantity. Some examples are the following:

1. *Observation:* No noticeable drawdown in the pressures in the model even after considerable withdrawal of fluid.
Cause: Rock compressibility is too high by an order of magnitude, causing the effects of changing saturation to be negligible as seen by inspection of the effective compressibility equation:

$$C_e = \frac{S_o C_o + S_w C_w + S_g C_g + C_f}{S_o} \tag{9.1}$$

Action: Use the correct rock compressibility value.
2. *Observation:* Water saturation appears to increase in model without any injection or influx of water.
Cause: Rock compressibility is too low, causing a free volume to develop in the pore space. In some models this free volume is filled with the immobile phase, which is usually water. As a result, an additional quantity appears in the water material balance.
Action: Correct the rock compressibility.

9.5 MODIFICATIONS INVOLVING RELATIVE PERMEABILITY DATA

Relative permeability is a very complex phenomenon, and it can be handled in many ways in the simulator. These different approaches affect the calculated performance to a greater or lesser degree, and it behooves the engineer to be familiar with the scheme used in the model with which he is working. As mentioned earlier there are two types of curves:

1. Imbibition
2. Drainage

These are used for different processes in the reservoir.

Relative permeability directly affects flow rate and consequently gas-oil ratios and water-oil ratios. Secondly, pressures are affected because of fluid movement incurred during the flow under the pressure gradients existing in the reservoir. The producing gas-oil ratio and the producing water-oil ratio are two criteria used in history matching which can be modified by relative permeability changes.

Consider the following. A field has been producing and the overall GOR (gas-oil ratio) is plotted versus cumulative production (Fig. 9.12).

The GOR is too high, indicating a too optimistic k_g/k_o curve. To reduce this high gas production the k_g/k_o curve is moved from left to right to decrease the quantity of gas flowing at a given liquid saturation (Fig. 9.13). It should be pointed out at this point that the lower gas production figure would also create a smaller pressure drop throughout the reservoir. This could be remedied if need be by a change in the quantity of fluid in the system.

Consider the opposite case of too low a GOR. A plot of GOR versus cumulative production is shown in Fig. 9.14. There is not enough gas flowing in the system at any given liquid saturation. To increase gas production the k_g/k_o curve is shifted from right to left to induce more favorable flow conditions to gas, as shown in Fig. 9.15. The higher gas production creates a secondary factor to be accounted for—i.e., lower reservoir pressures. The use of grid data modifications can be made to correct these discrepancies.

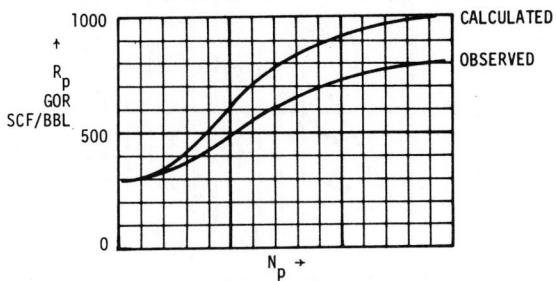

Figure 9.12: Producing gas/oil ratio data too high.

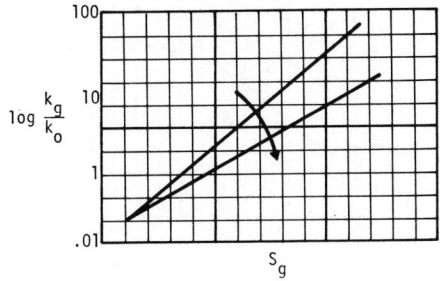

Figure 9.13: Shift down in k_g/k_o data.

Another example of the effect of relative permeability on gas-oil ratio can be seen from Fig. 9.16. Free gas starts flowing too early in the model; the curve, however, has the general shape of the observed curve. A shift in the critical gas saturation as shown in Fig. 9.17 delays the production of gas until the gas saturation reaches a higher value. This has a secondary effect of increasing reservoir pressures since the total gas production is decreased.

The opposite example is shown in Fig. 9.18, where the computed GOR lags below the measured GOR at a given saturation. The relative permeability –versus–gas curve can be modified by shifting the S_{g_c} to the right (Fig. 9.19). The effect of this is to make the gas flow at an earlier saturation.

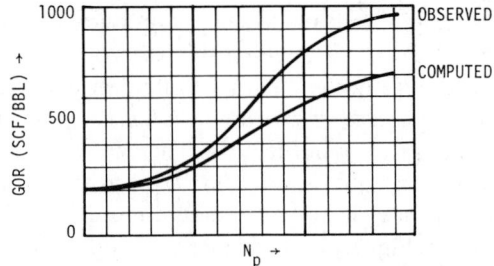

Figure 9.14: Producing gas/oil ratio too low.

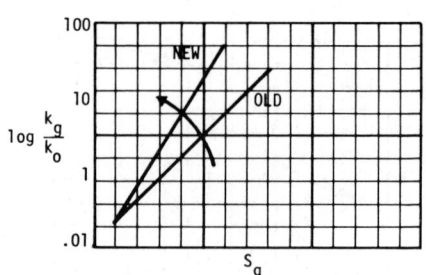

Figure 9.15: Shift up k_g/k_o data.

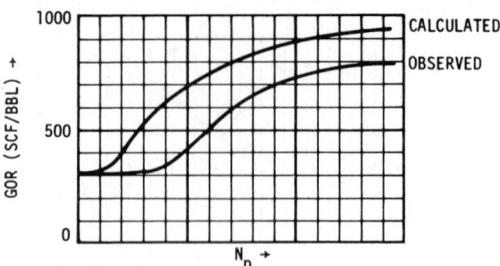

Figure 9.16: Model produces gas too early.

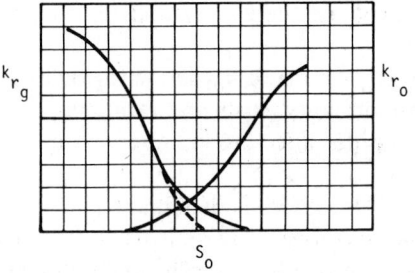

Figure 9.17: Shift in critical gas saturation.

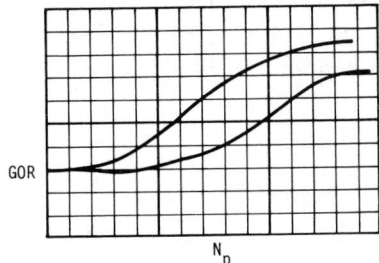

Figure 9.18: Model produces gas too late.

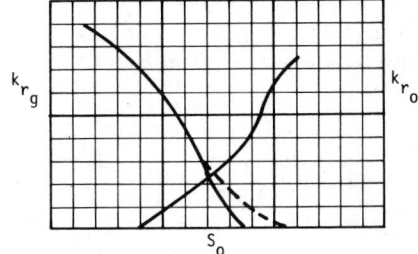

Figure 9.19: Shift in critical gas saturation.

9.6 PRESENCE OF COMMUNICATING AREAS

In some history runs it is possible to vary all the parameters governing the oil in place and still not obtain a match. An example is as follows.

In a given reservoir, even after varying the reservoir data within the ranges of credibility the following abnormalities are still unresolved:

1. Field observed pressures are still to high to be matched.
2. Field observed production rates are still too high compared to model figures.

The obvious question remains: what can be maintaining these pressures and rates? The answer is unrecognized reservoir energy.

The reservoir is getting energy in some form from *outside* the limits of the *presently defined area*. This source of reservoir energy can be either

1. additional undeveloped productive zones, or
2. unrecognized water drive influencing the reservoir.

The engineer can incorporate these conditions into his model by

1. adding additional acreage to his simulator by expanding his model in the area where the abnormality exists and rerunning; and
2. simulating an aquifer affecting his reservoir in the given region.

A careful analysis of these two is then made during several runs to define which process may be operative. If there is no evidence of increasing water cut, the abnormality is most likely due to undeveloped acreage.

It should be mentioned that the presence of communicating zones should be looked into as a last resort and the engineer should not flagrantly increase the productive acreage unless the evidence is overpowering. In these cases it is generally good practice to recommend offset drilling to prove up the acreage. Figure 9.20 indicates the presence of undeveloped sand in a reservoir system.

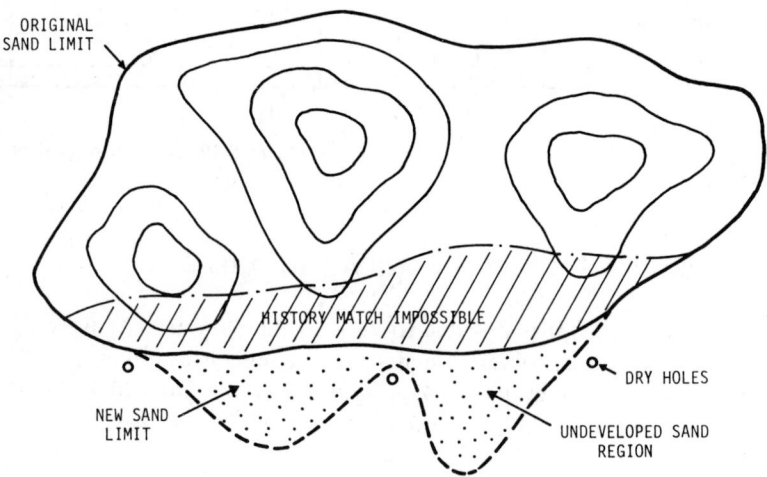

Figure 9.20: Communicating areas.

9.7 AUTOMATIC HISTORY-MATCHING METHODS

There have been several attempts to relieve the engineer of the burden of the history-matching procedure and to generate a history match automatically by a series of simulation runs. These methods are iterative in a general sense and usually couple some statistical analysis with optimization techniques to obtain the "best" combination of parameters to match the reservoir history. The large majority of automatic history-matching approaches can be envisaged as occurring in two steps:

1. Generate a representative set of simulation runs with known parameters randomly or otherwise.

2. Optimize the solution to get a best fit by linear programming– or search-type methods.

All the current methods pick parameters which form the history-matching criteria. These parameters are pressures, production rates, or producing ratios.

Jahns[1] uses a nonlinear regression approach to match reservoir pressures. The outline of his method is shown in Fig. 9.21.

The reservoir is zoned, and each zone has a descriptor for the transmissibility term (kh/μ) and the storage term (ϕch). The pressure behavior is then a function of those two parameters. These independent parameters are then varied in a formal manner, and by the use of regression analysis, the values that minimize the following relation are selected:

$$\text{Min}: \quad E = \sum_{i=1}^{n} W_i(P_i^{\text{obs}} - P_i^{\text{cal}})^2 \qquad (9.2)$$

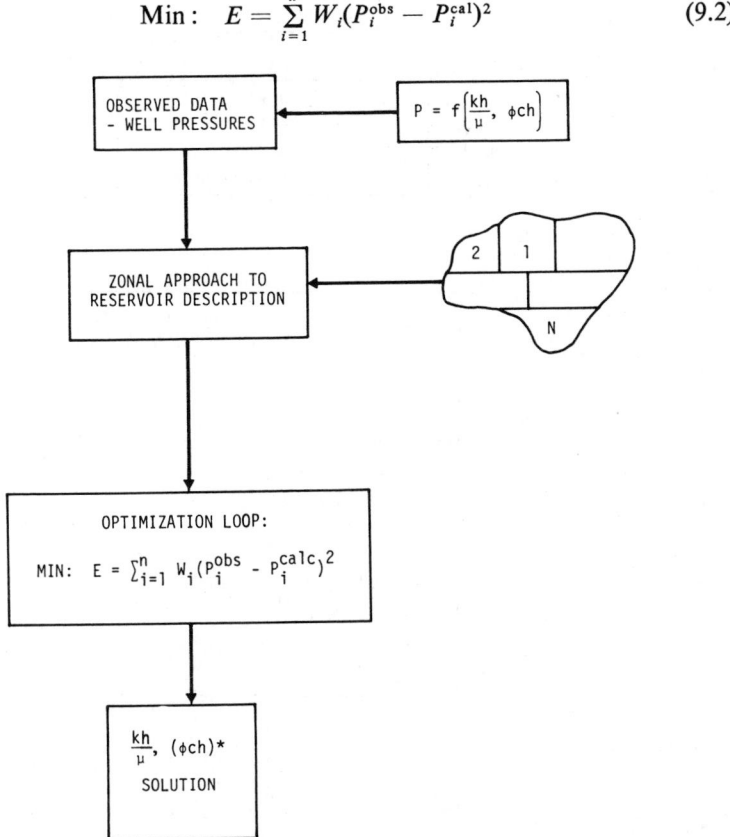

Figure 9.21: Jahns method.

This method can create several problems of its own and suffers from serious drawbacks. There is no guarantee of the minimum, and several different solutions can be obtained that indicate the existence of many minima. There is also no guarantee of convergence, and because of the unconstrained nature of the optimization there is a clear and present danger of negative transmissibility and storage values—a physically unrealistic situation. Most of the above problems can be resolved by careful scrutiny of result and making reruns where infeasibilities exist.

Coats[2] developed a method which combines least squares and linear programming to obtain a solution that is realistic in the sense that no negative or physically impossible parameter values result. The outline of his method is shown in Fig. 9.22.

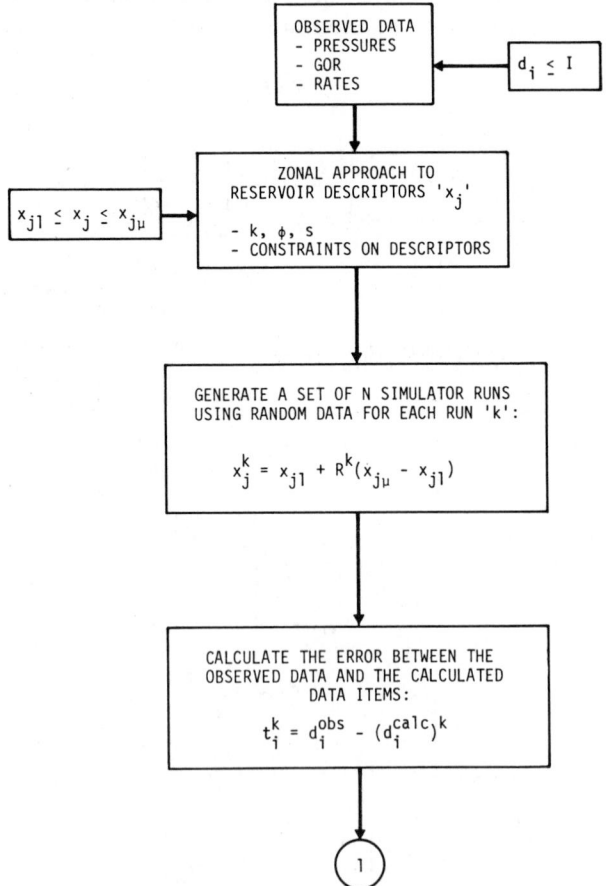

Figure 9.22: Coats method.

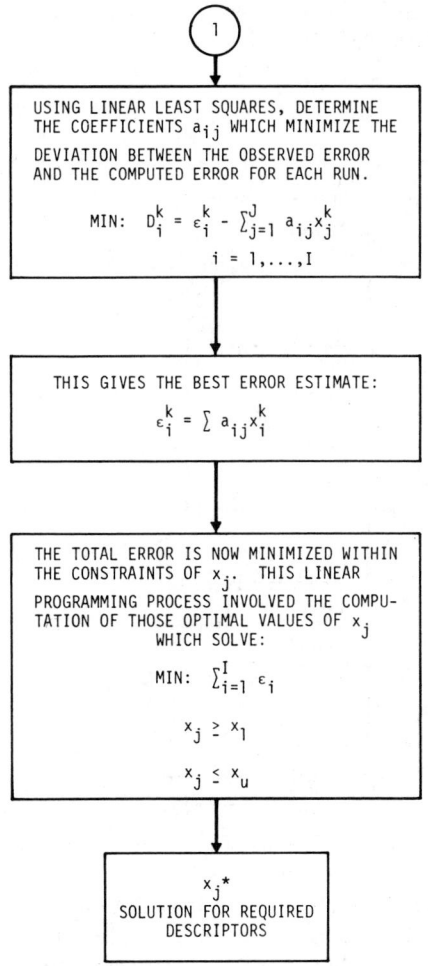

Figure 9.22 (Cont'd.)

This method has provided adequate results because of the constraints placed on the parameters, but it has been noted that some of the parameters are obtained at their upper levels. This seems to indicate that the assumptions made as to the linear nature of the problem are not valid as expected.

Slater and Durrer[3] use a gradient method as a search technique to find the "best" solution. Their method is outlined in Fig. 9.23.

Automatic history-matching methods currently leave something to be desired. The methods appear mathematically pure, but for several reasons to be enumerated they do not solve the problem.

The biggest drawback in these methods is their total objectivity. As the

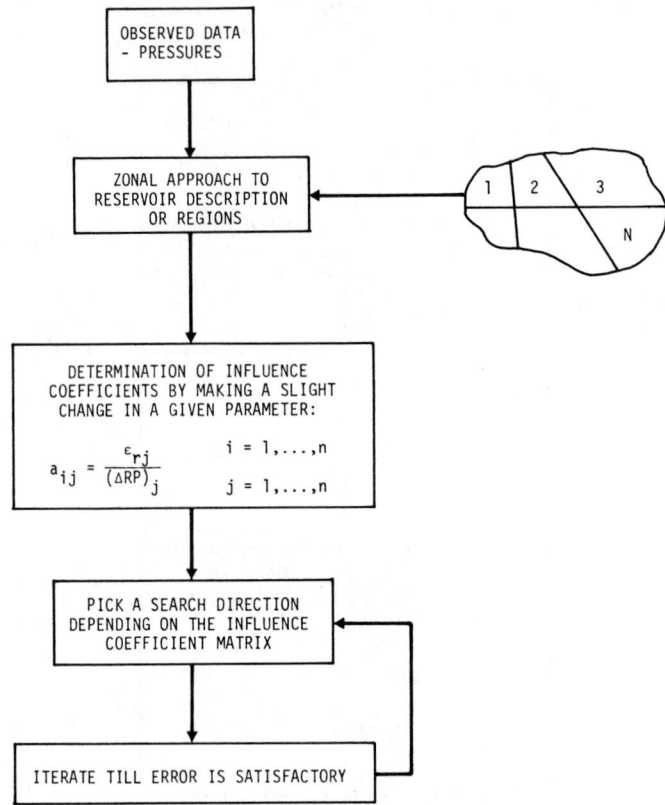

Figure 9.23: Slater and Durrer method.

engineer well knows, there are areas in the reservoir for which he has a good feel. His experience in working with the given field for some time has given him an intuitive approach in formulating the properties in that region which is far superior to that obtainable by automatic methods.

Secondly, the trial-and-error approach used in this history-matching section employs several variables at a time. In some studies all the data in every cell have to be reanalyzed and rebuilt in trying to match history. The recontouring process affects every single cell in the model directly. In the automatic history-matching mode the reservoir is divided into relatively large zones which are generally few in number, and within these zones rock and fluid parameters are grouped together to further blur the definition of the reservoir.

Furthermore, the automatic history-matching methods may produce data which are inherently infeasible. At this point the engineer has to enter the process to remove the infeasibility by starting at a new point. Even in the linear programming model which maintains feasibility because of the constrained optimization process, there is a tendency to obtain extremal values showing that the linear programming model was either at the upper or lower end of the constraint range.

In general, automatic history matching is a good tool for hypothetical reservoirs generated for academic interest, but they stumble badly in trying to predict the parameters in a "true-to-life" field. The search for a truly automatic procedure must center around including more parameters in the function descriptions of the field and solving these in a manner which allows feasibility.

The optimum procedure at the present stage of simulation development should combine the best of both worlds. The automatic approach could be used to get ballpark figures as a start; from this point the engineer will use his intuition to contour and develop the model as he best knows.

9.8 CASE STUDY: SIMULATION OF A SECONDARY RECOVERY PROJECT[4]

The Postle Reservoir is a 21,300-acre field in Texas County, Oklahoma. The Morrow Sand produces at a depth of 6100 ft and the area under study includes a gas cap region, an oil column and a water/oil contact zone. The study area is shown in Fig. 9.24. It consists of the Postle Upper Morrow and the Hough Morrow A Units. The Postle study is a classic example of the use of a simulator to "manage" a reservoir effectively. In this case study we shall look at the input requirements, the history-matching process, and the selection of the optimal operational design for secondary recovery underwater flood.

The basic input data—porosity, thickness, and water saturation—were obtained from the sonic and induction logs, and along with other geologic data a 56×73 grid was set up with only 2130 *actual* data cells being used in the model. A total of almost 10,000 data cards were required to set up the basic reservoir data file for rock and fluid properties.

The history match covered seven years of recorded history. The history-matching process involved a comparison of measured pressures, water-oil ratios, and gas-oil ratios with those calculated by the model. In the process of history matching it was apparent that a single relative permeability relationship was inadequate. Five different relative permeability curves had to be used to obtain a reasonable match in the whole field. These curves are shown in Figure 9.25. In addition, the critical gas saturation had to be modified

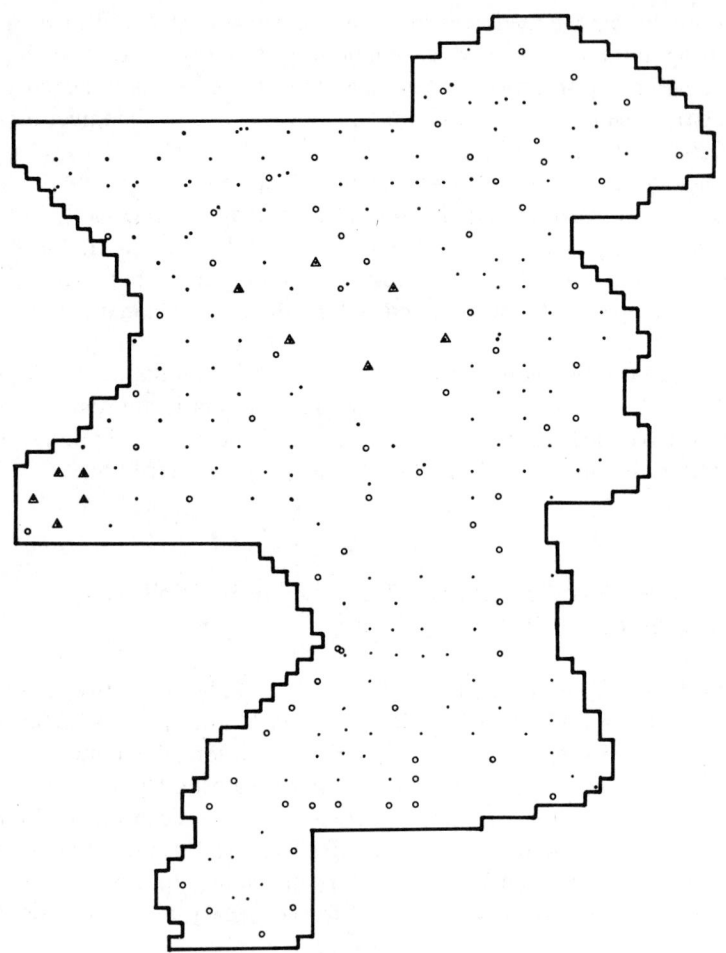

Figure 9.24: Postle Reservoir area.

in the gas cap area to obtain an acceptable pressure match and gas production rates.

The feasibility studies carried out on the Postle area involved varying the injection patterns and injection rates and pressure to optimize recovery. At first an optimum pattern was developed after analyzing a line drive, peripheral, and a modified peripheral flood pattern. The optimum plan resulted in improved sweep efficiency and about 15% greater ultimate recovery. The simulation study increased the number of injectors from 8 to 13 and the injection rate from 15,200 to 24,000 bbl water/day. This resulted in an increase in oil recovery rate.

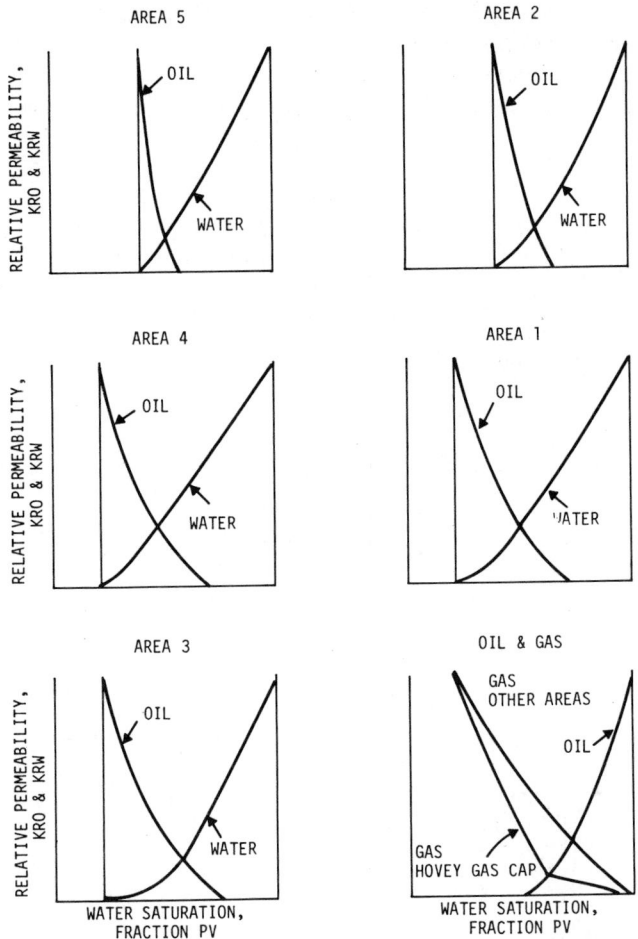

Figure 9.25: Postle simulation model studies relative permeability curves.

By examination of the sweep efficiency and by developing a movable oil plot $\phi h(S_o - S_{or})$ the location of additional reserves at any time can be pinpointed; one such map is shown in Fig. 9.26. Using these data new producer locations were selected and drilled to accelerate the economic recovery. Four of the six wells selected under this scheme were economic successes. Within the study itself 16 of the 18 producers studied were economic successes.

The improvement in reservoir management obtained by utilization of the simulation approach is shown in Figs. 9.27 and 9.28, where the difference between the *presimulation* study (classical waterflood approach) and the *simulation* study are apparent for the two areas within the Postle field. The cross-hatched and dotted areas are subdivided to show the contributions from the additional wells drilled or the effects of converting selected wells to injection.

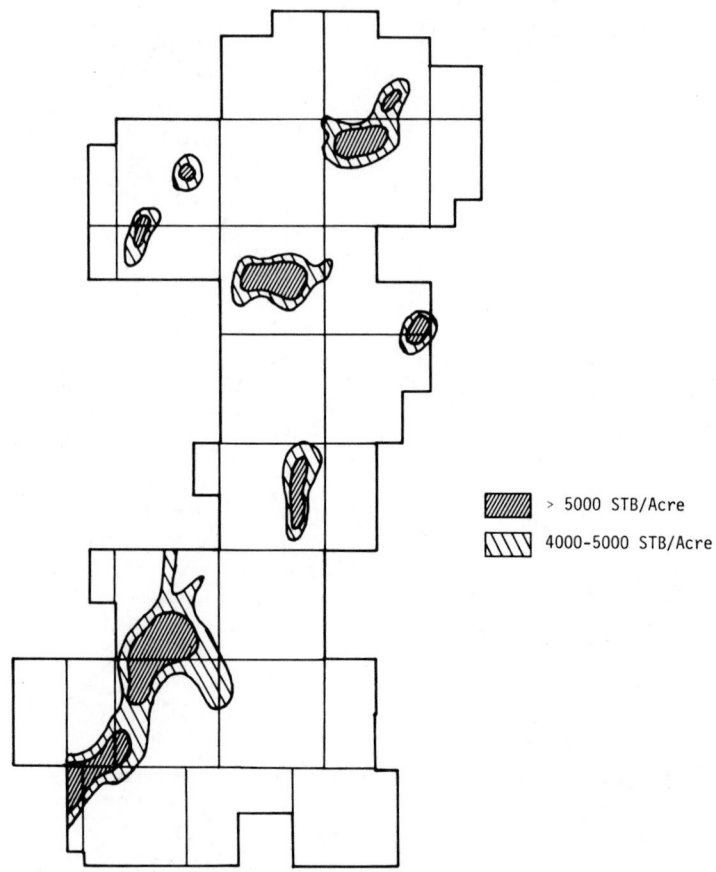

> 5000 STB/Acre

4000–5000 STB/Acre

Figure 9.26: Movable oil plot, $\phi h(S_o - S_{oirr})$.

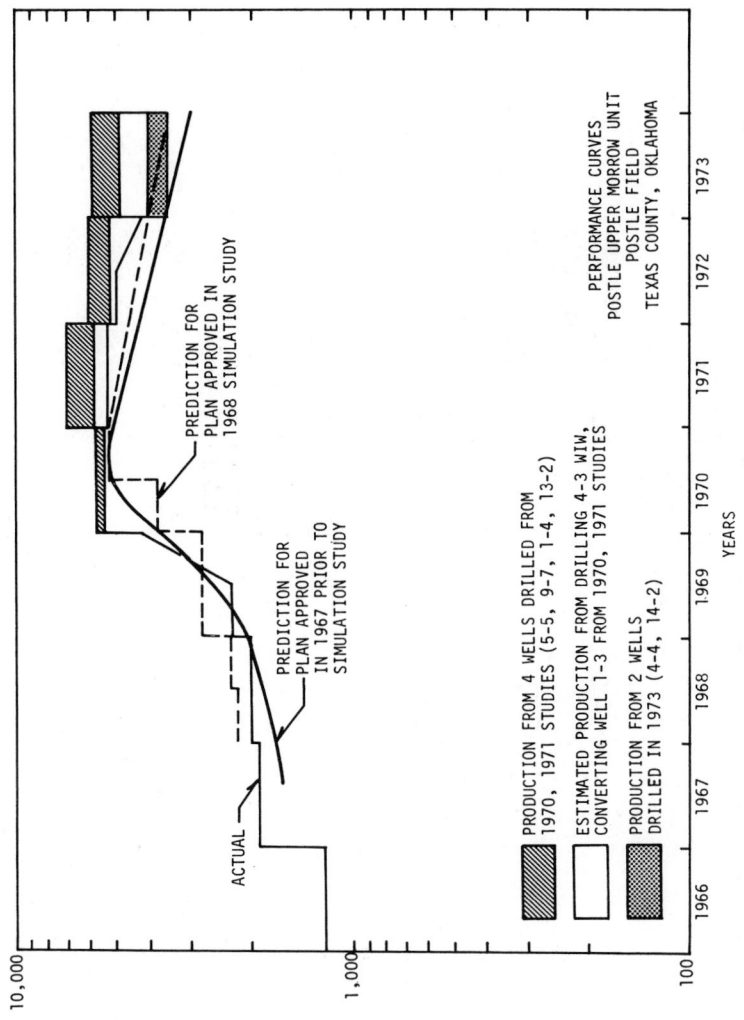

Figure 9.27: Postle performance curve.

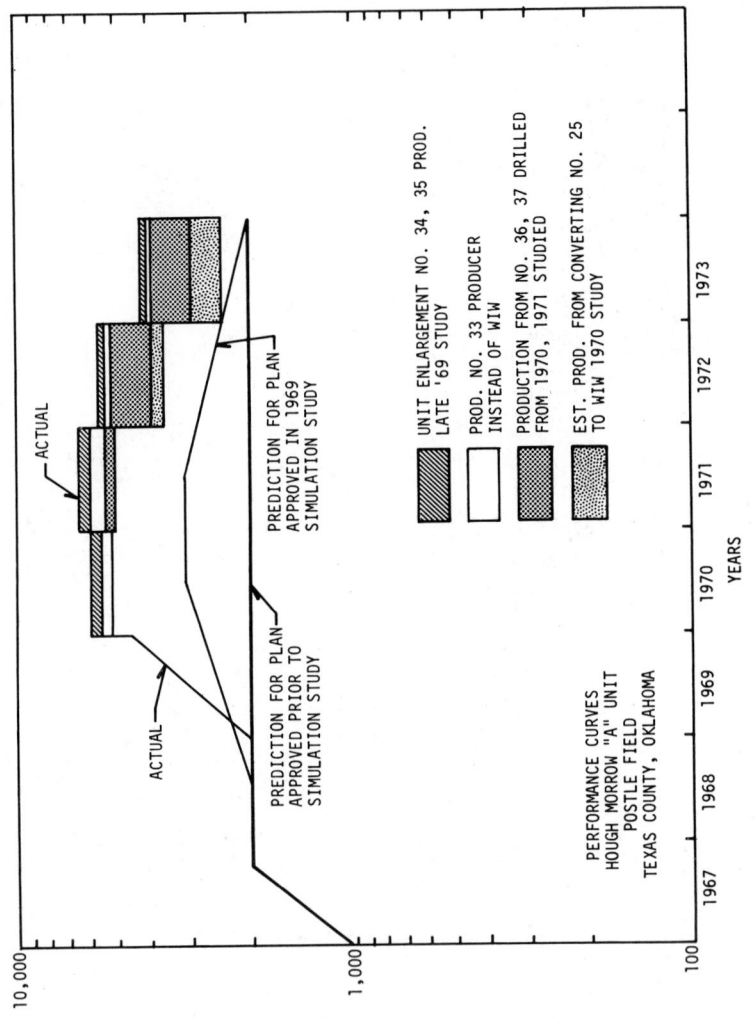

Figure 9.28: Hough Morrow "A" performance curve.

REFERENCES

1. H. L. JAHNS, "A Rapid Method for Obtaining a Two-dimensional Reservoir Description From Well Pressure Response Data," *Soc. Pet. Eng. J.* (Dec. 1966), 315–32; *Trans. AIME* **237**.

2. K. H., COATS, J. R. DEMPSEY, and J. H. HENDERSON, "A New Technique for Determining Reservoir Description From Field Performance Data," *Soc. Pet. Eng. J.* (March 1970), 66–74; *Trans. AIME* **249**.

3. G. E. SLATER, and E. J. DURRER, "Adjustment of Reservoir Simulation Models to Match Field Performance," *Soc. Pet. Eng. J.* (Sept. 1971), 295–305; *Trans. AIME* **251**.

4. J. P. WATSON, "Reservoir Simulation Model Studies of the Postle Morrow 'A' Sandstone," master's thesis, University of Oklahoma, 1974.

BIBLIOGRAPHY

BOBERG, T. C., E. G. WOODS, and W. J. MCDONALD, JR., "Application of Inverse Simulation to a Complex Multi-reservoir System," SPE 4626, 48th Annual Meeting, Las Vegas, Nev., Sept. 30–Oct. 3, 1973.

BREIT, V. S., K. A. BISHOP, D. W. GREEN, and E. E. TROMPETER, "A Technique for Assessing and Improving the Quality of Reservoir Parameter Estimates Used in Numerical Simulators," SPE 4546, 48th Annual Meeting, Las Vegas, Nev., Sept. 30–Oct. 3, 1973.

BRIGGS, J. E. and T. N. DIXON, "Some Practical Considerations in the Numerical Solution of Two-dimensional Reservoir Problems," *Soc. Pet. Eng. J.* (June 1968), 185–94; *Trans. AIME,* **243**.

CARTER, R. D., L. F. KEMP, JR., A. C. PIERCE, and D. L. WILLIAMS, "Performance Matching with Constraints," SPE 4260, Third Symposium on Numerical Simulation of Reservoir Performance Symposium, Houston, Texas, Jan. 10–12, 1973.

CHAVENT, C., M. DUPUY, and P. LEMONNIER, "History Matching by Use of Optimal Control Theory," SPE 4627, 48th Annual Meeting, Las Vegas, Nev., Sept. 30–Oct. 3, 1973.

CHEN, WEN H., GEORGE R. GAVALAS, and JOHN H. SEINFELD, "A New Algorithm for Automatic History Matching," SPE 4545, 48th Annual Meeting, Las Vegas, Nev., Sept. 30–Oct. 3, 1973.

COATS, K. H., J. R. DEMPSEY, and J. H. HENDERSON, "A New Technique for Determining Reservoir Description From Field Performance Data," *Soc. Pet, Eng. J.* (March 1970), 66–74; *Trans. AIME,* **249**.

HIRASAKI, GEORGE J., "Estimation of Reservoir Parameters by History Matching Oil Displacement by Water or Gas," SPE 4283, Third Symposium on Numerical Simulation of Reservoir Performance, Houston, Texas, Jan. 10–12, 1973.

JAHNS, H. O., "A Rapid Method for Obtaining a Two-dimensional Reservoir Description From Well Pressure Response Data," *Soc. Pet. Eng. J.* (Dec. 1966), 315–27; *Trans. AIME,* **237.**

JOYNER, H. D. and W. J. LOVINGFOSS, "Use of a Computer Model in Matching History and Predicting Performance of Low-permeability Gas Wells," *J. Pet. Tech.* (Dec. 1971), 1415–20.

KRUGER, W. D., "Determining Areal Permeability Distribution by Calculations," *J. Pet. Tech.* (July 1961), 691–96; *Trans. AIME,* **222.**

MANN, L. D. and G. A. JOHNSON, "Predicted Results of Numeric Grid Models Compared With Actual Field Performance, "*J. Pet. Tech.* (Nov. 1970), 1390–98.

SLATER, G. E., and E. J. DURRER, "Adjustment of Reservoir Simulation Models to Match Field Performance," *Soc. Pet. Eng. J.* (Sept. 1971), 295–305; *Trans. AIME,* **251.**

SOLORZANO, LUZBEL NAPOLEON, and SERGIO ENRIQUE ARREDONDO, "Method for Automatic History Matching of Reservoir Simulation Models," SPE 4594, 48th Annual Meeting, Las Vegas, Nev., Sept. 30–Oct. 3, 1973.

THOMAS, L. K. and L. J. HELLOMS, "A Nonlinear Automatic History Matching Technique for Reservoir Simulation Models," *Soc. Pet. Eng. J.* (Dec. 1972), 508–14; *Trans. AIME,* **253.**

YEH, WILLIAM W. G., "On the Optimal Identification of Parameters in a Parabolic System," SPE 4547, 48th Annual Meeting, Las Vegas, Nev., Sept. 30–Oct. 3, 1973.

10 The Well in the Simulator

10.1 INTRODUCTION

The well which penetrates the petroleum reservoir is our only window into the vast unknown to which we most diligently attribute rock and fluid properties, make predictions, and perform all the other engineering calculations that comprise the scope of reservoir engineering. The well in the simulator is equally important, since it is the location at which the disturbances are initiated in the system. The way in which the reservoir system responds to these perturbations represents its behavior, and this accurate representation is the primary goal in making a study. All our observations and measurements are made at the well; as yet there is no way to effectively measure parameters remote from the wellbore. The ways in which wells are handled in the simulator have a significant impact on the calculated response of the simulator. We shall explore the ways in which the well is treated in the simulator.

If we go back to the single-phase flow equations and include a production term, we get Eqs. (10.1), (10.2), and (10.3):

$$\frac{\partial}{\partial x}\left(\frac{k_o}{\mu_o B_o}\frac{\partial \Phi_o}{\partial x}\right) + \frac{\partial}{\partial y}\left(\frac{k_o}{\mu_o B_o}\frac{\partial \Phi_o}{\partial y}\right) - \frac{q_o}{B_o} = \frac{\partial}{\partial t}\left(\phi\frac{S_o}{B_o}\right) \qquad (10.1)$$

$$\frac{\partial}{\partial x}\left(\frac{k_g}{\mu_g B_g} + \frac{R_s k_o}{\mu_o B_o} + \frac{R_{sw} k_w}{\mu_w B_w}\right)\frac{\partial \Phi_g}{\partial x} - \frac{q_g}{B_g} = \frac{\partial}{\partial t}\left[\phi\left(\frac{S_g}{B_g} + \frac{R_{so} S_o}{B_o} + \frac{R_{sw} S_w}{B_w}\right)\right] \qquad (10.2)$$

(For brevity the gas equation is shown with x-derivatives only.)

$$\frac{\partial}{\partial x}\left(\frac{k_w}{\mu_w B_w}\frac{\partial \Phi_w}{\partial x}\right) + \frac{\partial}{\partial y}\left(\frac{k_w}{\mu_w B_w}\frac{\partial \Phi_w}{\partial y}\right) - \frac{q_w}{B_w} = \frac{\partial}{\partial t}\left(\phi\frac{S_w}{B_w}\right) \qquad (10.3)$$

The terms q_o, q_g, and q_w are the production terms which are the *volumetric* production rates for the given cell in the model. The production terms can be either positive or negative:

$$q > 0 \quad \text{Production well}$$
$$q = 0 \quad \text{Shut-in well}$$
$$q < 0 \quad \text{Injection well}$$

The term q represents a *source* or *sink* in that particular cell as shown in Fig. 10.1. The source or sink creates a *disturbance* within the system, and the fluids must readjust to the perturbation in accordance with the dynamics of flow in porous media. This readjustment causes changes in pressure throughout the system.

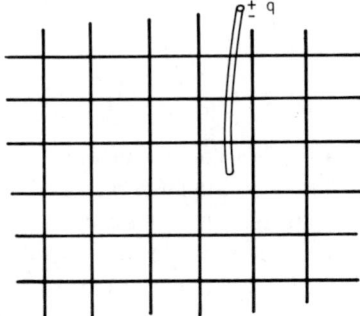

Figure 10.1: Well location in grid.

Equation (10.1), (10.2), and (10.3) must be solved as indicated earlier. The values of q_o, q_g, and q_w must either be (a) known a priori in order to set up the finite-difference calculations, or (b) included in implicit formulations of the finite-difference equations. The former scheme is defined as *explicit production*, while the latter is *implicit production*.

This operation in Fig. 10.2 may be an iterative scheme under some formulations.

There are two basic inner boundary conditions or wellbore conditions:

1. Constant pressure condition shown in Fig. 10.3.
2. Constant flow rate shown in Fig. 10.4.

In the first case, the pressure at the wellbore is set, and the flux into the wellbore is calculated and used in Eq. (10.1). In the second case, the flux across the inner boundary is prescribed, and the existing bottom hole pressure in the region of the well is calculated from the wellbore flow equation.

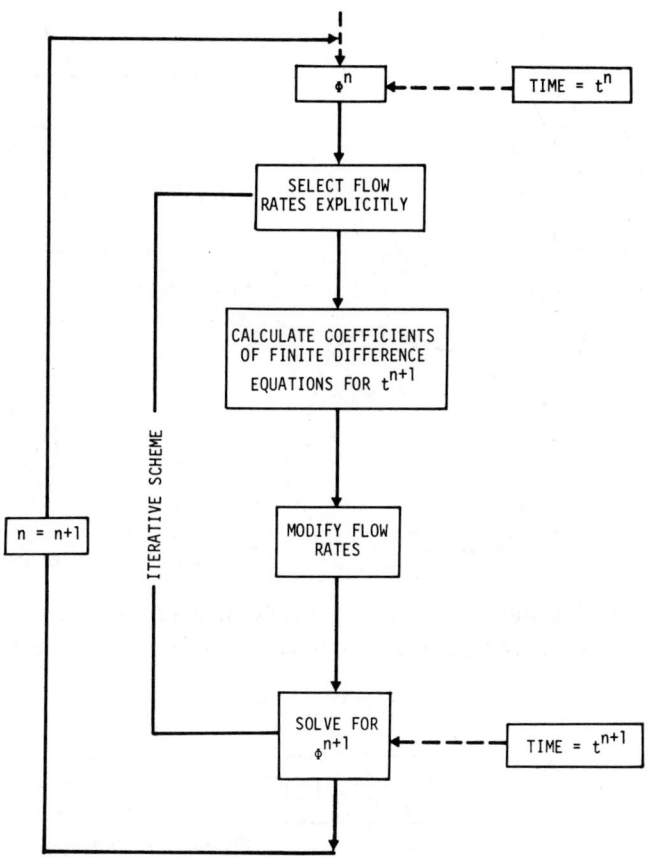

Figure 10.2: Solution process for Φ^{n+1} using explicit production terms.

Development of Production Term

Consider a well producing from a circular drainage area as shown in Fig. 10.5. The radial flow of a slightly compressible fluid under steady state is given by the integration of the velocity term from r_w to r_e:

$$q = \frac{7.07 k_{ro} k h (P_e - P_w)}{\mu_o B_o \ln(r_e/r_w)} \qquad (10.4)$$

In the presence of an impairment to flow—e.g., a skin factor S—this equation becomes:

$$q = \frac{7.07 k_{ro} k h (P_e - P_w)}{\mu_o B_o \ln(r_e/r_w) + S} \qquad (10.5)$$

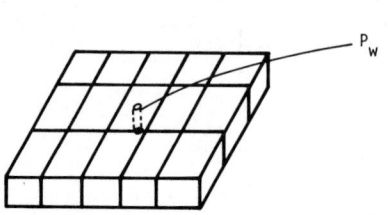

Figure 10.3: Constant P_w.

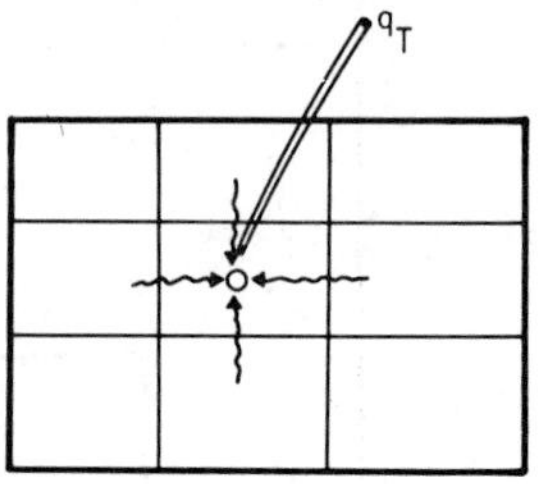

Figure 10.4: Constant flow rate.

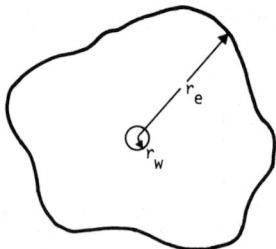

Figure 10.5: Drainage area.

In a reservoir simulator, however, the cells are not circular but rectangular or polygonal as seen in Fig. 10.6. A different formulation must be developed.

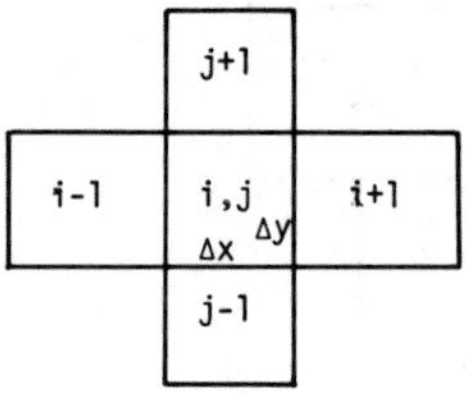

Figure 10.6: Polygonal areas in reservoir grid.

The radius r_e must be determined in terms of the cell dimensions Δx and Δy. The radius can be calculated for the equivalent circle:

$$\text{Area} = \Delta x \cdot \Delta y$$

$$= \pi r^2 \qquad (10.6)$$

$$r = \sqrt{\frac{\Delta x\, \Delta y}{\pi}}$$

The pressure, however, is not easily determined. The pressure obtained during the solution of the pressure equation is a material balance average pressure over the whole block. To account for the changing nature of the boundary pressure during the simulation we must modify our equation using the Jenkins-Aronofsky[1] radius-of-drainage concept in which a steady-state–type equation is used to define unsteady radial flow as long as the value of

r_e is suitably changed:

$$q = \frac{7.07 k_{ro} k h (\bar{P} - P_{wf})}{\mu_o B_o [\ln (r_e/r_w) - \frac{3}{4} + S]} \tag{10.7}$$

$$q = \frac{7.07 k_{ro} k h (\bar{P} - P_{wf})}{\mu_o B_o \left[\ln \left(\frac{\sqrt{\frac{\Delta x \, \Delta y}{\pi}}}{r_w} \right) - \frac{3}{4} + S \right]} \tag{10.8}$$

where $\bar{P}$ is the average pressure in the cell.

Now we can specify either the flowing bottom hole pressure or the production rate—but not both.

If P_{wf} is specified, the equations above are used as is to solve for q. If the value of flow is specified, the equation is rearranged to solve for the bottom hole pressure as shown below:

$$P_{wf} = \bar{P} - \frac{q_o \mu_o B_o [\ln (r_e/r_w) - \frac{3}{4} + S]}{7.07 k_{ro} k h} \tag{10.9}$$

The fluid saturations and *PVT* data are obtained from the conditions existing within the cell (see Fig. 10.7) at that particular time step.

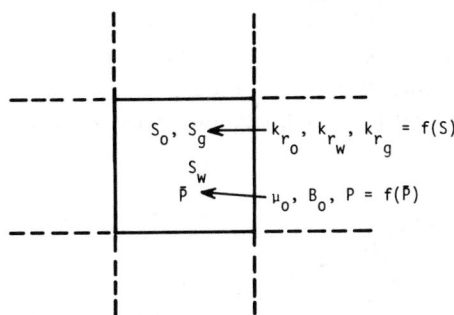

Figure 10.7: Evaluation of cell properties.

10.2 EXPLICIT PRODUCTION

Generally, the oil production rate is given and the production of the other phases is computed from relative permeability conditions which exist at that time in the production cell.

Given the oil rate at the new time level,

$$q_o^{n+1} = q_o \tag{10.10}$$

then the water rate is obtained from the mobility ratio term:

$$q_w^{n+1} = M_{wo}^n q_o^{n+1} \tag{10.11}$$

Similarly, the gas rate is:

$$q_g^{n+1} = M_{go}^n q_o^{n+1} + R_s^n q_o \tag{10.12}$$

where

$$M_{w_o} = \frac{k_{r_w}/\mu_w B_w}{k_{r_o}/\mu_o B_o} \tag{10.13}$$

= Mobility ratio of water to oil

$$M_{g_o} = \frac{k_{r_g}/\mu_g B_g}{k_{r_o}/\mu_o B_o} \tag{10.14}$$

= Mobility ratio of gas to oil

The rates shown are calculated using data evaluated at the *start of the time step*; thus, all parameters are known, and q can be *explicitly* calculated.

This explicit calculation is adequate for two-dimensional areal calculations where cell sizes are large or the flow rates are small. If the cells are small or flow rates very high, severe oscillations in saturation and consequently in flow rates can occur, as shown in Fig. 10.8. Note that the well goes on and off due to the severe oscillations in saturation which cause a fluctuation in mobilities because of their dependence on relative permeability. A method for combating this fluctuation will be formulated in the next section.

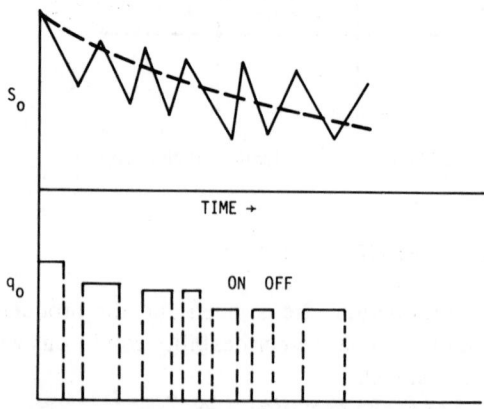

Figure 10.8: Oscillations in production rates and saturations.

Implicit Production

It was mentioned earlier that the use of explicit flow calculations can create inconsistencies in the model—namely, saturation oscillations and fluctuating flow rates. This can occur under any or all of the following conditions:

1. Small cells
2. Large time steps
3. Converging flow pattern
4. Extremely high production or injection rates

These inconsistencies can be largely eliminated by use of implicit production, which utilizes implicit mobilities in calculating the flow terms. One formulation put forward by Coats is as follows. Given the oil flow rate at a new time level $(n + 1)$,

$$q_o^{n+1} \tag{10.15}$$

the water rate can be evaluated using new mobility data:

$$q_w^{n+1} = M_{wo}^{n+(1/2)} q_o \tag{10.16}$$

and the gas rate can be similarly determined:

$$q_g^{n+1} = M_{go}^{n+(1/2)} q_o + R_s^n q_o \tag{10.17}$$

where $M_{wo}^{n+(1/2)}$ and $M_{go}^{n+(1/2)}$ are time-centered mobility values. These values are unknown at the start of time step, and they are evaluated as follows using old data and the derivative of the mobility curve:

$$M_{wo}^{n+(1/2)} = M_{wo}^n + 0.5 \frac{\partial M_{wo}}{\partial S_w} \Delta S_w \tag{10.18}$$

$$M_{go}^{n+(1/2)} = M_{go}^n + 0.5 \frac{\partial M_{go}}{\partial S_g} \Delta S_g \tag{10.19}$$

where $\partial M_{wo}/\partial S_w$ and $\partial M_{go}/\partial S_g$ are approximated from given relative permeability data by obtaining the chord slope:

$$\Delta S_w = (S_w^{n+1} - S_w^n) \tag{10.20}$$

$$\Delta S_g = (S_g^{n+1} - S_g^n) \tag{10.21}$$

$$M_{wo}^{n+(1/2)} = M_{wo}^n + 0.5 M_{wo}'(S_w^{n+1} - S_w^n) \tag{10.22}$$

$$M_{go}^{n+(1/2)} = M_{go}^n + 0.5 M_{go}'(S_g^{n+1} - S_g^n) \tag{10.23}$$

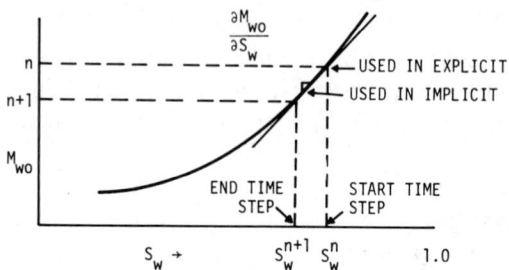

Figure 10.9: Implicit mobility used in *implicit production*.

The slopes are estimated from the relative permeability data as shown in Fig. 10.9.

Note that Eqs. (10.22) and (10.23) contain some unknown saturations S_w^{n+1} and S_g^{n+1}. The original equations (Eqs.10.16 and 10.17) can be simplified to separate two terms, one containing known parameters—i.e., n time level terms—and the other containing $(n + 1)$ time level terms. The M'_{wo}- and M'_{go}-values are obtained by evaluating the chord slopes, or the value of $M^{n+(1/2)}$ can be determined iteratively by a trial-and-error process beginning with the M^n-level as the initial guess. The use of this fully iterative scheme on the mobility term increases the running time of the model; except in coning models, it normally suffices to use M'_{wo} and M'_{go} evaluated at the n–time level. If $\Delta S_w = S_w^{n+1} - S_w^n$ is too large, note that the mobilities will lag too far behind.

Specification of Total Producing Rate q_T

When the total fluid rate is selected from a well, this quantity must be divided between the three phases flowing as follows:

$$q_T = q_o + q_w + q_g \tag{10.24}$$

where the individual phase flow rates are calculated as shown below:

Oil: $\qquad q_o = \dfrac{\alpha_o}{B_o} q_T \text{ STB/D} \tag{10.25}$

Water: $\quad q_w = \dfrac{\alpha_w}{B_w} q_T \text{ STB/D} \tag{10.26}$

Gas: $\qquad q_g = \dfrac{\alpha_g}{B_g} q_T + R_s q_o \text{ SCF/D} \tag{10.27}$

where

$$\alpha_o = \frac{k_{ro}/\mu_o}{k_{ro}/\mu_o + k_{rg}/\mu_g + k_{rw}/\mu_w} \tag{10.28}$$

and α_w and α_g are similarly defined.

Equations (10.25) through (10.27) are explicit equations. In some situations it will be necessary to use implicit productions. These are defined analogously to the case where oil flow rate was specified.

The oil production rate is:

$$q_o = \frac{q_T}{B_o}\left[\alpha_o^n + 0.5\frac{\partial\alpha_o^n}{\partial S_o}(S_o^{n+1} - S_o^n)\right] \qquad (10.29)$$

The water production rate is:

$$q_w = \frac{q_T}{B_w}\left[\alpha_w^n + 0.5\frac{\partial\alpha_w^n}{\partial S_w}(S_w^{n+1} - S_w^n)\right] \qquad (10.30)$$

The gas production rate is:

$$q_g = \frac{q_T}{B_g}\left[\alpha_g^n + 0.5\frac{\partial\alpha_g^n}{\partial S_g}(S_g^{n+1} - S_g^n)\right] + R_s q_o \qquad (10.31)$$

The partial derivative is approximated by the chord slope of the saturation mobility curve as done in Fig. 10.9:

$$\frac{\partial\alpha_o}{\partial S_o} = \frac{\alpha_o^{n+1} - \alpha_o^n}{S_o^{n+1} - S_o^n} \qquad (10.32)$$

$$\frac{\partial\alpha_w}{\partial S_w} = \frac{\alpha_w^{n+1} - \alpha_w^n}{S_w^{n+1} - S_w^n} \qquad (10.33)$$

$$\frac{\partial\alpha_g}{\partial S_g} = \frac{\alpha_g^{n+1} - \alpha_g^n}{S_g^{n+1} - S_g^n} \qquad (10.34)$$

The $(n + 1)$ and n time level values in these equations are separated out in a manner similar to that shown in Eq. (10.23).

10.3 GENERALIZED DEVELOPMENT FOR MULTICELL WELLS

The equations for computing the production rates for each fluid in multicell wells are developed for the generalized system shown in Fig. 10.10. The

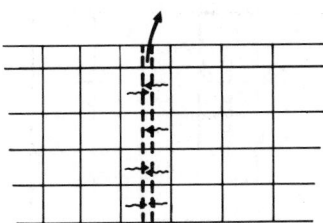

Figure 10.10: A well completed in several blocks.

total oil rates for a well completed in n blocks is the summation of the rates in each block:

$$q_o = \sum_{i=1}^{n} \left\{ \left[\frac{7.07h}{\ln(r_e/r_w) - \frac{3}{4} + S} \right] \frac{\lambda_o}{B_o} (P_e - P_{w_i}) \right\}_i \tag{10.35}$$

$$q_w = \sum_{i=1}^{n} \left\{ \left[\frac{7.07h}{\ln(r_e/r_w) - \frac{3}{4} + S} \right] \frac{\lambda_w}{B_w} (P_e - P_{w_i}) \right\}_i \tag{10.36}$$

$$q_g = 5.615 \sum_{i=1}^{n} \left\{ \left[\frac{7.07h}{\ln(r_e/r_w) - \frac{3}{4} + S} \right] \left(\frac{\lambda_o R_s + \lambda_g}{B_g} \right) (P_e - P_{w_i}) \right\}_i \tag{10.37}$$

where $\lambda_i = \left(\dfrac{kk_r}{\mu} \right)_i$,

$S =$ skin effect
$r_e =$ equivalent radius of the block in which the well is completed
$P_e =$ generally, the average pressure of the cell in which the well is located

If the cells are small—i.e., if the $\phi h \, \Delta V$-product is the same order of magnitude as the $q \, \Delta t$-product—the boundary pressure should include the effects of the adjoining cells. One method is shown below and uses a weighting factor F_i:

$$P_{e_i} = \bar{P}_i + F_i \frac{\left[\sum_{j=1}^{4} (h\lambda)_j (\bar{P}_j - \bar{P}) \right]_i}{\sum_{j=1}^{4} (h\lambda)_j} \tag{10.38}$$

where F_i is a factor by which the pressures of the surrounding cells are weighted before being added to the average pressure of the cell containing the well. The factor F_i lies between 0 and 1. Equation (10.38) in effect weights the pressure in each direction by the total mobility term in that direction as shown in Fig. 10.11. As F approaches 0, P_e approaches $\bar{P}_i$. As a rule of thumb, F lies between 0.5 and 0.7 for these small cells.

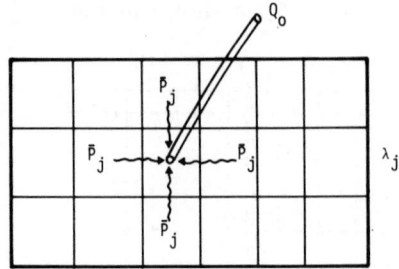

Figure 10.11: Averaging cell pressures.

Consider a well completed in n blocks as shown in Fig. 10.10. Then by Eq. (10.35):

$$q_{oT} = \sum_{i=1}^{N} \left\{ \left[\frac{7.07h}{\ln(r_e/r_w) + S} \right] \frac{\lambda_o}{B_o} (P_e - P_{w_i}) \right\} \qquad (10.35)$$

where q_g and q_w are defined by Eqs. (10.36) and (10.37).

There are two possible ways in which the flow rate can be specified:

1. Specify the total *oil* for the well—i.e., sum all n layers.
2. Specify the total *fluid* production for the well—i.e., sum the oil, gas, and water phases.

Case 1 may be dictated by an allowable constraint, and case 2 may be dictated by an equipment limitation. In either case, the equation can be formulated *explicitly* or *implicitly*.

Case 1: Total oil production—explicit and implicit scheme: The production rate from each layer is required to allow the calculation of the new pressures and saturations.

Given q_o^{n+1}—i.e., total oil production—then

$$q_w^{n+1} = M_{w_o}^n q_o^{n+1} \qquad (10.11)$$

$$q_g^{n+1} = M_{g_o}^n q_o^{n+1} + R_{s_o} q_o^{n+1} \qquad (10.12)$$

and using Eq. (10.35):

$$\text{For oil:} \qquad q_{oT} = \sum_{i=1}^{N} q_{o_i} \qquad (10.39)$$

$$\text{For water:} \qquad q_{wT} = \sum_{i=1}^{N} q_{w_i} \qquad (10.40)$$

$$\text{For gas:} \qquad q_{gT} = \sum_{i=1}^{N} q_{g_i} \qquad (10.41)$$

The ratio of q_{o_i} to q_{oT} is obtained from Eqs. (10.39) and (10.35):

$$\frac{q_{o_i}}{q_{oT}} = \frac{\left\{ \dfrac{7.07h\lambda_o(P_e - P_w)}{B_o[\ln(r_e/r_w) - \frac{3}{4} + S]} \right\}_i}{\sum\limits_{i=1}^{n} \left\{ \dfrac{7.07h\lambda_o(P_e - P_w)}{B_o[\ln(r_e/r_w) - \frac{3}{4} + S]} \right\}_i} \qquad (10.42)$$

Simplifying and assuming constant skin factors and pressure drawdown, we can write the individual oil rate from a layer:

$$q_{o_i}^{n+1} = q_{oT}^{n+1} \left[\frac{\lambda_{o_i}/B_{o_i}}{\sum\limits_{i=1}^{N} (\lambda_o/B_o)_i} \right]^n \qquad (10.43)$$

For the explicit scheme, the mobility values are all evaluated at the old time step values as shown by the superscript n or Eq. (10.43).

For the water produced in the ith layer, we can use the same relation developed earlier:

$$q_{w_i}^{n+1} = M_{wo_i}^n q_{o_i}^n$$

$$= \frac{(k_{rw}/\mu_w B_w)^n}{(k_{ro}/\mu_o B_o)^n} q_{o_i}^{n+1} \tag{10.44}$$

In the implicit case:

$$q_{w_i}^{n+1} = M_{wo}^{n+(1/2)} q_{o_i}^{n+1} \tag{10.45}$$

The mobility ratio $M_{wo}^{n+(1/2)}$ is evaluated from the slope of the mobility ratio saturation curve as shown in Fig. 10.9:

$$q_{w_i}^{n+1} = M_{wo}^n + 0.5 \frac{\partial M_{wo}}{\partial S_w}(S_w^{n+1} - S_w^n)q_{o_i}^{n+1} \tag{10.46}$$

By a similar development the gas phase production from each layer is determined explicitly using the calculated oil value:

$$q_{g_i}^{n+1} = M_{go_i}^n q_{o_i}^{n+1} + R_{s_i}^n q_{o_i}^{n+1} \tag{10.47}$$

and implicitly use is made of the new mobility ratio data:

$$q_{g_i}^{n+1} = M_{go_i}^{n+(1/2)} q_{o_i}^{n+1} + R_{s_i}^n q_{o_i}^{n+1} \tag{10.48}$$

where

$$M_{go_i}^{n+(1/2)} = M_{go_i}^n q_{o_i}^{n+1} + 0.5 \frac{\partial M_{go}}{\partial S_g}(S_g^{n+1} - S_g^n) \tag{10.49}$$

Therefore, the implicit production of gas from the ith layer is:

$$q_{g_i} = M_{go_i}^n + 0.5 \frac{\partial M_{go}}{\partial S_g}(S_g^{n+1} - S_g^n)q_{o_i}^{n+1} + R_{s_i}^n q_{o_i}^{n+1} \tag{10.50}$$

Equations (10.43) through (10.50) indicate the expressions used to evaluate the individual rates of each phase from each layer under either an explicit or implicit scheme.

Case 2: Total fluid production specified for the well—explicit and implicit:
Given q_T^{n+1} from the well, this is produced in three phases and from n layers. Therefore, if the total production of each phase is found, the determination of phase production from each layer is exactly as before.

Let α_{o_T}, α_{g_T}, α_{w_T} be the total mobility ratio of oil, gas, and water, respectively, for all layers. Therefore,

$$\alpha_{o_T} = \sum_{i=1}^{n} \left(\frac{k_{ro}/\mu_o}{\dfrac{k_{ro}}{\mu_o} + \dfrac{k_{rw}}{\mu_w} + \dfrac{k_{rg}}{\mu_g}} \right)_i \tag{10.51}$$

$$\alpha_{w_T} = \sum_{i=1}^{n} \left(\frac{k_{rw}/\mu_w}{\dfrac{k_{ro}}{\mu_o} + \dfrac{k_{rw}}{\mu_w} + \dfrac{k_{rg}}{\mu_g}} \right)_i \tag{10.52}$$

$$\alpha_{g_T} = \sum_{i=1}^{n} \left(\frac{k_{rg}/\mu_g}{\dfrac{k_{ro}}{\mu_o} + \dfrac{k_{rw}}{\mu_w} + \dfrac{k_{rg}}{\mu_g}} \right)_i \tag{10.53}$$

Then the fraction of the total production q_T that will be oil is:

$$q_{o_T} = \left(\frac{\alpha_{o_T}}{\alpha_{o_T} + \alpha_{w_T} + \alpha_{g_T}} \right) q_T \tag{10.54}$$

Similarly,

$$q_{w_T} = \left(\frac{\alpha_{w_T}}{\alpha_{o_T} + \alpha_{w_T} + \alpha_{g_T}} \right) q_T \tag{10.55}$$

Then the gas production can be determined by subtraction:

$$q_{g_T} = q_T - (q_{o_T} + q_{w_T}) \tag{10.56}$$

The individual layer phase rates can be determined by prorating the production to each zone based on its fractional contribution to the total mobility of that phase in all n layers. Consider the oil phase. The total oil production is known from Eq. (10.54). The individual layer oil production is obtained with an equation identical to Eq. (10.43):

$$q_{o_i}^{n+1} = q_{o_T}^{n+1} \frac{(\lambda_o/B_o)_i}{\sum\limits_{i=1}^{n} (\lambda_o/B_o)_i} \tag{10.43}$$

To include variations in pressures in each layer, the above procedures must be modified to include the pressure differential term of Eq. (10.35). The distributing term will now be of the following type:

$$q_{o_i}^{n+1} = q_{o_T}^{n+1} \frac{[(\lambda_o/B_o)(P_e - P_{w_f})]_i}{\sum\limits_{i=1}^{n} [(\lambda_o/B_p)(P_e - P_{w_i})]_i} \tag{10.57}$$

This gives a more accurate measure of the deliverability of a layer where the flowing pressures and average pressures differ significantly from layer to layer.

10.4 SELECTION OF RUN PARAMETERS

Time Step Selection

Most models come equipped with instructions on time step selection. The engineer, however, should be aware of the mechanics of the process so that his selection is based on sound judgment. Too large a time step leads to oscillations in fluid saturations within the cell containing the well and subsequent instability, while too short a time step, though stable, leads to excessive running times.

In the IMPES formulation of the simulation model, we have seen that saturations lag behind pressures; the former are calculated explicitly, while the latter are calculated implicitly.

Consider the two-dimensional cell with its associated oil, gas, and water content as shown in Fig. 10.12. The following discussion will give a physical

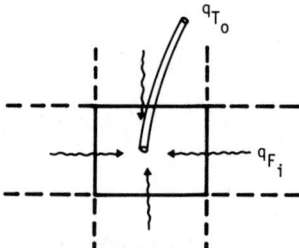

Figure 10.12: Production location.

feel for the way in which the time step affects the model behavior, particularly in the cell containing a well. This well is producing at an oil rate q_{T_o}, and this rate is maintained during a time increment Δt. The total barrels oil production is thus

$$N_p = q_{T_o} \Delta t \tag{10.58}$$

The total oil in place in the cell is, in barrels,

$$N_o = \phi \, \Delta x \, \Delta y \, h S_o \tag{10.59}$$

The flux into the cell in any time step is

$$Q_F = \Delta t \sum_{i=1}^{4} q_{F_i} \tag{10.60}$$

Thus, the oil in this cell at the end of a time step is, by a material balance,

$$N = N_o + Q_F - N_p \tag{10.61}$$

Now at any time the oil remaining in a cell is

$$N < 0 \quad \text{if} \quad N_p > (N_o + Q_F) \tag{10.62}$$

$$N > 0 \quad \text{if} \quad N_p < (N_o + Q_F) \tag{10.63}$$

Thus, the value of N and the subsequent oil saturation is a function of

1. N_p, or more directly, the time step size; and
2. Q_F, or indirectly, the pressure gradients and fluid saturations between the well cell and contiguous cells.

Generally, Q_{F_i} is difficult or impossible to evaluate a priori, and as such the engineer has very little control over it. The objective here now is to maintain N as large as possible consistent with fast model execution.

In the past, one useful criterion was to "select a time step such that there is less than 10–20% change in the fluid saturation in the block containing the well." For a many-celled model of varying cell dimensions, this can be expressed as:

$$\Delta t_{\max} = \min \left[\left(\frac{0.1\phi \, \Delta x \, \Delta y \, hS_o}{q_{T_o} 5.615} \right)_{i=1,N \, \text{cell}} \right]$$

Thus, the maximum time increment is less than the time required to produce one-tenth a pore volume of the smallest cell containing a well, or in the case of significant variation in production rates, it is based on the smallest (oil pore volume–production rate) ratio, since the smallest cell may not be depleting at the fastest rate.

This rule does not prevent the engineer from doing some time step optimizing on his model study. It is recommended that a series of short runs be made with several different time steps and the computed pressure, saturation, and flow results compared. That value of Δt which seems to produce instability can be used as an operating upper limit for that model.

There are now simulators that incorporate automatic time step selection schemes. These involve the determination in some way of a critical time step which would impose a ceiling on the step size used at a particular point in time in the simulation study. The approach is similar in principle to the variable-step integration process of ordinary differential equations. It costs more in computing overhead per step, but larger steps are taken when the model permits, and as a result the overall running time can be decreased.

It is difficult to calculate directly the critical time step value for a given reservoir model because of the nonlinearity of the system and its time-varying parameters such as mobility and flow rates. However, Todd, O'Dell, and Hirasaki[2] have derived some time step limitations based on a two-phase system of oil and gas. Their development assumes an IMPES model with relative permeability and capillary pressure lagging behind the pressure calculation. The technique involves the familiar expansion of the conservation laws, and by a matrix eigenvalue analysis they determine the limitations imposed by the capillary pressure and relative permeability terms. Their results express the time step limitation as constrained by both critical values.

$$\Delta t_{\max} \leq \frac{1}{1/\Delta t_{cp} + 1/\Delta t_{RP}} \tag{10.64}$$

where the capillary pressure limitation is

$$\Delta t_{cp} = \frac{\phi \, \Delta x \, \Delta y \, \Delta z}{\dfrac{\partial P_c}{\partial S_g} \displaystyle\sum_{i=1}^{N} (T_o f_g)_i} \tag{10.65}$$

and the relative permeability limitation is

$$\Delta t_{RP} = \frac{\phi \, \Delta x \, \Delta y \, \Delta z}{4 \dfrac{\partial f_g}{\partial S_g} \displaystyle\sum_{i=1}^{M} |q|_i} \tag{10.66}$$

where

$T_o =$ transmissibilities between cells

$|q| =$ absolute value of the net flow in a direction i

$M =$ number of directions, maximum of 3

$N =$ number of block faces, maximum of 6

Note that Eqs. (10.65) and (10.66) require computations of derivatives of capillary and fractional flow terms in addition to fluid flux. Obviously, these must be obtained from old values unless an iterative scheme is used. There is some computational overhead involved in determining the critical time step size. The criterion indicated by Eq. (10.64) must not be violated by *any* cell in the model; to ensure the sanctity of this limitation, a search must be carried out encompassing every cell in the simulator. If the saturations are not changing rapidly, then it is conceivable that the computation of Eq. (10.64) and the attendant search should be made every k time steps, where k is an arbitrary integer. The selection of k will, of course, be arbitrary and be based on the experience of the engineer with his model.

The problems associated with time step size are manifested in several ways in a simulation study. They have been mentioned earlier in this chapter and are responsible for these phenomena:

1. Instabilities in fluid production, showing up as oscillations
2. Negative or inconsistent saturations.

All of the above problems can be remedied by the use of smaller step sizes; however, this is an implicitly more costly approach. Another approach is to modify the operation of the model to attempt to cure the above problems. To remove the oscillations in production rates, the use of implicit production terms as indicated in Sec. 10.2 is recommended. The saturation problem has been looked at by several investigators, and the minimization of this tendency can be achieved by the use of implicit saturation techniques. By its very nature, the process requires an iterative procedure in which the coefficients obtained from relative permeability, pressure, and capillary terms are updated constantly during a time step until convergence is achieved. This approach can be optimized and convergence speeded up if the new fluid- and saturation-dependent parameters are obtained by an extrapolation technique.

Flow Rate Selection

The flow rate at a well is based on one of two factors:

1. a prescribed bottom hole pressure, or
2. a prescribed withdrawal rate of oil.

The engineer can set a definite bottom hole pressure if he knows *in advance* that the fluid level within the production string will remain at some pumped-off point. The pressure would then be calculated from the hydrostatic gradient.

The usual situation, however, involves the setting of a production rate. This production rate can be dictated by any of the following:

1. Allowable production
2. Profit plan considerations
3. Pressure maintenance requirements
4. Gas-oil ratio control
5. Lift equipment capacity or limitations.

The production rate during a prediction run is set on:

1. Individual well total
2. Lease total
3. Unit total

If the individual well rate is set, the simulator will produce this quantity of fluid from the well as long as reservoir conditions are adequate. A better way to set the production or injection rate is to set a lease total and allow the model to prorate this total between the producers or injection wells. This is illustrated by the following.

Company A wishes to produce Q_m bbl/day of oil from a lease containing N producing wells. The engineer can attempt to calculate the individual possible production rates for each well, a tedious process which is error-prone, or he can set Q_m bbl/day of oil as the input lease total limit and allow the simulator to produce those quantities from each well consistent with its productivity and with any other limitations due to equipment or gas-oil ratios that may exist. The simulator will then perform the operations in Fig. 10.13.

This factor is only constant during a given time step and can vary with each time step.

This procedure forces the lease total to sum to Q_m and by examination of the individual well rates the wells which are faltering in production can be set aside for stimulation or some workover.

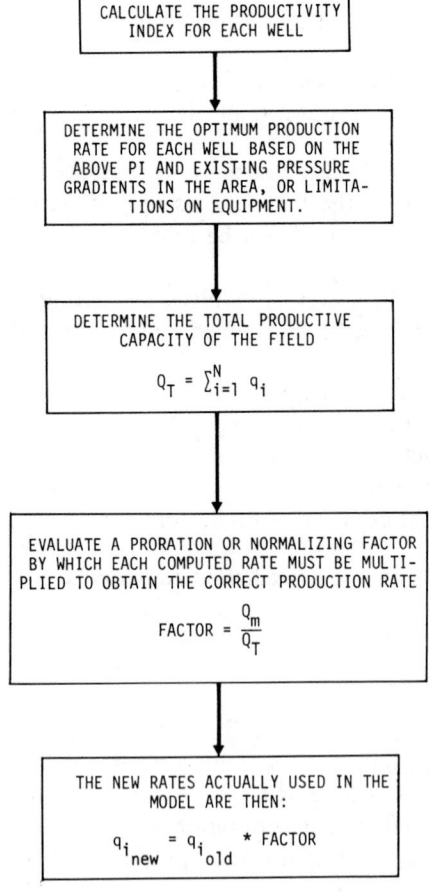

Figure 10.13: Allocation of production rates.

The same process is applicable in the case of water or gas injection when the correct injection figure for each well is not known.

10.5 MIGRATION OF FLUIDS WITHIN THE RESERVOIR

During the life of a reservoir millions of barrels of fluid are moved through the reservoir from one area to another. It is sometimes critical to know exactly how much fluid is moving across a given line or out of a given area. This usually occurs when a lease line runs through the model or the ownership varies from one area to another in the region under study.

Determination of Migration

Consider the following situation illustrated in Fig. 10.14. Company *A* owns 100% of lease *A* and 80% of lease *B*, which offsets lease *A*. Production operations in lease *B* are such that a lowering of pressure results and fluid moves from region *A* to region *B*. This fluid can be determined as follows:

1. Identify the faces of those blocks which make up the boundary.
2. Sum the fluid flux across these faces *algebraically* during every time step to obtain the cumulative migration.
3. It is possible to determine and output the direction and magnitude of migration across each face. This gives the engineer an indication as to the location of trouble spots. It is possible that migration could be going in different directions across the lease boundary as shown in Fig. 10.15.

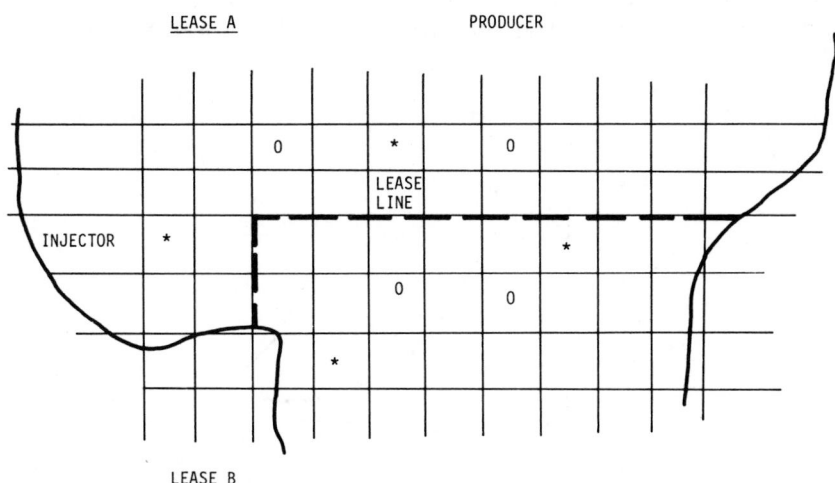

Figure 10.14: Adjacent leases with common oil reservoir.

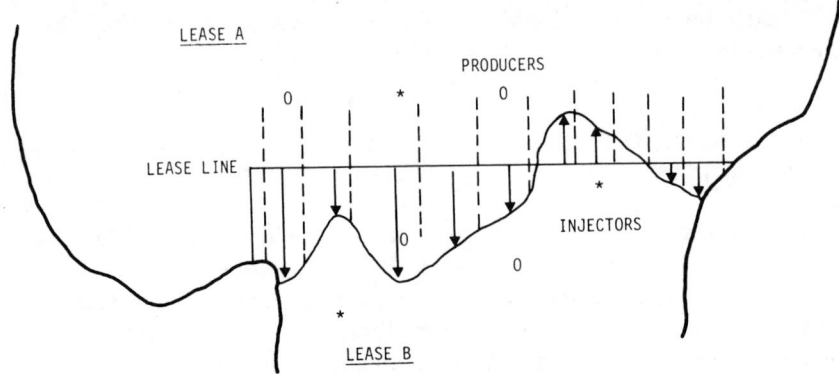

Figure 10.15: *Y*-direction migration.

The rate of migration is easily determined at every point on the boundary, since the transmissibility between each block and its neighbor is known from rock and fluid data and also from the existing pressures in each of these blocks. These rates can be printed out at every time step and the flow "vector" plotted. This "vector" gives the magnitude and direction of the movement and enables the engineer to see more readily where his lease is losing or gaining fluid.

Monitoring of Migration

The easiest method of monitoring the migration across lease lines or out of productive zones is by an analysis of the cumulative migration curve:

1. Determine the net migration by summing the fluid fluxes across the faces which delineate the boundary.
2. Plot the cumulative migration versus time, as shown in Fig. 10.16.

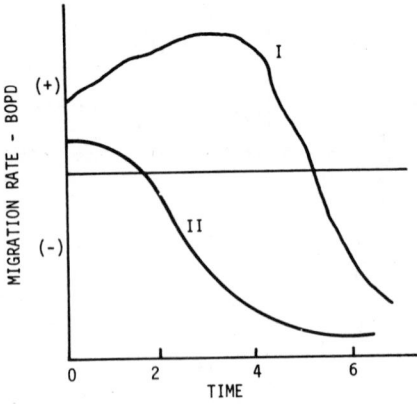

Figure 10.16: Migration: Direction and rate: $A \longrightarrow B = (+)$; $B \longrightarrow A = (-)$.

3. At selected intervals plot the flow rate vectors at the lease line and determine the locations where the migration occurs.

It is clear that a loss of fluid from a given lease is an economic drawback; not only is the loss detrimental, but it could be aggravated if it occurs early in the project life. The more oil within a given area, the more expansive the energy and the higher the production rates. The ideal ethical case is to have zero migration, but since this is generally unattainable, it is better to lose the reservoir fluid to migration very late in the life of the project. In this way the present-worth equivalent of that oil moving into your neighbor's lease is smallest.

Controlling Migration

The objective in controlling migration is to minimize the loss across a boundary and ideally to maintain an algebraic zero balance. This is easier said than done. The engineer is immediately faced with the question: Should I design the system to lose now but regain the fluid later, or should I gain some now and hope not to lose it all later? This is a real problem, and it is tied up with the fact that migration may go in either or both directions depending on the pressure gradients and fluid mobilities.

The mechanics of maintaining a balance on migration are necessarily heuristic. The engineer has two degrees of freedom in his trial-and-error approach. He can vary locations of injectors and producers and/or he can vary production and injection rates. To implement this he locates injectors and producers close to the lease boundary and monitors the migration pattern. This procedure is repeated until a minimum migration occurs when plotted on a graph like that in Fig. 10.17.

In some circumstances the company involved is unable to drill more wells, for any of several reasons. In this situation it is usually possible to

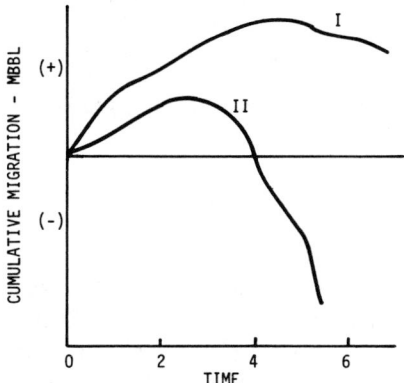

Figure 10.17: Cumulative migration.

convert one or more producers to injectors and modify the injection pattern to create an overpressured zone. This, in effect, constitutes a "pressure hill" up which the oil will not be able to migrate, and the oil will be lost. This approach again must be iterated to minimize migration.

10.6 FLOOD PATTERN—SWEPT AREA

In any secondary project it is beneficial to determine the location of the flood front at any given time in the project life. This knowledge allows the engineer to optimize his project economics by

1. selecting new drilling locations in the unswept area;
2. modifying injection or production rates to contact as much as possible of the oil in place; and
3. sequencing the conversion of producers to injectors to increase flooding efficiency.

To analyze the simulator output and determine the flood front location, we must review the fractional flow equation.

Fractional Flow Equation

The Buckley-Leverett theory is used to develop the concept of fractional flow. In any displacement process there is more than one mobile phase, and the total flow rate is distributed between these phases according to the fractional flow equation (Eq. 10.67).

For water displacing oil the complete equation in a linear system is:

$$f_w = \frac{1 - \dfrac{k_o}{\mu_o q_t}\left[\dfrac{\partial P_c}{\partial x} + g(\Delta \rho)\sin \alpha\right]}{1 + \dfrac{k_o}{k_w}\dfrac{\mu_w}{\mu_o}} \qquad (10.67)$$

Equation (10.67) is simplified by neglecting gravity and capillary effects:

$$f_w = \frac{1}{1 + \dfrac{k_o}{k_w}\dfrac{\mu_w}{\mu_o}} \qquad (10.68)$$

Equation (10.68), which shows the fractional flow of water to be a function of k_o, k_w, μ_w, and μ_o, indicates that f_w will be a *strong* function of fluid saturation, since relative permeability changes significantly with saturation. The changes in μ_w and μ_o are not sufficiently large to effect changes in f_w. Now:

$$q_T = q_o + q_w$$
$$q_w = f_w q_w \qquad (10.69)$$
$$q_o = (1 - f_w)q_T$$

The fractional flow curve (Fig. 10.18) is developed by evaluating the fractional flow equation at varying saturations of the displacing phase. This curve is used to obtain the displacing phase saturation at the flood front as follows:

1. Locate S_{w_i}, the initial water saturation at start of flood.
2. Draw a tangent from S_{w_i} to the f_w-curve.
3. Drop a vertical line from the point of tangency to the S_w-axis.
4. The point on the horizontal axis is the water saturation at the front, S_{w_f}.

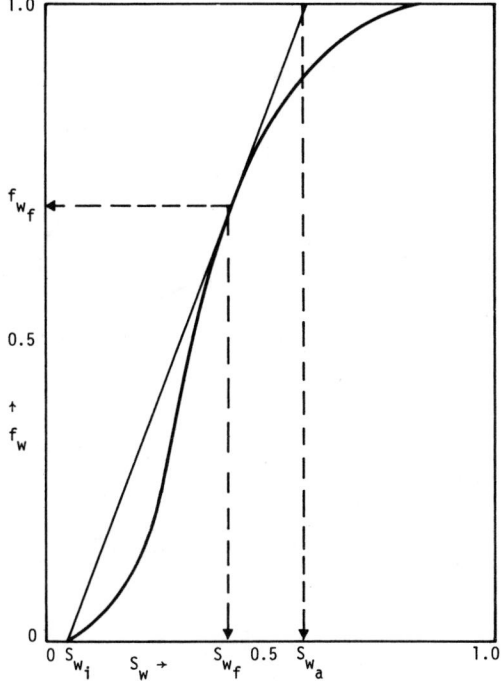

Figure 10.18: Fractional flow curve.

The output of saturations from a simulator is usually a discretized map consisting of the S_w-, S_g-, or S_o-values at each grid point or in some cases a contour map of the saturations. In order to use this information effectively, the engineer must locate the flood front on this map of fluid saturations.

Even though there is some blurring of the saturations at the front, the cell sizes of the model are such that the derived flood front is still usually representative.

Contouring the Front Location

From the fractional flow curve we have obtained the saturation at the front, S_{w_f}, as described earlier. With this value of S_{w_f} the engineer must now interpolate between his data to connect all points with a saturation equal to S_{w_f}. All areas behind this contour constitute swept area, and all ahead of this constitute unswept area. This does not imply that the areas behind the flood are watered out; there is still a lot of oil to be produced down to an uneconomic water-oil ratio. Wells in this area will continue to produce oil at increasing water cut until the economic limit is reached. Figure 10.19 shows the location of the saturation fronts at different periods of time during a simulation study.

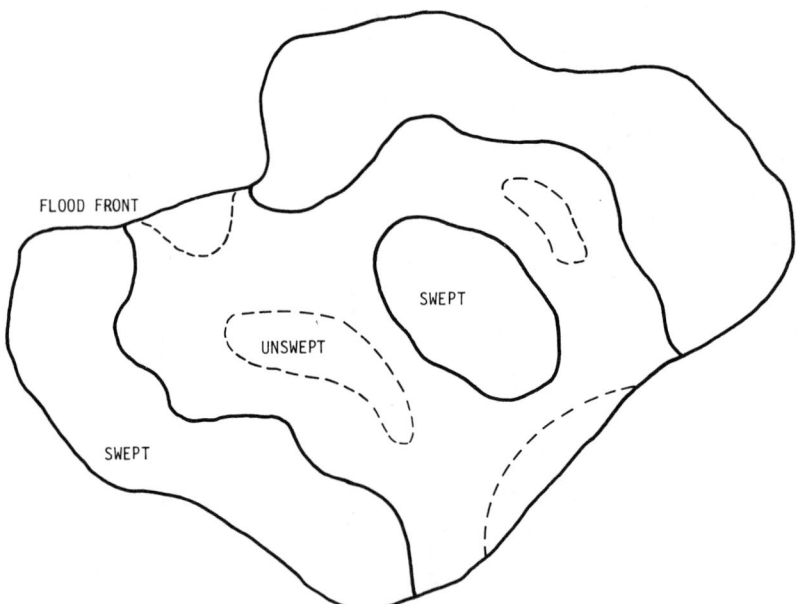

Figure 10.19: Flood front location after 5 years _____, after 10 years _ _ _ _ _ .

Selection of New Locations

By inspection of the flood front map the engineer can determine which areas have not yet been flooded and which are economically feasible areas for new wells. Several parameters must be analyzed to confirm a new location; among these are:

1. Unswept location
2. Oil reservoir volume
3. Reservoir pressure.

The presence of an unswept location is not sufficient to warrant a new well. There must be economically recoverable reserves in an area, and these are dictated by sufficient reservoir volume (reserves) and a high enough reservoir pressure (productivity) to deliver this oil if that location is tapped. The $\phi h S_o$-maps are essential at this time in making a decision on new well locations. A composite map made by superimposing Fig. 10.19 on Fig. 10.20 gives a clearer picture of the possible areas for infill drilling.

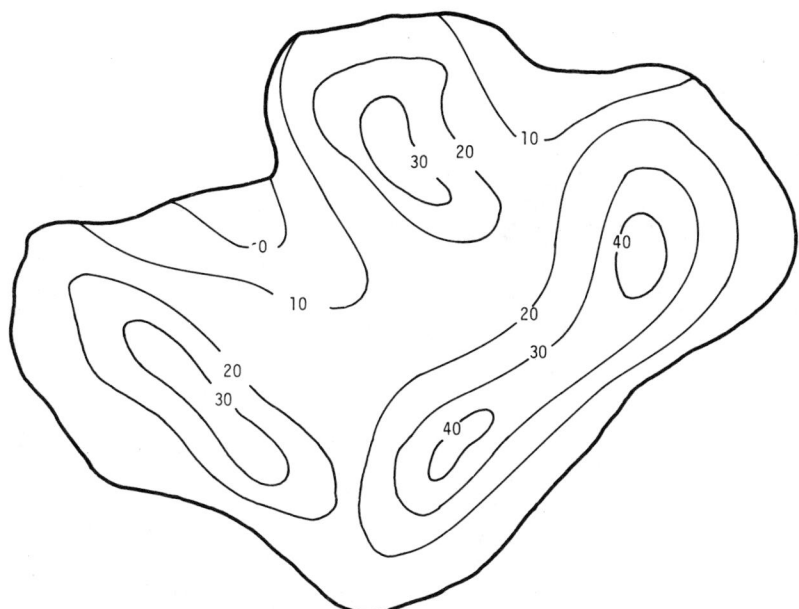

Figure 10.20: Net hydrocarbon map ($\phi h S_o$) at 5 years.

After selecting the location or locations at which to drill the new wells, the next decision is to determine in what sequence and at what times these additional wells should be drilled. There is no analytical way in which this can be obtained, and a trial-and-error approach is best in this case. The engineer picks the best new well and drills this location first. The other wells are drilled up as production declines or as ranked by the engineer.

10.7   CASE STUDY: TWO-PHASE PRODUCTION FROM A LAYERED SYSTEM

The explicit production of two phases from a three-layered reservoir is to be determined over a given period of time (see Fig. 10.21). Given a fixed total withdrawal rate, the individual layer production of oil and water was required. Data included PVT properties and the relative permeability curve in Fig. 10.22. The water saturation in each layer is changing due to a combination of influx and water injection. The calculated water saturations in the wellbore area are shown in Fig. 10.23. In a simulator it will be known because it is calculated prior to determing the flow rates. The equations used are:

Total:
$$q_{oT} = \sum_{i=1}^{N} q_{oi}$$

$$q_{wT} = \sum_{i=1}^{N} q_{wi}$$

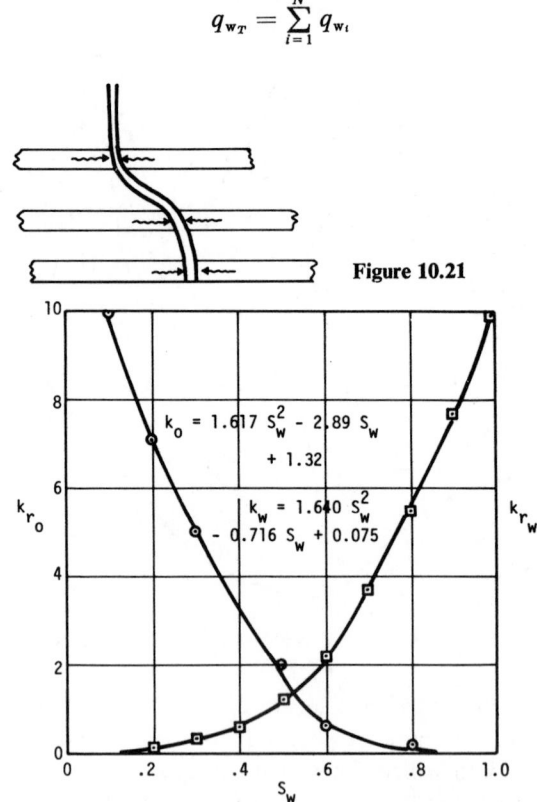

Figure 10.21

Figure 10.22: Relative permeability data.

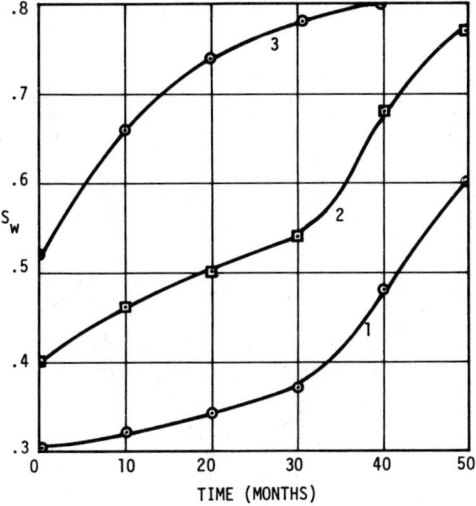

Figure 10.23: Calculated water saturations.

The individual rates are:

$$\text{Oil:} \qquad q_{o_i} = \frac{\left(\dfrac{k_o}{\mu_o B_o}\right)_i}{\displaystyle\sum_{i=1}^{N} \left(\dfrac{k_o}{\mu_o B_o}\right)_i}$$

$$\text{Water:} \qquad q_{w_i} = q_{o_i} \frac{M_{w_i}}{M_{o_i}}$$

where

$$M_{o_i} = \left(\frac{k_o}{\mu_o B_o}\right)_i$$

$$M_{w_i} = \left(\frac{k_w}{\mu_w B_w}\right)_i$$

Data:

$$\mu_o = 1.12 \qquad\qquad B_o = 1.452$$
$$\mu_w = 0.84 \qquad\qquad B_w = 1.003$$
$$q_{o_T} = 300 - 4t$$

The program to compute the flow rates involves a least-squares approximation for the determination of the appropriate relative permeability values. The results are shown in Fig. 10.24. The results obtained by using least

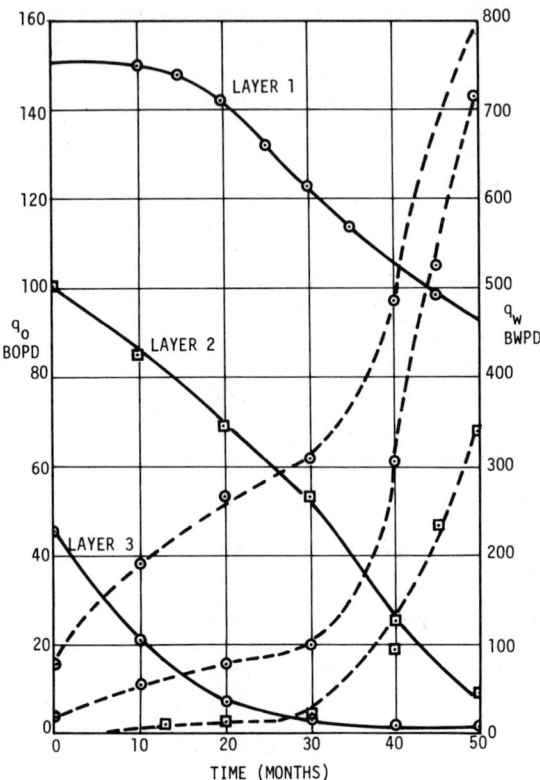

Figure 10.24: Oil and water production.

squares, linear interpolation, and a grangian interpolation are shown in Figs. 10.25, 10.26, and 10.27, respectively.

Comments: Note the decrease in production of oil as the layers become increasingly water-saturated. Also, the rate of change of oil production in each layer varies significantly, as indicated by the changing slopes in Fig. 10.24. This behavior is very much a function of the relative permeability relationship, and the results would be even more pronounced if a different relative permeability curve were used for each layer.

QC AND QW RESULTS, USING LEAST SQUARES CURVE FOR KO AND KW

```
***************************************************************************
*  TIME  *                   QO                   *                   QW                   *
*   IN   * * * * * * * * * * * * * * * * * * * * * * * * * * * * * * * * * *
*  DAYS  *  LAYER 1  *  LAYER 2  *  LAYER 3  *  LAYER 1  *  LAYER 2  *  LAYER 3  *
*  *     * * * * * * * * * * * * * * * * * * * * * * * * * * * * * * * * * *
*        *           *           *           *           *           *           *
*   0    * 146.026   *  99.195   *  54.779   *   4.382   *  27.380   *  78.040   *
*   1    * 143.500   *  96.727   *  55.773   *   6.002   *  31.197   *  79.456   *
*   2    * 145.168   *  93.625   *  53.207   *   6.072   *  35.098   *  85.898   *
*   3    * 142.129   *  95.034   *  50.837   *   7.929   *  35.626   *  92.973   *
*   4    * 143.786   *  91.896   *  48.318   *   8.021   *  39.860   * 100.093   *
*   5    * 146.984   *  89.697   *  43.319   *   8.199   *  44.839   * 115.226   *
*   6    * 147.888   *  90.249   *  37.863   *   8.250   *  45.115   * 129.681   *
*   7    * 148.617   *  90.694   *  32.689   *   8.290   *  45.337   * 144.898   *
*   8    * 150.145   *  87.390   *  30.465   *   8.376   *  50.176   * 154.041   *
*   9    * 150.615   *  87.664   *  25.721   *   8.402   *  50.333   * 170.458   *
*  10    * 153.324   *  85.016   *  21.660   *   8.553   *  55.901   * 190.534   *
*  11    * 152.201   *  84.393   *  19.406   *   8.490   *  55.492   * 197.875   *
*  12    * 153.473   *  80.971   *  17.556   *   8.561   *  60.820   * 208.535   *
*  13    * 148.858   *  81.476   *  17.666   *  10.731   *  61.200   * 209.838   *
*  14    * 150.073   *  78.056   *  15.871   *  10.819   *  66.835   * 220.887   *
*  15    * 148.733   *  77.359   *  13.909   *  10.722   *  66.237   * 228.365   *
*  16    * 147.309   *  76.618   *  12.073   *  10.620   *  65.603   * 235.737   *
*  17    * 148.276   *  73.185   *  10.539   *  10.689   *  71.305   * 247.102   *
*  18    * 146.675   *  72.395   *   8.930   *  10.574   *  70.535   * 254.342   *
*  19    * 142.166   *  72.848   *   8.986   *  12.935   *  70.976   * 255.932   *
*  20    * 142.960   *  69.416   *   7.624   *  13.007   *  76.846   * 267.583   *
*  21    * 140.360   *  68.154   *   7.486   *  12.771   *  75.449   * 262.718   *
*  22    * 138.587   *  67.293   *   6.120   *  12.609   *  74.496   * 269.500   *
*  23    * 138.336   *  63.555   *   6.109   *  12.587   *  79.849   * 269.012   *
*  24    * 134.708   *  64.296   *   4.996   *  15.187   *  80.781   * 282.546   *
*  25    * 132.067   *  63.035   *   4.898   *  14.889   *  79.197   * 277.006   *
*  26    * 132.441   *  59.714   *   3.844   *  14.931   *  85.071   * 288.202   *
*  27    * 128.090   *  60.045   *   3.866   *  17.630   *  85.542   * 289.796   *
*  28    * 127.696   *  56.450   *   3.854   *  17.576   *  91.133   * 288.906   *
*  29    * 124.035   *  57.052   *   2.912   *  20.593   *  92.106   * 302.731   *
*  30    * 123.588   *  53.510   *   2.902   *  20.519   *  97.861   * 301.639   *
*  31    * 119.295   *  53.788   *   2.917   *  23.653   *  98.369   * 303.204   *
*  32    * 118.781   *  50.315   *   2.904   *  23.551   * 104.230   * 301.896   *
*  33    * 117.347   *  48.571   *   2.082   *  27.551   * 113.991   * 322.074   *
*  34    * 115.297   *  46.569   *   2.134   *  31.825   * 123.872   * 330.106   *
*  35    * 113.242   *  44.569   *   2.189   *  36.524   * 134.460   * 338.531   *
*  36    * 113.402   *  40.307   *   2.291   *  42.512   * 156.929   * 354.312   *
*  37    * 114.250   *  36.271   *   1.479   *  49.556   * 183.399   * 386.697   *
*  38    * 114.427   *  32.021   *   1.552   *  57.201   * 212.212   * 405.620   *
*  39    * 113.413   *  28.895   *   1.693   *  74.573   * 254.176   * 442.457   *
*  40    * 112.479   *  25.664   *   1.857   *  96.309   * 304.846   * 485.327   *
*  41    * 109.284   *  24.710   *   2.006   * 120.982   * 343.910   * 524.374   *
*  42    * 106.089   *  23.733   *   2.179   * 151.138   * 389.658   * 569.538   *
*  43    * 104.093   *  21.640   *   2.267   * 168.049   * 422.558   * 592.584   *
*  44    * 102.084   *  19.554   *   2.362   * 186.696   * 458.458   * 617.386   *
*  45    *  98.966   *  18.435   *   2.599   * 232.261   * 525.051   * 679.521   *
*  46    *  94.524   *  18.822   *   2.654   * 251.429   * 536.084   * 693.800   *
*  47    *  92.553   *  16.663   *   2.785   * 279.220   * 584.798   * 727.931   *
*  48    *  90.568   *  14.505   *   2.927   * 310.199   * 638.712   * 765.240   *
*  49    *  89.518   *  11.589   *   2.893   * 306.603   * 655.420   * 756.366   *
*  50    *  87.415   *   9.542   *   3.044   * 340.336   * 715.313   * 795.666   *
***************************************************************************
```

Figure 10.25

QC ANC GW RESULTS, FINDING KC AND KW BY LINEAR INTERPCLATICN

```
*******************************************************************************
* TIME *            *         QO            *              *         QW
* IN   * * * * * *  * * * * * * * * * * * *  * * * * * * *  * * * * * * * * * *
* DAYS *  LAYER 1  *  LAYER 2  *  LAYER 3  *  LAYER 1  *  LAYER 2  *  LAYER 3
* *   * *  * * * *  * * * * * *  * * * * *  * * * * * *  * * * * *  * * * * *
*      *            *           *           *            *           *
*   0  *  142.586  *   59.810   *   57.605  *   16.513   *   33.026  *   77.061
*   1  *  140.195  *   97.414   *   58.391  *   18.412   *   36.825  *   78.113
*   2  *  141.337  *   94.419   *   56.243  *   18.562   *   40.500  *   84.374
*   3  *  138.405  *   95.411   *   54.184  *   20.463   *   40.925  *   90.945
*   4  *  139.623  *   92.389   *   51.987  *   20.643   *   44.726  *   97.479
*   5  *  142.270  *   90.205   *   47.524  *   21.034   *   49.079  *  111.013
*   6  *  143.021  *   90.681   *   42.298  *   21.145   *   49.338  *  123.346
*   7  *  143.479  *   90.972   *   37.549  *   21.213   *   49.496  *  138.472
*   8  *  144.615  *   87.692   *   35.692  *   21.381   *   53.452  *  148.477
*   9  *  144.784  *   87.795   *   31.421  *   21.406   *   53.514  *  166.485
*  10  *  147.229  *   85.205   *   27.566  *   21.767   *   58.046  *  187.439
*  11  *  146.197  *   84.608   *   25.196  *   21.615   *   57.639  *  195.131
*  12  *  147.497  *   81.280   *   23.223  *   21.807   *   61.786  *  205.953
*  13  *  143.198  *   81.513   *   23.289  *   23.692   *   61.963  *  206.542
*  14  *  144.557  *   78.156   *   21.286  *   23.916   *   66.230  *  217.701
*  15  *  143.495  *   77.582   *   18.922  *   23.741   *   65.744  *  225.233
*  16  *  141.849  *   76.692   *   17.458  *   23.468   *   64.990  *  233.481
*  17  *  142.649  *   73.049   *   16.303  *   23.601   *   68.987  *  245.689
*  18  *  140.951  *   72.179   *   14.870  *   23.320   *   68.166  *  253.528
*  19  *  136.699  *   72.388   *   14.913  *   25.186   *   68.363  *  254.263
*  20  *  137.500  *   68.750   *   13.750  *   25.334   *   72.383  *  266.610
*  21  *  135.000  *   67.500   *   13.500  *   24.873   *   71.067  *  261.762
*  22  *  133.257  *   66.629   *   12.114  *   24.552   *   70.149  *  268.905
*  23  *  132.446  *   63.514   *   12.041  *   24.403   *   75.532  *  267.268
*  24  *  129.018  *   64.054   *   10.929  *   26.368   *   76.174  *  280.087
*  25  *  126.488  *   62.798   *   10.714  *   25.851   *   74.681  *  274.595
*  26  *  126.404  *   60.079   *    9.517  *   25.834   *   80.372  *  284.745
*  27  *  122.236  *   60.224   *    9.540  *   27.622   *   80.565  *  285.431
*  28  *  121.386  *   57.140   *    9.474  *   27.430   *   85.719  *  283.445
*  29  *  117.987  *   57.649   *    8.364  *   29.404   *   86.483  *  296.349
*  30  *  117.133  *   54.563   *    8.303  *   29.192   *   91.581  *  294.205
*  31  *  112.973  *   54.703   *    8.324  *   30.988   *   91.815  *  294.956
*  32  *  112.110  *   51.629   *    8.261  *   30.751   *   96.808  *  292.703
*  33  *  110.486  *   50.249   *    7.265  *   33.304   *  105.170  *  310.836
*  34  *  108.098  *   48.490   *    7.412  *   35.769   *  113.268  *  317.150
*  35  *  105.933  *   46.523   *    7.544  *   40.045   *  121.349  *  322.788
*  36  *  105.741  *   42.427   *    7.833  *   45.356   *  138.587  *  335.129
*  37  *  105.754  *   39.445   *    6.801  *   51.196   *  164.089  *  360.996
*  38  *  105.010  *   35.943   *    7.048  *   57.134   *  190.447  *  374.093
*  39  *  103.074  *   33.347   *    7.579  *   70.219   *  226.749  *  402.296
*  40  *  101.294  *   30.471   *    8.235  *   85.837   *  270.229  *  437.135
*  41  *   97.459  *   29.681   *    8.860  *  102.609   *  303.552  *  470.292
*  42  *   94.553  *   28.085   *    9.362  *  126.490   *  334.295  *  496.926
*  43  *   91.836  *   26.647   *    9.517  *  137.769   *  356.364  *  505.155
*  44  *   89.125  *   25.188   *    9.688  *  149.591   *  379.587  *  514.219
*  45  *   85.128  *   24.615   *   10.256  *  178.173   *  419.696  *  544.417
*  46  *   80.942  *   24.747   *   10.311  *  189.074   *  421.935  *  547.321
*  47  *   78.189  *   23.245   *   10.566  *  203.946   *  450.721  *  560.853
*  48  *   75.437  *   21.709   *   10.854  *  219.985   *  481.873  *  576.152
*  49  *   74.133  *   19.200   *   10.667  *  216.183   *  492.074  *  566.194
*  50  *   71.429  *   17.582   *   10.989  *  233.322   *  526.034  *  583.305
*******************************************************************************
```

Figure 10.26

302

GC ANC GW RESULTS, FINDING KO AND KW BY LAGRANGIAN INTERPOLATION

```
***************************************************************************
TIME  *                      QO                 *                 QW
IN    * * * * * * * * * * * * * * * * * * * * * * * * * * * * * * * * * * *
DAYS  *  LAYER 1  *  LAYER 2  *  LAYER 3  *  LAYER 1  *  LAYER 2  *  LAYER 3  *
*   * * * * * * * * * * * * * * * * * * * * * * * * * * * * * * * * *
      *           *           *           *           *           *
 0    *  142.875  *  100.012  *   57.113  *   16.547  *   33.093  *   75.038  *
 1    *  140.460  *   97.467  *   58.073  *   18.026  *   36.165  *   76.298  *
 2    *  141.885  *   94.266  *   55.849  *   18.208  *   39.248  *   82.010  *
 3    *  138.826  *   95.409  *   53.764  *   19.733  *   39.723  *   88.282  *
 4    *  140.226  *   92.180  *   51.594  *   19.932  *   43.086  *   94.791  *
 5    *  142.935  *   89.791  *   47.274  *   20.317  *   47.133  *  108.992  *
 6    *  143.489  *   90.139  *   42.372  *   20.396  *   47.316  *  123.078  *
 7    *  144.009  *   90.466  *   37.525  *   20.470  *   47.487  *  138.771  *
 8    *  145.352  *   87.206  *   35.401  *   20.667  *   51.417  *  148.365  *
 9    *  145.795  *   87.448  *   30.757  *   20.724  *   51.560  *  166.070  *
10    *  148.312  *   84.867  *   26.820  *   21.082  *   56.208  *  187.413  *
11    *  147.216  *   84.240  *   24.545  *   20.926  *   55.792  *  195.492  *
12    *  148.342  *   80.921  *   22.737  *   21.086  *   60.199  *  206.712  *
13    *  143.934  *   81.239  *   22.827  *   22.664  *   60.435  *  207.526  *
14    *  144.975  *   77.951  *   21.074  *   22.828  *   65.125  *  219.039  *
15    *  143.667  *   77.248  *   19.084  *   22.622  *   64.538  *  227.160  *
16    *  142.235  *   76.479  *   17.286  *   22.396  *   63.895  *  234.726  *
17    *  142.968  *   73.186  *   15.846  *   22.512  *   68.650  *  246.035  *
18    *  141.384  *   72.375  *   14.242  *   22.262  *   67.889  *  253.530  *
19    *  137.078  *   72.630  *   14.292  *   23.923  *   68.129  *  254.425  *
20    *  137.627  *   69.384  *   12.989  *   24.019  *   73.050  *  265.990  *
21    *  135.125  *   68.122  *   12.753  *   23.582  *   71.722  *  261.154  *
22    *  133.402  *   67.254  *   11.344  *   23.281  *   70.807  *  268.293  *
23    *  132.846  *   63.857  *   11.297  *   23.184  *   75.095  *  267.175  *
24    *  129.383  *   64.404  *   10.213  *   25.048  *   75.738  *  280.232  *
25    *  126.846  *   63.141  *   10.013  *   24.557  *   74.252  *  274.737  *
26    *  126.891  *   60.184  *    8.925  *   24.565  *   79.073  *  285.655  *
27    *  122.737  *   60.318  *    8.944  *   26.385  *   79.249  *  286.292  *
28    *  122.018  *   57.090  *    8.892  *   26.231  *   83.832  *  284.614  *
29    *  118.557  *   57.516  *    7.926  *   28.335  *   84.458  *  297.860  *
30    *  117.779  *   54.347  *    7.874  *   28.149  *   89.238  *  295.905  *
31    *  113.679  *   54.434  *    7.887  *   30.242  *   89.380  *  296.377  *
32    *  112.839  *   51.333  *    7.829  *   30.019  *   94.311  *  294.186  *
33    *  111.143  *   49.829  *    7.029  *   32.955  *  102.509  *  312.470  *
34    *  108.798  *   48.048  *    7.154  *   36.000  *  110.776  *  318.035  *
35    *  106.354  *   46.349  *    7.296  *   39.462  *  119.876  *  324.371  *
36    *  105.662  *   42.767  *    7.571  *   43.992  *  139.699  *  336.579  *
37    *  105.887  *   39.220  *    6.893  *   49.493  *  164.371  *  365.901  *
38    *  105.417  *   35.401  *    7.181  *   55.335  *  191.146  *  381.194  *
39    *  103.537  *   32.720  *    7.743  *   68.573  *  228.640  *  411.019  *
40    *  101.648  *   29.981  *    8.371  *   84.923  *  272.569  *  444.329  *
41    *   97.887  *   29.214  *    8.899  *  103.060  *  303.643  *  472.356  *
42    *   94.268  *   28.299  *    9.433  *  123.854  *  336.838  *  500.705  *
43    *   91.850  *   26.496  *    9.654  *  134.875  *  359.789  *  512.426  *
44    *   89.394  *   24.727  *    9.878  *  146.786  *  383.920  *  524.346  *
45    *   85.625  *   23.878  *   10.496  *  176.150  *  425.080  *  557.156  *
46    *   81.475  *   23.983  *   10.542  *  187.843  *  426.935  *  559.587  *
47    *   78.956  *   22.243  *   10.802  *  204.207  *  455.501  *  573.359  *
48    *   76.389  *   20.541  *   11.070  *  221.887  *  485.793  *  587.627  *
49    *   75.038  *   18.087  *   10.875  *  217.965  *  496.283  *  577.240  *
50    *   72.368  *   16.499  *   11.133  *  236.389  *  528.094  *  590.973  *
***************************************************************************
```

Figure 10.27

REFERENCES

1. R. Jenkins, and J. S. Aronofsky, "Unsteady Radial Flow of Gas Through Porous Media," *J. Appl. Mech.* (June 1953), 210.

2. M. R. Todd, D. M. O'Dell, and G. J. Hirasaki, "Methods for Increased Accuracy in Numerical Reservoir Simulators," *Soc. Pet. Eng. J.* (Dec. 1972), 515–30; *Trans. AIME*, **253**.

BIBLIOGRAPHY

Akbar, Ali M., M. D. Arnold, and A. Herbert Harvey, "Numerical Simulation of Individual Wells in a Field Simulation Model," SPE 4073, 47th Annual Meeting, San Antonio, Texas, Oct. 8–10, 1973.

Byrne, William B., Jr., and Richard A. Morse, "The Effects of Various Reservoir and Well Parameters on Water Coning Performance," SPE 4287, Third Symposium on Numerical Simulation of Reservoir Performance, Houston, Texas, Jan. 10–12, 1973.

Chappelear, J. E. and W. L. Rogers, "Some Practical Considerations in the Construction of a Semi-implicit Simulator," SPE 4276, Third Symposium on Numerical Simulation of Reservoir Performance, Houston, Texas, Jan. 10–12, 1973.

Coats, K. H., "An Analysis for Simulating Reservoir Performance Under Pressure Maintenance by Gas and/or Water Injection," *Soc. Pet. Eng. J.* (Dec. 1968), 331–40.

Coats, K. H. and J. G. Richardson, "Calculation of Water Displacement by Gas in Development of Aquifer Storage," *Soc. Pet. Eng. J.* (June 1967), 105–12; *Trans. AIME*, **240**.

Cone, Clifford, "Case History of the University Block 9 (Wolfcamp) Field-A Gas-Water Injection Secondary Recovery Project," *J. Pet. Tech.* (Dec. 1970), 1485–91.

Cottin, Rene H., "Application of a Multiphase Coning Model to Optimize Completion and Production of Thin Oil Columns Lying Between Gas Cap and Water Zone," SPE 4632, 48th Annual Meeting, Las Vegas, Nev., Sept. 30–Oct. 3, 1973.

Craig, F. F., Jr., "Effect of Reservoir Description on Performance Predictions," *J. Pet. Tech.* (Oct. 1970), 1239–45.

Dorsey, J. B., L. D. Jones, and A. Bencheikh, "Numerical Simulation of the Zarzaitine Devonian F4 Reservoir, Algeria," SPE 4281, Third Symposium on Numerical Simulation of Reservoir Performance, Houston, Texas, Jan. 10–12, 1973.

Field, M. B., I. M. Wytrychowski, and J. K. Patterson, "A Numerical Simulation of Kaybob South Gas Cycling Projects," *J. Pet. Tech.* (Oct. 1971), 1253–62.

FIELD, M. B., J. W. GIVENS, and D. S. PAXMAN, "Kaybob South—Reservoir Simulation of a Gas Cycling Project With Bottom Water Drive," *J. Pet. Tech.* (April 1970), 481–92; *Trans. AIME,* **249.**

FRANCISCO, CHACON H., "Optimum Well Spacing for the Gas Reservoirs of the Mexican Republic," SPE 4579, 48th Annual Meeting, Las Vegas, Nev., Sept. 30–Oct. 3, 1973.

GIVENS, JAMES W., "A Practical Two-dimensional Model for Simulating Dry Gas Reservoirs with Bottom Water Drive," First Symposium on Numerical Simulation of Reservoir Performance, Dallas, Texas, April 22–23, 1968.

GRAUE, D. J. and R. A. FILGATE, "A Study of Water Coning in the Kaybob South Beaverhill Lake Field," SPE 3623, 46th Annual Meeting, New Orleans, Oct. 3–6, 1971.

HALL, H. N., "Predicting Gravity-Drainage Performance Using a Three-dimensional Model," *J. Pet. Tech.* (May 1968), 517–24; *Trans. AIME,* **243.**

McCARTY, D. G. and E. C. BARFIELD, "The Use of High-speed Computers for Predicting Flood-out Patterns," *Trans. AIME,* **213,** 139–45.

McCREARY, J. G., "A Simple Method for Controlling Gas Percolation in Numerical Simulation of Solution Gas Drive Reservoirs," *Soc. Pet. Eng. J.* (Mar. 1971), 85–91; *Trans. AIME,* **251.**

McCULLOCH, R. C., J. R. LANGTON, and A. SPIVAK, "Simulation of High Relief Reservoirs, Rainbow Field, Alberta, Canada," *J. Pet. Tech.* (Nov. 1969), 1399–408.

McNEILL, W. E., JR., and J. E. GARRETT, "Predicting Optimum Shut-in of Wells in Peripheral and Line Drive Waterfloods," *J. Pet. Tech.* (April 1971), 497–505; *Trans. AIME,* **251.**

MERCHANT, A. R., M. D. ARNOLD, and A. HERBERT HARVEY, "A Technique for Improving Material Balance Accuracy in Reservoir Simulation Models," SPE 4548, 48th Annual Meeting, Las Vegas, Nev., Sept. 30–Oct. 3, 1973.

NOLEN, J. S., and D. W. BERRY, "Tests of the Stability and Time-step Sensitivity of Semi-implicit Reservoir Simulation Techniques," *Soc. Pet. Eng. J.* (June 1972), 253–66; *Trans. AIME,* **253.**

SANDREA, R. J. and S. M. FAROUQ ALI, "The Effects of Isolated Permeability Interferences on the Sweep Efficiency and Conductivity of a Five-spot Network," *Soc. Pet. Eng. J.* (March 1967), 20–30; *Trans. AIME,* **240.**

SPIVAK, A. and K. H. COATS, "Numerical Simulation of Coning Using Implicit Production Terms," *Soc. Pet. Eng. J.* (Sept. 1970), 254–67; *Trans. AIME,* **249.**

THOMPSON, FRED R., and RICHARD A. THACHUK, "Compositional Simulation of a Gas Cycling Project, Bonnie Glen D-3A Pool, Alberta, Canada," SPE 4280, Third Symposium on Numerical Simulation of Reservoir Performance, Houston, Texas, Jan. 10–12, 1973.

TODD, M. R. and W. J. LONGSTAFF, "The Development, Testing, and Application of a Numerical Simulator for Predicting Miscible Flood Performance," *J. Pet. Tech.* (July 1972), 874–82; *Trans. AIME,* **253.**

TODD, M. R., P. M. O'DELL, and G. J. HIRASAKI, "Methods for Increased Accuracy in Numerical Reservoir Simulators," *Soc. Pet. Eng. J.* (Dec. 1972), 515–30; *Trans. AIME,* **253**.

VAN POOLLEN, H. K., E. A. BREITENBACH, and D. H. THURNAU, "Treatment of Individual Wells and Grids in Reservoir Modeling," *Soc. Pet. Eng. J.* (Dec. 1968), 341–46.

WEAVER, R. H., "Simulation of Waterflood Behavior in a Reservoir Previously Invaded by Water," *J. Pet. Tech.* (Aug. 1972), 909–15.

YAZDI, MOJTABA and PAUL B. CRAWFORD, "Performance of a Simulated Fractured Matrix Reservoir Subject to a Bottom-water Drive," SPE 4573, 48th Annual Meeting, Las Vegas, Nev., Sept. 30–Oct. 3, 1973.

11 Optimization and Simulation

11.1 CONCEPTS IN OPTIMIZATION

Introduction

The engineer or manager is faced with many complex problems in which the outcome is strongly affected by the type of decision made. The total concept of decision making has become so involved due to the interaction of many influences, parameters, and previous decisions that the manager can no longer hold onto his intuitive judgment, no matter how good he is. He can no longer comprehend the totality of a project as complex as, for example, an offshore development study and make a singlehanded intuitive decision, no matter how good his data. In any project there are innumerable ways in which resources, be they men, machines, or money, can be allocated. This allocation process produces a fixed outcome for a given allocation scheme; the instinctive desire is to obtain the best results, where *best* is defined within the limits of those constraints acting at the time. The objective, therefore, is to reach an *optimum* decision—one which cannot be improved upon.

Optimization, in general, deals with the allocation of limited resources with the objective in mind of extremizing (maximizing or minimizing) some outcome; this outcome is called the *objective function*. It is the basic decision criterion.

Process of Analysis

In order to make a decision, the engineer must first understand the problem. A firm understanding of the factors which affect the problem will ensure that

307

a more complete model of the problem can be built, and the subsequent results are then more realistic and will allow more effective implementation of the model. In an effort to specify the parameters involved the engineer has to quantify his data; this quantitative approach has been used intuitively by good engineers for decades, but, as mentioned earlier, the complexity of today's problems calls for a structured approach to the subject. Wagner[1] divides the process into the following subphases:

1. *Formulating the problem:* This initial process defines all the elements of the problem, labels the parameters as controllable or not, and sets the limitations and the objectives.

2. *Model building:* This involves the definition or derivation of mathematical formulas and the determination of interrelationships or interdependencies; it also delimits the domain of the problem in space and time.

3. *Analysis:* This forms the computational step of the problem. The model is solved by some algorithm and the objective determined along with other pertinent information. If the model is too complex, this step may be impossible to perform by present technological standards; on the other hand, the too-simple model leads to trivial results. A necessary adjunct to this step is to determine the sensitivity of the results to model inputs, a process which is essential in determining acceptability of the results.

4. *Implementation of results:* This involves the use of the findings of the study in the field and the subsequent feedback of this data into the initial model to produce a more up-to-date version. The longer one works with a given project, the more information he obtains from it and the more realistic is his forecast. It is a philosophical pity that we only know everything about a project from a post mortem examination of it.

Optimization Methods

There are several methods available for the optimal solution of engineering problems, and several excellent texts[2-5] have been written on each of these methods. Some methods are:

1. Linear programming
2. Dynamic programming
3. Nonlinear programming
4. Network analysis
5. Gradient methods
6. Geometric programming
7. Graph theory
8. Integer programming

11.2 LINEAR PROGRAMMING

By far the most widely used technique in problem optimization is that of linear programming. As the name implies, all the parameters are assumed to be linearly interrelated, and although this is not the case in nature, some not too restrictive assumptions can usually be made to linearize the parameters. The linear programming model has a characteristic format which consists of

1. an objective function, and
2. a set of constraints,

both of which must be linear. The objective function is the cost or profit function; it is a combination of the system parameters which produce a given outcome—e.g.:

$$z = 2x_1 + 3x_2 \tag{11.1}$$

is a function z described by the linear combination of the two parameters x_1 and x_2. The associated coefficients may be the profit associated with the sale of each x_i, or it may be the cost associated with the manufacture of each x_i. In the former case we would try to maximize z, whereas in the latter we would try to minimize z. The parameters x_1 and x_2 are not available in infinite quantities, else there would be no need to optimize. They are constrained with certain limits; these limits can be due to supply, cost, work schedule, or several other conditions. The net result is that there are conditions imposed on these variables. This set forms the constraint equations—e.g.:

$$\begin{aligned}
x_1 + 2x_2 &\leq 10 \\
4x_1 + 3x_2 &\leq 20 \\
x_1 &\leq 18 \\
x_2 &\leq 12 \\
6x_1 + 10x_2 &\leq 40
\end{aligned} \tag{11.2}$$

These form a set of linear inequalities. Associated with the above constraints are the nonnegativity constraints:

$$x_1 \geq 0, \qquad x_2 \geq 0 \tag{11.3}$$

The complete format for the linear programming can be generalized to the following form:

$$\text{Maximize:} \quad Z = \sum_{i=1}^{N} c_i x_i \tag{11.4}$$

subject to:

$$a_{11}x_1 + a_{12}x_2 + \ldots + a_{12}x_n \leq b_1$$

$$\vdots$$ (11.5)

$$a_{n1}x_1 + \ldots + a_{nn}x_n \leq b_n$$

$$x_i \geq 0$$

This system is readily solved by the simplex[3] algorithm. Some examples of linear programming models follow.[2]

Diet Problem: The objective is to determine the quantities of food that should be eaten to meet certain nutritional requirements at minimum cost.

Data

Vitamin	Gal Milk	Lb Beef	Doz Eggs	Min. Daily Req.
A	1	1	10	1 mg
B	100	10	10	50 mg
C	10	100	10	10 mg
Cost	$1.00	$1.10	$0.50	

If

$$x_M = \text{gal of milk}$$
$$x_B = \text{lb of beef}$$
$$x_E = \text{doz eggs}$$

we can set the problem up as a minimization of costs:

$$\text{Minimize} \quad Z = 1.0x_M + 1.10x_B + 0.50x_E$$

subject to

$$1x_M + 1x_B + 10x_E \geq 1 \qquad \text{Vitamin A}$$
$$100x_M + 10x_B + 10x_E \geq 50 \quad \text{Vitamin B}$$
$$10x_M + 100x_B + 10x_E \geq 10 \quad \text{Vitamin C}$$
$$x_M, x_B, x_E \geq 0$$

Oil Storage Problem: Crude oil flows from two separate oil fields through different pipelines into a common storage battery. The stored crude is then transported through a single line to a refinery (see Fig. 11.1).

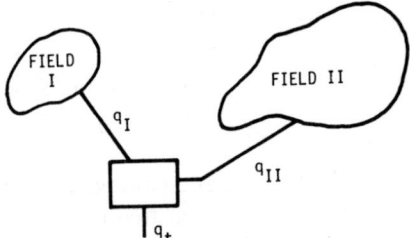

Figure 11.1: Oil storage problem.

The limitations on the operations are as follows: maximum capacity q_I is 4000 bbl/day, q_{II} is 3000 bbl/day. The limiting gas-oil ratio (GOR) is 2000 SCF/STB; GOR from Field I is 1000 SCF/STB and from Field II 4000 SCF/STB. The storage area has a capacity of 8500 bbl and contains 5000 bbl at the start of the day. The current refinery demand in q_t is 2000 bbl/day. The net value of the crude oil is \$2/bbl from Field I and \$3/bbl from Field II. Determine the production rates for each field to maximize the income for the day.

We can set the problem up as a maximization of profit Z. Thus, the objective function is:

$$\text{Maximize:} \quad Z = 2q_I + 3q_{II} \tag{a}$$

The constraints are then developed:

$$\text{pipeline capacity:} \quad q_I \leq 4000$$
$$q_{II} \leq 3000 \tag{b}$$
$$\text{storage volume:} \quad V_s \leq 8500$$

$$V_s = V_{init} + q_I + q_{II} - q_t \quad \text{by material balance}$$
$$= 5000 + q_I + q_{II} - 2000$$

Thus,

$$q_I + q_{II} = V_s - 5000 + 2000$$
$$= 8500 - 5000 + 2000 = 5500$$

But this is a limiting capacity; thus,

$$q_I + q_{II} \leq 5500 \tag{c}$$

gas-oil ratio: Total gas produced/total oil $= \text{GOR}_T$

$$\text{GOR}_T \leq 2000$$

Thus,

$$\frac{1000q_I + 4000q_{II}}{q_I + q_{II}} \leq \text{GOR}_T = 2000$$

This can be simplified by:

$$1000q_I + 4000q_{II} \leq 2000q_I + 2000q_I$$
$$-q_I + 2q_{II} \leq 0 \tag{d}$$

Equations (a), (b), (c), and (d) together form the linear programming model, which can be compactly written as:

$$\text{Maximize:} \quad Z = 2q_I + 3q_{II}$$

subject to

$$q_I \leq 4000$$
$$q_{II} \leq 3000$$
$$q_I + q_{II} \leq 5500$$
$$-q_I + 2q_{II} \leq 0$$
$$q_I \geq 0$$
$$q_{II} \geq 0$$

Since this is a two-parameter problem, it is possible to solve it graphically, and the student should verify that:

$$q_I = 3667 \text{ bbl/day}$$
$$q_{II} = 1833 \text{ bbl/day}$$
$$z = \$12,835$$

Linear Inequalities

Most of the constraints in the linear programming models are inequalities. A linear inequality can best be visualized from a geometrical interpretation.

Consider a two-parameter case:

$$2x_1 + 3x_2 \leq 12 \tag{11.6}$$

This can be represented as illustrated in Fig. 11.2. The line represents the function $(2x_1 + 3x_2)$; since we had an inequality, the area in which we are interested lies below the given line. Thus, a point $(\bar{x}_1, \bar{x}_2)$ within the cross-hatched area satisfies this inequality. If, on the other hand, the inequality sign were reversed—

$$2x_1 + 3x_2 \geq 12 \tag{11.7}$$

then the area in Fig. 11.3 would represent it. Just as in numerical analysis we

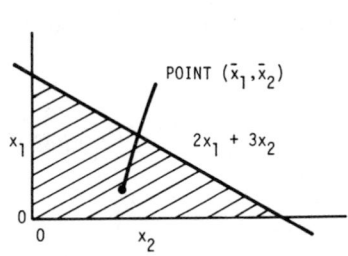

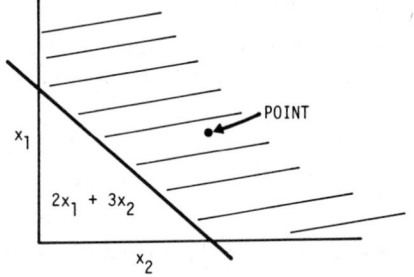

Figure 11.2: Linear inequalities. Figure 11.3: Linear inequalities.

have simultaneous equations, we can have simultaneous inequalities. Consider:

$$2x_1 + 3x_2 \leq 12$$
$$x_1 \leq 5$$
$$x_2 \leq 3 \qquad (11.8)$$
$$x_2 \geq 1$$
$$x_1 \geq 1$$

Figure 11.4 is a geometrical representation of the above set of inequalities. The cross-hatched polygon now satisfies this set of conditions. This geometric

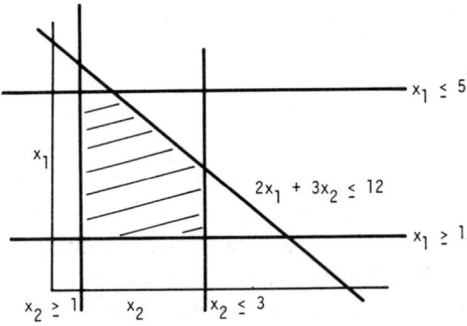

Figure 11.4: Set of inequalities.

representation gives us some idea of the solution space in two dimensions. The solution to our optimization problem lies somewhere within that polygon. From inspection it is clear that there is an infinite number of $(\bar{x}_1, \bar{x}_2)$-values within that polygon; we shall need a method of finding the correct one in a finite number of steps.

Slack Variables

Slack variables are nonnegative variables introduced into the constraint inequality to make it an equality—e.g.:

$$x_1 + 3x_2 + x_3 \leq 7$$
$$2x_1 + 4x_2 + 3x_3 \leq 12$$

(11.9)

Introducing the slack variables x_4 and x_5, we obtain:

$$x_1 + 3x_2 + x_3 + x_4 = 7$$
$$2x_1 + 4x_2 + 3x_3 + x_5 = 12$$

(11.10)

The linear equalities are now used in the solution of the linear programming model.

EXAMPLE [1] A GRAPHICAL EXAMPLE: An organization has two products in its line and must determine how to allocate its resources optimally. The data for the problem is given below:

Hours/Unit

Product	Foundry	Machine Room	Finishing	Profit per Unit
x	6	3	4	2
y	6	6	2	3
Total Available Hours	120	72	60	—

Each of the three work stations has a specified hour/unit production time, and each has a total number of available hours. The objective is to determine how many units each of x and y to produce in order to maximize the profit received from the sale.

Formulation of Objective Function: The profit received is:

$$z = 2x + 3y$$

The constraints are based on the machine usage and total time limitations:

$$6x + 6y \leq 120 \quad \text{Foundry}$$
$$3x + 6y \leq 72 \quad \text{Machine room}$$
$$4x + 2y \leq 60 \quad \text{Finishing}$$
$$x, y \geq 0$$

These simultaneous linear inequalities determine the set of constraints governing the problem. Since there are only two unknowns, this problem can be solved graphically as shown in Fig. 11.5.

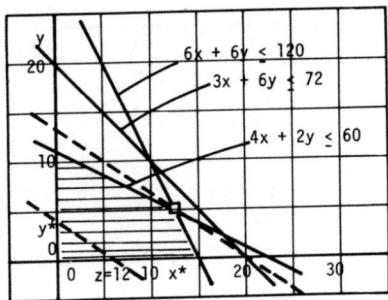

Figure 11.5: Graphical solution for two variables.

Procedure in Two Unknowns

1. Graph each constraint and label the feasible region.
2. Choose any profit value and plot that profit line—e.g.:

$$z = 12$$
$$\text{Plot:} \quad 2x + 3y = 12$$

3. Move parallel to this line, since *all profit lines will have the same slope* but different intercepts.
4. The line which passes through the highest feasible point is the optimal line.
5. The values of each variable are read off the axes.

$$\text{Optimal:} \quad x^* = 16$$
$$y^* = 5$$
$$z^* = 2(16) + 3(5) = 47$$

Simplex Method

The simplex method is an algorithm first developed by Dantzig[3] to solve linear programming problems algebraically. It is obvious that for more than two variables the graphical methods break down, and we must resort to computational techniques. To fully understand the algorithm the student should have some background in matrix solution of simultaneous linear equations, and for this reason we shall look at it only from a qualitative point of view. From Fig. 11.6 we see that there are an infinite number of feasible solutions that satisfy the constraint set; Dantzig, however, proved that the actual number of combinations that enter into the solution is finite, and these form

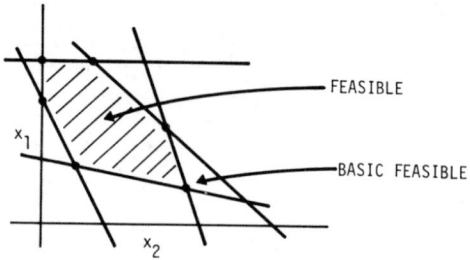

Figure 11.6: Basic feasible solutions.

what are called the basic feasible solutions. These solutions are those points at the intersections of the constraint equations. In the example below this distinction is illustrated.

The simplex method looks only at these basic feasible solutions, and after finding a solution it determines whether the objective function can be improved by using another variable. This process continually improves the objective function until the optimal is reached.

The simplex algorithm can be verbalized as follows for a general linear programming model.

1. Select m variables that constitute a basic feasible solution.
2. Eliminate these from the objective function
3. Check the objective function to determine if there is a zero variable in the basic feasible solution which will cause an increase in the objective function if it were made positive. (I.e., is the objective function improvable?)
4. Determine how large this new variable can be made until one of the original m basic feasible variables becomes zero.
5. Eliminate the variable in step 4 which becomes zero.
6. Enter the variable in step 4 as part of the next trial and solve for a new set of m variables.
7. Begin the process again at step 3.
8. If the objective function cannot be improved, the optimum solution

consists of the current value of the m basic feasible variables and the current value of the objective function.

The algorithm is readily available in several computer codes.[6,7]

11.3 APPLICATIONS IN PETROLEUM ENGINEERING

At the exploration and operations end of the petroleum industry the level of utilization of optimization techniques lags far behind the current state of the art of optimization methods. Optimization techniques have been used rather infrequently in the past in these areas, but since the late 1960s companies have begun to integrate into their decision-making processes some formalized optimization techniques. These techniques usually are directed toward developmental problems: where, when, and how many wells should be drilled; secondary recovery problems; when to start injection and how much. There are problems involved in the gathering and distribution of gas, in pipeline design, and in gas storage.

The most frequent optimization tool used is linear programming, and the engineer is generally faced with two subproblems:

1. *Formulation of the linear programming model of his system:* This usually involves some simulation to obtain the raw data followed by some superposition technique or similar process to show interrelationships.

2. *Solution of the linear programming model:* This is usually a very straightforward step in which the engineer uses a library program generally available at the computer center. These include MPS (Mathematical Programming System), LPS (Linear Programming System), or some similar readily available system.

The first of these two subproblems is more difficult and time-consuming. Some of the techniques involved will be developed in the next section, and a case history will point out the uses.

The Simulator as a Tool

In this section the simulator is used as an instrument to obtain the necessary raw data. In most cases the engineer has to determine the interaction of several parameters which affect the outcome of his decision. He does not have the time or the resources to make a single run for every theoretically possible combination of parameters, so he has to develop a set of correlations which he can later use in his optimization processes. In effect, he uses the simulator to build a data base incorporating the range of possibilities of these parameters, and subsequently he searches these data items in a very selective fashion to determine his best operating decision.

Secondly, the simulator can be used in an "interactive" mode. It can perform as a single entity in an iterative scheme whereby the engineer is making decisions based on a given set of constraints, and after each decision the simulator is rerun to determine some information necessary for the final decision. This interactive approach by its very nature poses problems of expense and time and may be prohibitive except for the very simplified models.

In addition to the simulator, other additional tools are required for the formulation of an optimization model. Let us look at some of these.

Superposition

In order to develop the full use of the simulator as a tool, we must first look at the concept of superposition and its application to many petroleum engineering problems.

Consider a well producing in a reservoir system with a production curve as shown in Fig. 11.7. The pressure drop at this well at any time $t > t_2$ can be obtained by superposition of the rates at different times. For a single well as shown in Fig. 11.8:

$$P_D = \frac{2\pi kh}{q\mu}(P_i - P) \tag{11.11}$$

Therefore, a shut-in well is obtained by superposing a producing well and an injection well of the same flow capacity. The pressure equation for our example in Fig. 11.7 can therefore be written for $t > t_2$ as:

$$
\begin{aligned}
\frac{2\pi kh}{\mu}(P_i - P_w) = \;& q_1 P_D(1, t_D) && \text{Production from } t = 0 \text{ at } q_1 \\
& - q_1 P_D[1, (t - t_1)_D] && \text{Injection from } t = t_1 \text{ at } -q_1 \\
& + q_2 P_D[1, (t - t_1)_D] && \text{Production from } t = t_1 \text{ at } q_2 \\
& - q_2 P_F[1, (t - t_2)_D] && \text{Injection from } t = t_2 \text{ at } -q_2 \\
& + q_3 P_D[1, (t - t_2)_D] && \text{Production from } t = t_2 \text{ at } q_3
\end{aligned} \tag{11.12}
$$

Equation (11.12) can be abbreviated by collecting terms as follows:

$$\frac{2\pi kh}{\mu}(P_i - P_w) = q_1 P_D(1, t_D) + (q_2 - q_1)P_D[1, (t - t_1)_D]$$

$$+ (q_3 - q_2)P_D[1, (t - t_2)_D] \tag{11.13}$$

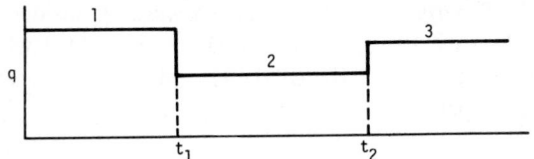

Figure 11.7: Variable flow rate.

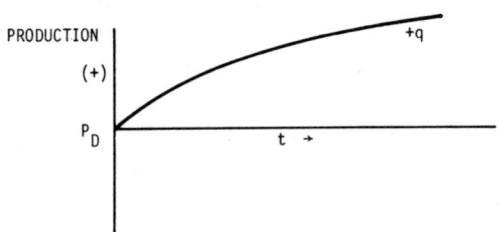

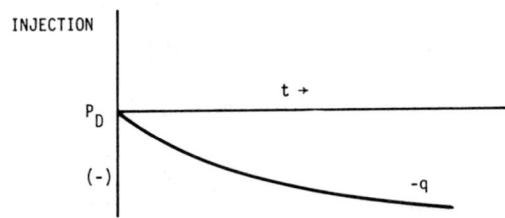

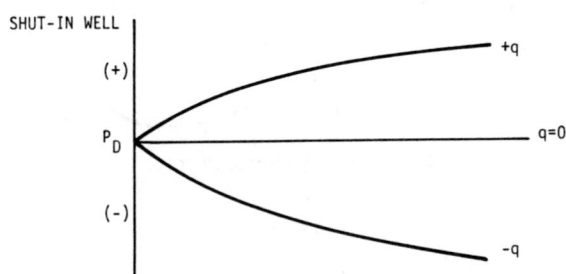

Figure 11.8: Dimensionless pressure drop.

or in general form:

$$\frac{2\pi k h}{\mu}(P_i - P_w) = q_1\left\{\underbrace{P_D(1, t_D)}_{\substack{\text{Initial}\\\text{flow rate}}} + \underbrace{\sum_{i=2}^{N}\left(\frac{q_i - q_{i-1}}{q_1}\right)P_D[1, (t - t_{i-1})_D]}_{\substack{\text{Subsequent}\\\text{flow rates}}}\right\} \quad (11.14)$$

Interference—Superposition in Time and Space: Consider two wells in a reservoir at a distance r_{12} apart as shown in Fig. 11.9 with production schedules as shown in Fig. 11.10. At the end of $t = t_1$ (*Note:* $q_i^j = i$th well rate in the jth time period):

For well 1:

$$\frac{2\pi kh}{\mu}(P_i - P_{w_1}) = \underbrace{q_1^1 P_D(1, t_{1_D})}_{\substack{\text{Well on}\\\text{itself}}} + \underbrace{q_2^1 P_D(r_{12}, t_{1_D})}_{\text{Interference}} \qquad (11.15)$$

For well 2:

$$\frac{2\pi kh}{\mu}(P_i - P_{w_2}) = \underbrace{q_2^1 P_D(1, t_{1_D})}_{\substack{\text{Well on}\\\text{itself}}} + \underbrace{q_1^2 P_D(r_{12}, t_{1_D})}_{\text{Interference}} \qquad (11.16)$$

At the end of $t = t_2$:
For well 1:

$$\frac{2\pi kh}{\mu}(P_i - P_w) = q_1^1 p_D(1, t_{2_D}) + (q_1^2 - q_1^1)P_D[1, (t_2 - t_1)_D]$$

$$+ q_2^1 P_D(r_{12}, t_2) + (q_2^2 - q_2^1)P_D[1, t_2 - t_1)_D] \qquad (11.17)$$

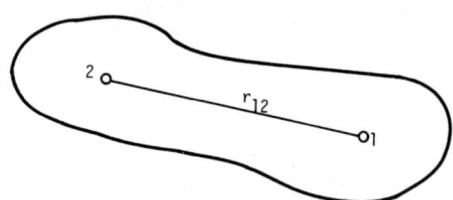

Figure 11.9: Well interference.

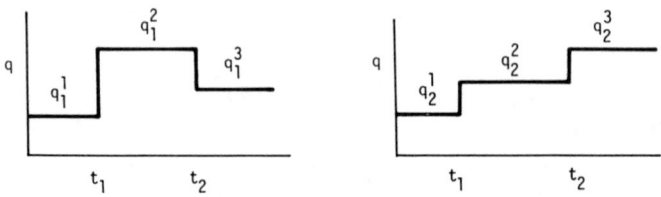

Figure 11.10: Well schedules.

where the first pair of terms includes the effect of well 1 on itself. A similar expression can be written for well 2.

EXAMPLE OF SUPERPOSITION IN TIME   For a well in a square (see Fig. 11.11), for which the P_D-values are available, the general equation in engineering units is:

$$P_i - P_w = \frac{141.2q\mu}{kh} P_D(r_D, t_D)$$

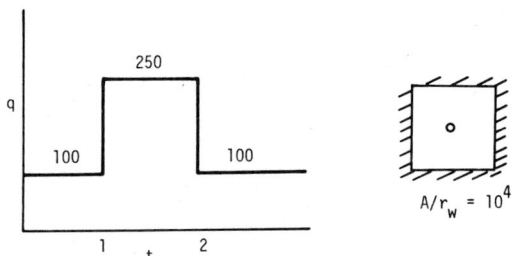

Figure 11.11: Flow rates.

At $t > t_2$:

$$P_i - P_w = \frac{141.2q_1\mu}{kh}\left\{ P_D(1, t_D) + \frac{q_2 - q_1}{q_1} P_D[1, (t_3 - t_1)_D]\right.$$

$$\left. + \frac{q_3 - q_2}{q_1} P_D[1, (t_3 - t_2)_D]\right\}$$

$$P_i - P_w = \frac{141.2q_1\mu}{kh}\left\{ P_D(1, 5000) + \frac{250 - 100}{100} P_D[1, (5000 - 200)]\right.$$

$$\left. + \frac{200 - 250}{100} P_D[1, (5000 - 2000)]\right\}$$

From Fig. 11.12 evaluate P_D:

$$= \frac{141.2 * 100 * 1.0}{100 * 10}\left[6.4 + (2.5 * 6.2) - \left(\frac{1}{2} * 5.2\right)\right]$$

$$= \frac{141.2 * 100}{1000} (18.3)$$

$$= 258 \text{ psi}   \text{ Pressure drop at well}$$

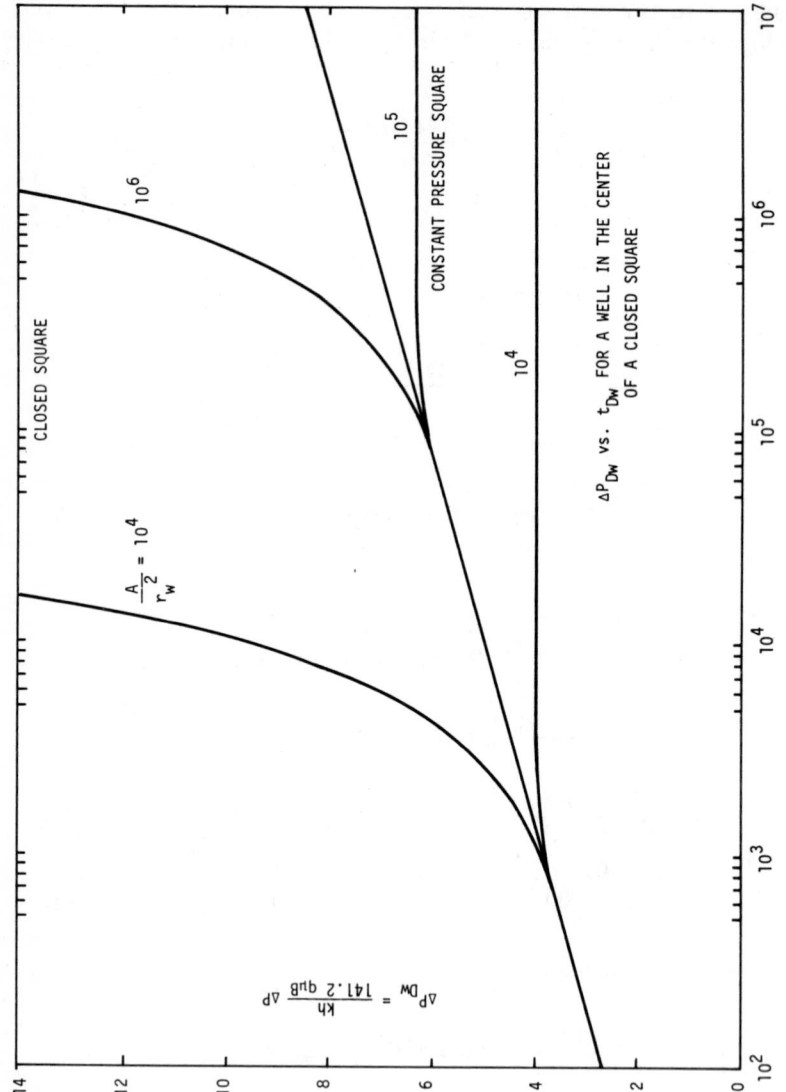

Figure 11.12: P_{Dw} for a well in a square.

322

11.4 APPLICATION TO GAS STORAGE OPTIMIZATION

The system under consideration is shown in Fig. 11.13. It involves the optimization of a gas storage facility adjacent to a large metropolitan area. The objective of the study is to meet the demand for gas during the peak season by

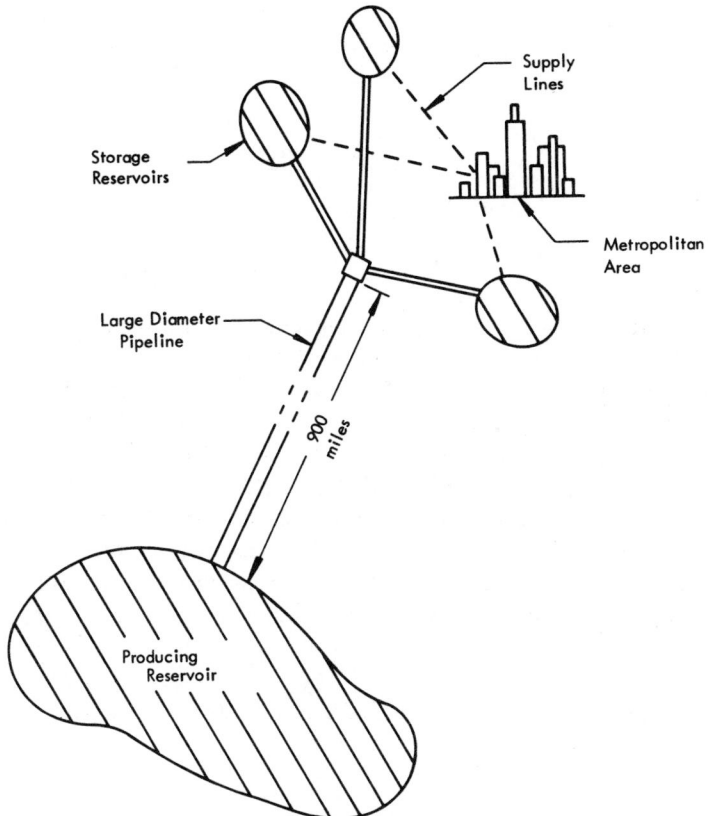

Figure 11.13: Gas system.

an optimal selection of producing rates and schedules. The dynamic nature of this problem is reflected in the optimization technique adopted. The solution process involves the use of superposition in time and space to develop the constraints for the linear model. Several authors, including Wattenbarger[8] and Coats,[9] have addressed this problem in recent years. Using Eqs. (11.13) through (11.17) we can write for any given pairs of wells at time t_1:

Well 1: $\qquad \dfrac{2\pi kh}{\mu}(P_i - P_{w_1}) = q_1^1 P_D(0, t_1) + q_2^1 P_D(r_{12}, t_1)$

$\left. \right\}$ (11.18)

Well 2: $\qquad \dfrac{2\pi kh}{\mu}(P_i - P_{w_2}) = q_2^1 P_D(0, t_1) + q_1^1 P_D(r_{12}, t_1)$

where r_{12} is the distance between wells.

At the end of $t = t_2$:

Well 1:

$$\dfrac{2\pi kh}{\mu}(P_i - P_{w_1}) = q_1^1 p_D(0, t_2) + (q_1^2 - q_1^1)P_D[r_0, (t_2 - t_1)]$$

$$+ q_2^1 P_D(r_{12}, t_2) + (q_2^2 - q_2^1)P_D[r_2, (t_2 - t_1)]$$

Well 2:

$$\dfrac{2\pi kh}{\mu}(P_i - P_{w_2}) = q_2^1 P_D(0, t_2) + (q_2^2 - q_2^1)P_D[0, (t_2 - t_1)]$$

$$+ q_1^1 P_D(r_{12}, t_2) + (q_1^2 - q_1^1)P_D[r_{12}, (t_2 - t_1)]$$

$\left. \right\}$ (11.19)

In each of the above expressions the first set of terms is the effect of the well on itself, the second set the effect of the other well on it.

This principle can be applied for as many wells as needed by extending the number of terms involved. An extension to three wells over three time steps is shown in Table 11.1. The first row of this tableau shows the flow rates for each well in each time period, while successive rows show the superposition of the interference effects of each well on itself and on the other wells.

This expanded form can be compacted by resorting to matrix and vector notation. Figure 11.14 illustrates the compact form of the superposition principle. It is basically a matrix multiplication:

$\vec{q}^1$	$\vec{q}^2$	$\vec{q}^3$	$\vec{q}^4$		$\vec{q}^n$	
A_{11}	0	0	0	...	0	$= \Delta P(1)$
A_{21}	A_{22}	0	0			$= \Delta P(2)$
A_{31}	A_{32}	A_{33}	0			.
A_{41}	A_{42}	A_{43}	A_{44}			
.	.	.	.	...	0	.
.	.	.	.			.
.	.	.	.			.
A_{n1}	A_{n2}	A_{n3}	A_{n4}	...	A_{nn}	$= \Delta P(n)$

Figure 11.14: Matrix of interference effect coefficients.

where $A_{i,j}$ is a matrix consisting of the appropriate $P_D(r, t)$-values as defined earlier. These matrices are referred to as influence matrices.

TABLE 11.1

Influence Matrix for 3 Wells in 3 Time Steps

q_1^1	q_2^1	q_3^1	q_1^2	q_2^2	q_3^2	q_1^3	q_2^3	q_3^3	Time steps
$P_D(r_{11}, t_1)$	$P_D(r_{12}, t_1)$	$P_D(r_{13}, t_1)$	0						
$P_D(r_{12}, t_1)$	$P_D(r_{22}, t_1)$	$P_D(r_{23}, t_1)$	0						I
$P_D(r_{13}, t_1)$	$P_D(r_{32}, t_1)$	$P_D(r_{33}, t_1)$	0						
$P_D(r_{11}, t_2)$ $-P_D(r_{11}, t_2 - t_1)$	$P_D(r_{12}, t_2)$ $-P_D(r_{12}, t_2 - t_1)$	$P_D(r_{13}, t_2)$ $-P_D(r_{13}, t_2 - t_1)$	$P_D(r_{11}, t_2 - t_1)$	$P_D(r_{12}, t_2 - t_1)$	$P_D(r_{13}, t_2 - t_1)$	0			
$P_D(r_{12}, t_2)$ $-P_D(r_{12}, t_2 - t_1)$	$P_D(r_{22}, t_2)$ $-P_D(r_{22}, t_2 - t_1)$	$P_D(r_{23}, t_2)$ $-P_D(r_{23}, t_2 - t_1)$	$P_D(r_{12}, t_2 - t_1)$	$P_D(r_{22}, t_2 - t_1)$	$P_D(r_{23}, t_2 - t_1)$	0			II
$P_D(r_{13}, t_2)$ $-P_D(r_{13}, t_2 - t_1)$	$P_D(r_{32}, t_2)$ $-P_D(r_{32}, t_2 - t_1)$	$P_D(r_{33}, t_2)$ $-P_D(r_{33}, t_2 - t_1)$	$P_D(r_{13}, t_2 - t_1)$	$P_D(r_{32}, t_2 - t_1)$	$P_D(r_{33}, t_2 - t_1)$	0			
$P_D(r_{11}, t_3)$ $-P_D(r_{11}, t_3 - t_1)$	$P_D(r_{12}, t_3)$ $-P_D(r_{12}, t_3 - t_1)$	$P_D(r_{13}, t_3)$ $-P_D(r_{13}, t_3 - t_1)$	$P_D(r_{11}, t_3 - t_1)$ $-P_D(r_{11}, t_3 - t_2)$	$P_D(r_{12}, t_3 - t_1)$ $-P_D(r_{12}, t_3 - t_2)$	$P_D(r_{13}, t_3 - t_1)$ $-P_D(r_{13}, t_3 - t_2)$	$P_D(r_{11}, t_3 - t_2)$	$P_D(r_{12}, t_3 - t_2)$	$P_D(r_{13}, t_3 - t_2)$	
$P_D(r_{12}, t_3)$ $-P_D(r_{12}, t_3 - t_1)$	$P_D(r_{22}, t_3)$ $-P_D(r_{22}, t_3 - t_1)$	$P_D(r_{23}, t_3)$ $-P_D(r_{23}, t_3 - t_1)$	$P_D(r_{12}, t_3 - t_1)$ $-P_D(r_{12}, t_3 - t_2)$	$P_D(r_{22}, t_3 - t_1)$ $-P_D(r_{22}, t_3 - t_2)$	$P_D(r_{23}, t_3 - t_1)$ $-P_D(r_{23}, t_3 - t_2)$	$P_D(r_{12}, t_3 - t_2)$	$P_D(r_{22}, t_3 - t_2)$	$P_D(r_{23}, t_3 - t_2)$	III
$P_D(r_{13}, t_3)$ $-P_D(r_{13}, t_3 - t_1)$	$P_D(r_{32}, t_3)$ $-P_D(r_{32}, t_3 - t_1)$	$P_D(r_{33}, t_3)$ $-P_D(r_{33}, t_3 - t_1)$	$P_D(r_{13}, t_3 - t_1)$ $-P_D(r_{13}, t_3 - t_2)$	$P_D(r_{32}, t_3 - t_1)$ $-P_D(r_{32}, t_3 - t_2)$	$P_D(r_{33}, t_3 - t_1)$ $-P_D(r_{33}, t_3 - t_2)$	$P_D(r_{13}, t_3 - t_2)$	$P_D(r_{32}, t_3 - t_2)$	$P_D(r_{33}, t_3 - t_2)$	

To formulate the optimization model, the following must be defined:

1. Decision variables
2. Constraints on the variables
3. Objective function to be extremized.

Decision Variables: The level of production at each time step for each well will decide whether the well should be produced. These are defined by the following:

$$q_i^k = \text{Rate of production of gas in the time step } k \text{ from well } i$$

where $i = 1, 2, \ldots, n$
    $k = 1, 2, \ldots, m$

Constraints: These are based on the physical limitations of the system. First, a well cannot physically produce gas below a certain minimum pressure and still enter the gathering system. This minimum is usually the inlet pressure of the compressors in the system or the minimum pressure experienced in the gathering line.

Mathematically, the pressure at any time must not fall below a certain value which may or may not change with time:

$$P_i^{(t)} \geq P_{\min}^{(t)} \tag{11.20}$$

Then the maximum pressure drop at any time is:

$$\Delta P_i^{(t)} \leq P_0 - P_{\min}^{(t)} \tag{11.21}$$

where P_0 is the original pressure value and $\Delta P_i^{(t)}$ is the total drop experienced at point i. This total drop can be expressed in terms of production rates by using the matrix operation as outlined previously and reading across the rows of the tableau.

Secondly, the demand should not be exceeded at any time step:

$$\sum_{i=1}^{N} q_i^k \leq D^k \tag{11.22}$$

Objective Function: The end result of the planning process is to maximize the withdrawals over all the time periods, since failure to meet this demand results in penalties for breach of contract. The objective function is therefore:

$$\text{Maximize: } \sum_{k=1}^{m} \sum_{i=1}^{n} q_i^k \tag{11.23}$$

The optimization model is:

$$\text{Maximize:} \quad \sum_{k=1}^{m} \sum_{i=1}^{n} q_i^k \tag{11.24}$$

subject to

$$\sum_{i=1}^{n} q_i^k \leq D^k \quad \text{Demand} \tag{11.25}$$

$$\sum_{i=1}^{n} A_{ki} q^{-i} \leq \Delta p_i^k \quad \begin{array}{l} \text{Drawdown} \\ \text{limit} \end{array} \tag{11.26}$$

A linear programming model can be written for one time step with four wells—e.g.:

$$\text{Maximize:} \quad \sum_{k=1}^{1} \sum_{i=1}^{4} q_i^k \tag{11.27}$$

subject to

$$
\begin{aligned}
a_{11}q_1 + a_{12}q_2 + a_{13}q_3 + a_{14}q_4 &\leq P_0 - P_{\min(1)} \\
a_{21}q_1 + a_{22}q_2 + a_{23}q_3 + a_{24}q_4 &\leq P_0 - P_{\min(2)} \\
a_{31}q_1 + a_{32}q_2 + a_{33}q_3 + a_{34}q_4 &\leq P_0 - P_{\min(3)} \\
a_{41}q_1 + a_{42}q_2 + a_{43}q_3 + a_{44}q_4 &\leq P_0 - P_{\min(4)} \\
q_1 + q_2 + q_3 + q_4 &\leq D^1
\end{aligned}
\tag{11.28}
$$

The linear programming tableau of this model is shown in Fig. 11.15.

Row	q_1	q_2	q_3	q_4	Slack					Right-hand side
1	a_{11}	a_{12}	a_{13}	a_{14}	1	0	0	0	0	P_1^*
2	a_{21}	a_{22}	a_{23}	a_{24}	0	1	0	0	0	P_2^*
3	a_{31}	a_{32}	a_{33}	a_{34}	0	0	1	0	0	P_3^*
4	a_{41}	a_{42}	a_{43}	a_{44}	0	0	0	1	0	P_4^*
5	1	1	1	1	0	0	0	0	1	D_1
0	1	1	1	1	0	0	0	0	0	Z

Figure 11.15: Linear programming tableau.

This tableau expands as more wells are added and the time steps increase. It very rapidly becomes a large-scale linear programming model.

The model is shown in Fig. 11.16 in block form with the identity matrices omitted for convenience. The optimization model has several interesting features which can be used to advantage in obtaining an efficient solution.

q_1	q_2	q_3	q_4		q_n	Right-hand side
A_{11} A_{21} A_{31} A_{41} . . A_{m1}	9 A_{22} A_{32} A_{42} . . . A_{m2}	 A_{33} A_{43} . . . A_{m3}	 A_{44} . . . A_{m4}	...	0 A_{mn}	$\leq P_o - P_{\min}(1)$ $\leq P_o - P_{\min}(2)$ $\leq P_o - P_{\min}(3)$ $\leq P_o - P_{\min}(4)$ . . $\leq P_o - P_{\min}(n)$
$11\cdots1$ 0	0 $11\cdots1$ 0	0 $11\cdots1$ 0	0 $11\cdots1$ 0	...	0 $11\cdots1$	$\leq D_1$ $\leq D_2$ $\leq D_3$ $\leq D_4$: $\leq D_m$
$11\cdots1$	$11\cdots1$	$11\cdots1$	$11\cdots1$	...	$11\cdots1$	

Figure 11.16: Linear programming model.

11.5 CASE STUDY: OPTIMIZATION OF GAS STORAGE RESERVOIR

The reservoir data for the given gas storage is shown in Figs. 11.17, 11.18, and 11.19. These include the isopach map, the porosity data, and the permeability data. The demand schedule over a planning horizon is shown in Fig. 11.20. The objective is to determine the operating schedule to meet the

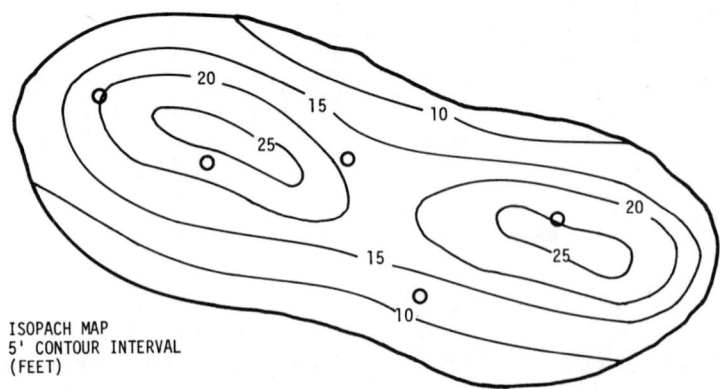

ISOPACH MAP
5' CONTOUR INTERVAL
(FEET)

Figure 11.17: Isopach map of gas reservoir showing well locations.

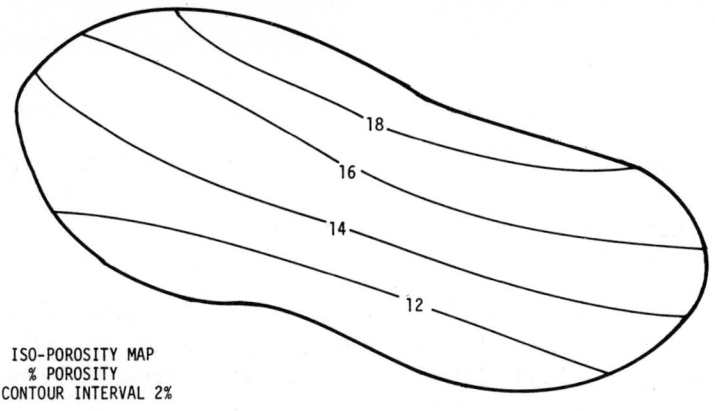

ISO-POROSITY MAP
% POROSITY
CONTOUR INTERVAL 2%

Figure 11.18: Isoporosity map of gas reservoir.

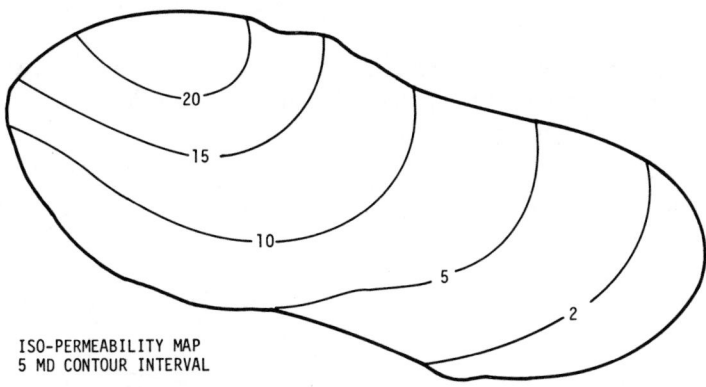

ISO-PERMEABILITY MAP
5 MD CONTOUR INTERVAL

Figure 11.19: Isopermeability map of gas reservoir.

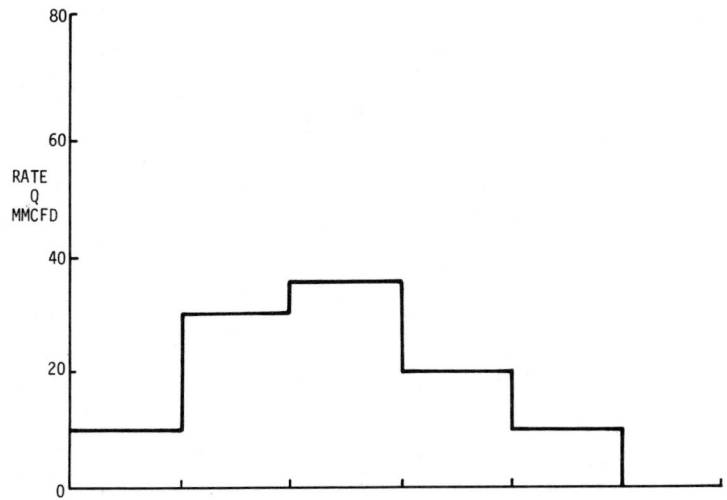

Figure 11.20: Gas demand curve.

demand. The following items should be determined over five time periods:

1. Scheduling of wells (in what sequence should wells be turned on?)
2. Production by well
3. "Slack" in the demand curve, if any
4. Effect on scheduling of changing the minimum pressure restriction

The first phase of this study consists of determining the influence functions of the wells on one another. This entails running a set of simulation studies with one well producing and all the others shut in. The dimensionless potential drop is computed continuously for each well as follows:

$$\Delta M_D = \frac{M_i - M_j}{M_i}$$

where M_i is the real gas potential at the original reservoir condition and M_j is the real gas potential at any time j. The interference or influence data for the five wells is shown in Figs. 11.21 through 11.25.

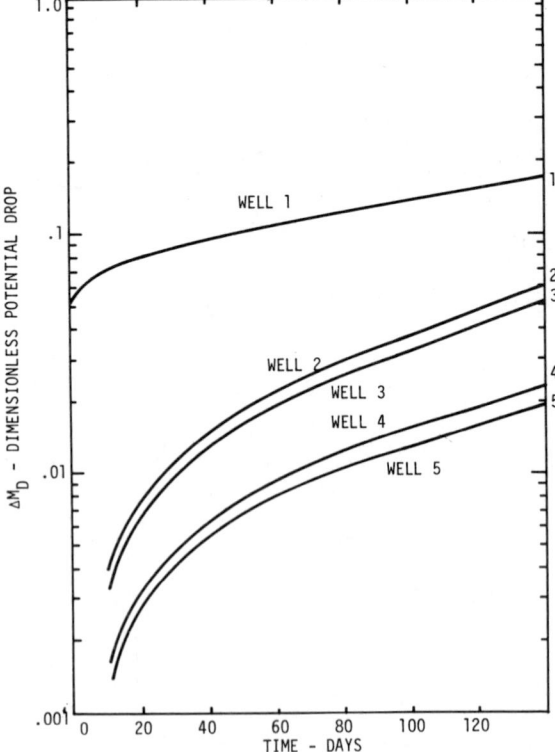

Figure 11.21: Influence function for gas reservoir—Well #1.

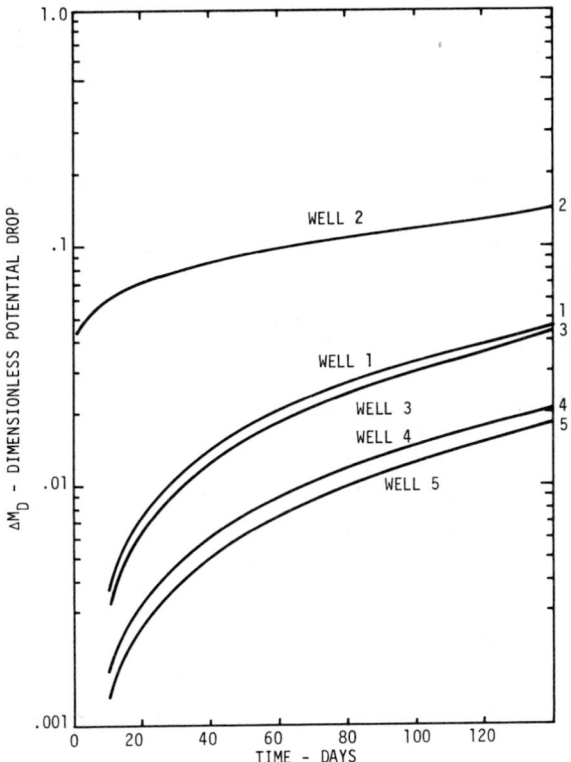

Figure 11.22: Influence function for gas reservoir—Well #2.

The second phase consists of setting up the tableau of coefficients for the optimization model, as shown in Fig. 11.26, and solving this optimization model with a suitable linear programming code. An example of the data is shown in Table 11.2. The general shape of the tableau for the five time periods is shown in Fig. 11.26 with a key to the matrix elements in Table 11.3. The elements of this tableau are the activity coefficients which are generated by superposition as indicated in Sec. 11.4. The format and calculations are identical to that shown in Table 11.1. The actual program is shown in Fig. 11.27, which is a set of computer code for the IBM MPS/360 system. This is a mathematical programming system which allows linear and separable programming to be done using a straightforward compiler-type program.

The results are shown in Table 11.4, where the activity column indicates the production by well in each of the five time steps. Note that only two wells are turned on in step 1, four in step 2, and all five in steps 3, 4, and 5.

Table 11.5 indicates the slack or overcapacity in each variable. Note that there is excess deliverability only in the first two steps or 40 days in several

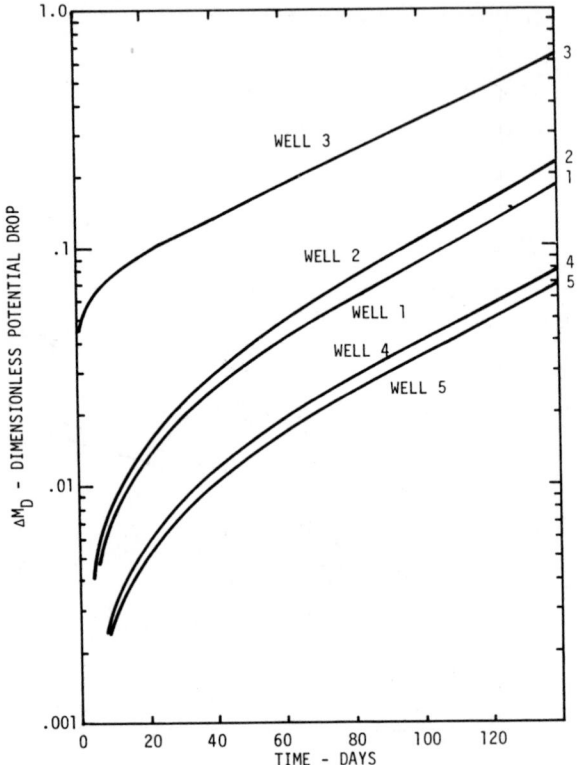

Figure 11.23: Influence function for gas reservoir—Well #3.

wells, but after that period all the pressure constraints are at their upper limits.

The actual deliverability curve and the projected demand curve are shown in Fig. 11.28. The engineer can now vary the right-hand side of his optimization tableau to try to find what line pressure is required to meet the demand if at all possible. This is done by increasing the right-hand-side constant within limits dictated by engineering design.

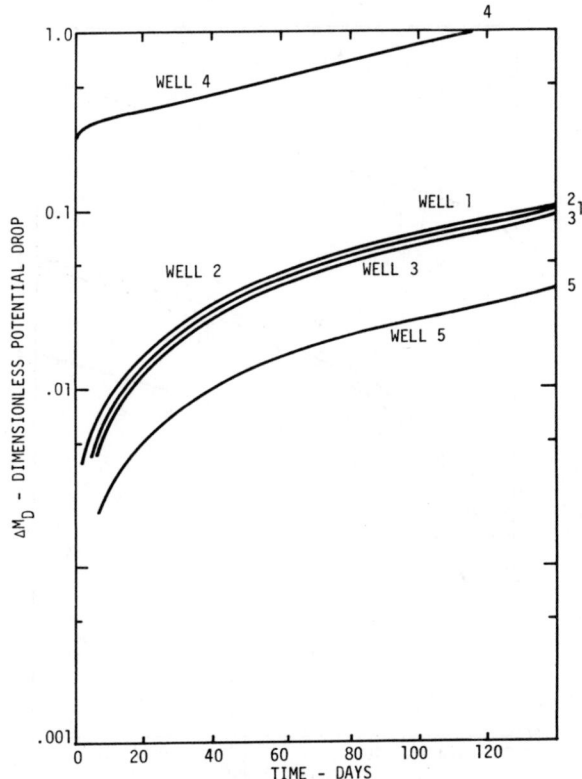

Figure 11.24: Influence function for gas reservoir—Well #4.

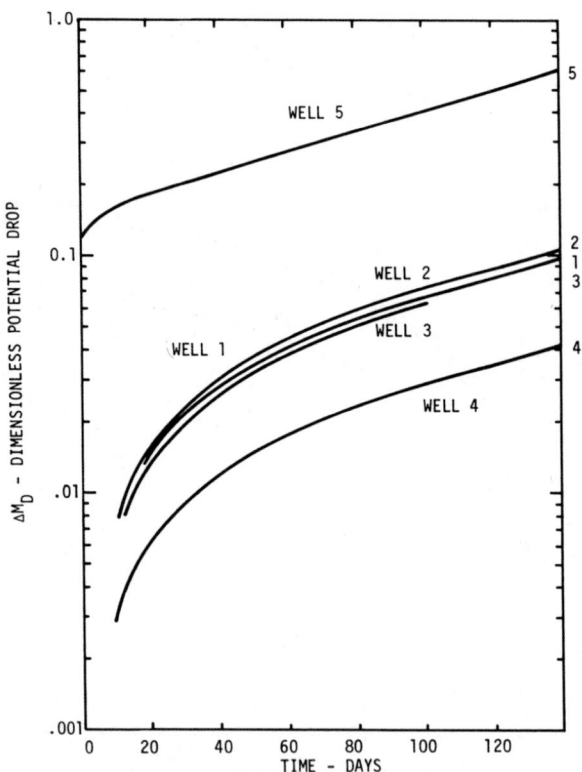

Figure 11.25: Influence function for gas reservoir—Well #5.

```
                              C C C C C C C C C C C C C C C C M
          C C C C C C C C C C 1 1 1 1 1 1 1 1 1 2 2 2 2 2 P
          1 2 3 4 5 6 7 8 9 0 1 2 3 4 5 6 7 8 9 0 1 2 3 4 5 D

R1     L  U V V V                                             T
R2     L  V U V V V                                           T
R3     L  V V U V V                                           T
R4     L  V V V T V                                           T
R5     L  V V V V T                                           T
R6     L  U V V V V U V V V                                   T
R7     L  V U V V U V U V V                                   T
R8     L  V V U V V V V U V V                                 T
R9     L  V V V U V V V T V                                   T
R10    L  V U V U V V V T                                     T
R11    L  U V V V V U V V V V U V V V                         T
R12    L  V U V V V U V V U V V V                             T
R13    L  V V U V V V V U V V V U V V                         T
R14    L  V V V U V V V V U V V V T V                         T
R15    L  V V V V U V V V U V V V V T                         T
R16    L  U V V V V U V V V U V V V V U V V V                 T
R17    L  V V V V V U V V V V U V V U V U V V V               T
R18    L  V V U V V V V U V V V V U V V V U V V               T
R19    L  V V V U V V V V U V V V V U V V V T V               T
R20    L  V V V V U V V V V T V U V V U V V V V T             T
R21    L  U V V V V U V V V V U V V V V U V V V V U V V V T   T
R22    L  V U V V V V V V V V U V V V V U V V U V U V V V T   T
R23    L  V V U V V V V U V V V U V V V V U V V V V U V V T V T
R24    L  V V V U V V V V U V V V V U V V V V U V V V V V T V T
R25    L  V V V V U V V V U V V V V U V U V V V U V V V V T T T
D1     L  1 1 1 1 1                                           A
D2     L            1 1 1 1 1                                 B
D3     L                      1 1 1 1 1                       B
D4     L                                1 1 1 1 1             B
D5     L                                          1 1 1 1 1   A
CBJFUN N  1 1 1 1 1 1 1 1 1 1 1 1 1 1 1 1 1 1 1 1 1 1 1 1 1
```

Figure 11.26: Symbolic simplex tableau for optimization model.

TABLE 11.2
Sample Data Input for Optimization Model

```
                GAS PRODUCTION OPTIMIZATION USING SUPERPOSITION

        C6          R14          .00700      R15          .00600
        C6          R16          .01400      R17          .00700
        C6          R18          .00680      R19          .00700
        C6          R20          .00700      R21          .01100
        C6          R22          .00600      R23          .00820
        C6          R24          .00700      R25          .00700
        C6          D2          1.00000      OBJFUN      1.00000
        C7          R6           .00740      R7           .07100
        C7          R8           .00660      R9           .00700
        C7          R10          .00750      R11          .00600
        C7          R12          .01600      R13          .00370
        C7          R14          .00800      R15          .01050
        C7          R16          .00700      R17          .01400
        C7          R18          .00700      R19          .00800
        C7          R20          .00600      R21          .00600
        C7          R22          .01000      R23          .00900
        C7          R24          .00750      R25          .00950
        C7          D2          1.00000      OBJFUN      1.00000
        C8          R6           .00640      R7           .00660
        C8          R8           .07800      R9           .00600
        C8          R10          .00700      R11          .00660
        C8          R12          .00840      R13          .01700
        C8          R14          .00660      R15          .00550
        C8          R16          .00680      R17          .00700
        C8          R18          .01900      R19          .00600
        C8          R20          .00650      R21          .00820
        C8          R22          .00900      R23          .03300
        C8          R24          .00650      R25          .00700
        C8          D2          1.00000      OBJFUN      1.00000
        C9          R6           .00700      R7           .00700
        C9          R8           .00600      R9           .16500
        C9          R10          .00300      R11          .00700
        C9          R12          .00800      R13          .00660
        C9          R14          .01500      R15          .00250
        C9          R16          .00700      R17          .00800
        C9          R18          .00600      R19          .01700
        C9          R20          .00240      R21          .00700
        C9          R22          .00750      R23          .00650
        C9          R24          .02000      R25          .00260
        C9          D2          1.00000      OBJFUN      1.00000
        C10         R6           .00700      R7           .00750
        C10         R8           .00700      R9           .00300
        C10         R10          .33000      R11          .00600
        C10         R12          .01050      R13          .00550
        C10         R14          .00250      R15          .03500
        C10         R16          .00700      R17          .00600
        C10         R18          .00650      R19          .00240
        C10         R20          .96000      R21          .00700
        C10         R22          .00950      R23          .00700
        C10         R24          .00260      R25          .04000
        C10         D2          1.00000      OBJFUN      1.00000
        C11         R11          .07800      R12          .00750
        C11         R13          .00640      R14          .00700
        C11         R15          .00700      R16          .01600
```

TABLE 11.3

Key to Matrix Elements

```
              GAS PRODUCTION OPTIMIZATION USING SUPERPOSITION

                              SUMMARY OF MATRIX

       SYMBOL                    RANGE                COUNT (INCL. RHS)

         Z              LESS THAN       .000001

         Y        .000001    THRU       .000009

         X        .000010               .000099

         W        .000100               .000999

         V        .001000               .009999          294

         U        .010000               .099999           70

         T        .100000               .999999           36

         1       1.000000              1.000000           50

         A       1.000001             10.000000            2

         B      10.000001            100.000000            3

         C     100.000001          1,000.000000

         D   1,000.000001         10,000.000000

         E  10,000.000001        100,000.000000

         F 100,000.000001      1,000,000.000000

         G        GREATER THAN  1,000,000.000000
```

```
             CONTROL PROGRAM COMPILER - MPS/360 V2-M7

0001                PROGRAM
0002                TITLE ('GAS PRODUCTION OPTIMIZATION USING SUPERPOSITION')
0003                INITIALZ
0062                MOVE (XCATA, 'GASOPTDATA')
0063                MOVE (XPBNAME, 'GASDATSET')
0064                CONVERT ('SUMMARY')
0065                BCDOUT
0066                SETUP ('MAX')
0067                PICTURE
0068                MOVE (XCBJ,'CBJFUN')
0069                MOVE (XRHS,'MPD')
0070                PRIMAL
0071                SOLUTION
0072                EXIT
0073                PEND
```

Figure 11.27: Program code for IBM MPS/360 system.

TABLE 11.4

Solution of Optimization Model Showing Flow Rates in Five Time Steps

GAS PRODUCTION OPTIMIZATION USING SUPERPOSITION - SECTION 2: COLUMNS

NUMBER	COLUMN	AT	ACTIVITY	INPUT COST	LOWER LIMIT	UPPER LIMIT	REDUCED COST
32	C1	BS	4.36709	1.00000	.	NONE	.02359-
33	C2	LL	.	1.00000	.	NONE	.29909-
34	C3	LL	.	1.00000	.	NONE	.
35	C4	BS	5.63291	1.00000	.	NONE	.00199-
36	C5	LL	.	1.00000	.	NONE	.
37	C6	BS	9.22486	1.00000	.	NONE	.
38	C7	BS	10.74396	1.00000	.	NONE	.
39	C8	BS	3.53695	1.00000	.	NONE	.
40	C9	BS	4.14505	1.00000	.	NONE	.
41	C10	LL	.	1.00000	.	NONE	1.25902-
42	C11	BS	5.98379	1.00000	.	NONE	.
43	C12	BS	6.43827	1.00000	.	NONE	.
44	C13	BS	7.67806	1.00000	.	NONE	.
45	C14	BS	2.78461	1.00000	.	NONE	.
46	C15	BS	1.71412	1.00000	.	NONE	.
47	C16	BS	4.02791	1.00000	.	NONE	.
48	C17	BS	4.53906	1.00000	.	NONE	.
49	C18	BS	4.62180	1.00000	.	NONE	.
50	C19	BS	1.72351	1.00000	.	NONE	.
51	C20	BS	1.33442	1.00000	.	NONE	.
52	C21	BS	3.10229	1.00000	.	NONE	.
53	C22	BS	3.28680	1.00000	.	NONE	.
54	C23	BS	1.50243	1.00000	.	NONE	.
55	C24	BS	1.20319	1.00000	.	NONE	.
56	C25	BS	.90529	1.00000	.	NONE	.

TABLE 11.5
Slack Activity in Time Periods

GAS PRODUCTION OPTIMIZATION USING SUPERPOSITION - SECTION 1: ROWS

NUMBER	ROW	AT	ACTIVITY	SLACK ACTIVITY	LOWER LIMIT	UPPER LIMIT	DUAL ACTIVITY
1	R1	BS	.38006	.57994	NONE	.96000	.
2	R2	BS	.07175	.88825	NONE	.96000	.
3	R3	BS	.06175	.89825	NONE	.96000	.19495-
4	R4	UL	.96000		NONE	.96000	.
5	R5	BS	.04747	.91253	NONE	.96000	.
6	R6	UL	.96000	.	NONE	.96000	3.63266-
7	R7	UL	.96000		NONE	.96000	3.82748-
8	R8	BS	.49670	.46330	NONE	.96000	.
9	R9	UL	.96000		NONE	.96000	1.84573-
10	R10	BS	.22263	.73737	NONE	.96000	.
11	R11	UL	.96000		NONE	.96000	4.63298-
12	R12	UL	.96000		NONE	.96000	5.07518-
13	R13	UL	.96000		NONE	.96000	4.45309-
14	R14	UL	.96000		NONE	.96000	2.45648-
15	R15	UL	.96000		NONE	.96000	1.18250-
16	R16	UL	.96000		NONE	.96000	7.01798-
17	R17	UL	.96000		NONE	.96000	7.98295-
18	R18	UL	.96000		NONE	.96000	6.86043-
19	R19	UL	.96000	.	NONE	.96000	3.51960-
20	R20	UL	.96000	.	NONE	.96000	1.65186-
21	R21	UL	.96000	.	NONE	.96000	8.91876-
22	R22	UL	.96000	.	NONE	.96000	9.87161-
23	R23	UL	.96000	.	NONE	.96000	9.10029-
24	R24	UL	.96000	.	NONE	.96000	4.11257-
25	R25	UL	.96000	.	NONE	.96000	1.99450-
26	D1	UL	10.00000	.	NONE	10.00000	.24049-
27	D2	BS	27.65083	2.34917	NONE	30.00000	.
28	D3	BS	24.59885	35.40115	NONE	60.00000	.
29	D4	BS	16.24670	3.75330	NONE	20.00000	.
30	D5	UL	10.00000	.	NONE	10.00000	.12931-
31	OBJFUN	BS	88.49638	88.49638-	NONE	NONE	1.00000

339

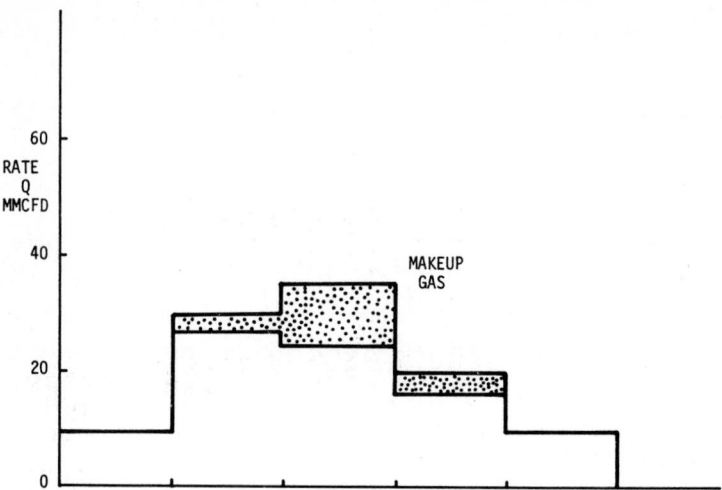

Figure 11.28: Predicted deliverability curve (millions of cubic feet per day).

11.6 APPLICATION TO MULTIRESERVOIR SYSTEMS

The optimization of several oil reservoir systems over a planning horizon requires the engineer to look at the optimum development of the reservoir and all the attendant peripheral facilities, pipelines, compressor, plants, etc. The optimization process can be expressed by the sequence of operations shown in Fig. 11.29. Note that model formulations for the reservoirs and the

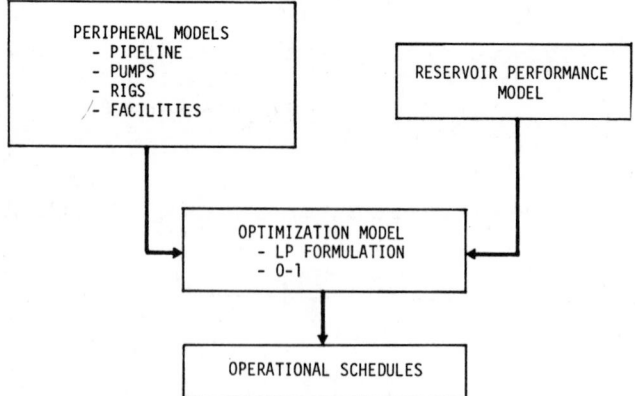

Figure 11.29: Optimization of multireservoir systems.

accessory systems (gathering systems and the like) actually take place in parallel.

Several investigators[9, 10, 11, 12, ...,19] have looked at this problem in more or less detail, and their formulations differ primarily in the handling of the reservoir and accessory models. The reason is obvious, since the reservoir behavior can be approximated in many different ways; the solution of the linear programming model, however, is rather straightforward, though not insignificant.

Aronofsky and Lee[10] used a Van Everdingen and Hurst[11] type of pressure function to explicitly calculate production rate effects on pressure in a reservoir model. This work was later expanded by Aronofsky and Williams[12] to multireservoir systems. Bohanon[13] used an explicit cumulative decline curve method in which the rate versus cumulative curve was linear. He extended his model to examine expansion in pipeline systems and other facilities.

There are several subsystems which enter into the development of a realistic optimization model for an oil field system. These areas can all be treated in different ways by different groups, and they may even be omitted on option if the engineer is looking for a simpler model. The total model, however, should include the following functions in some form:

1. Deliverability function
2. Pipeline function
3. Cost functions
4. Objective function

We shall now investigate the manner in which the more important of these functions enters the optimization model.

Deliverability

The deliverability function involves several important parameters: flow rate q, time t, pressure P, and, secondarily, cumulative production R_T. These variables can be and have been used in several different combinations. If we consider that there is a lower limit on pressure P^* below which it is no longer feasible to produce a well or a reservoir, we could write a constraint of the following type:

$$P_j = P_0 - \frac{q\mu}{2\pi kh} P_{D_j} \geq P^* \qquad (11.29)$$

where the pressure at a point j is determined from the original pressure P_0, the flow rate q, and the dimensionless pressure function P_D. This approach was used by Lee and Aronofsky.[10] It assumes a linearity in the response of pressure and a homogeneity in rock and fluid throughout the reservoir system.

The pressure calculated at each well is constrained not to drop below some limit. The defining constraint for this variable will be of the type:

$$P_j \geq P^* \tag{11.30}$$

In the case of more than one well or more than one reservoir interfering with one another, superposition in time and space must be used to obtain the individual pressure terms. This was pursued in Sec. 11.3, and here we shall simply state the results for a two-well case:

$$P_{1,j} = P_0 - \frac{\mu}{2\pi kh}[q_1^1 P_D(1,j) + q_2^1 P_D(r_{12},j)] \tag{11.31}$$

$$P_{2,j} = P_0 - \frac{\mu}{2\pi kh}[q_2^1 P_D(1,j) + q_1^2 P_D(r_{12},j)] \tag{11.32}$$

These are just Eqs. (11.15) and (11.16) rewritten for time j. This set of equations is an expansion of Eq. (11.29) to include interference effects, but it is fundamentally the solution of the diffussivity equation for a radial system. This concept of superposition can be improved in practice and applied to real-life problems using the influence matrix formulation of Sec. 11.3. The influence matrix concept considers the reservoir heterogeneities and variations in fluid properties more accurately. However, there is still some approximation involved, since the linearity of the superposition principle is assumed when multiplying rates by pressure drops to obtain total pressure drop. The typical form of the influence matrix system is as follows:

$$a_{11}q_1 + a_{12}q_2 + \ldots + a_{1n}q_n \leq P_1^*$$
$$a_{21}q_1 + a_{22}q_2 + \ldots + a_{2n}q_n \leq P_2^*$$
$$\vdots \tag{11.33}$$
$$a_{n1}q_1 + a_{n2}q_2 + \ldots + a_{nn}q_n \leq P_n^*$$

where a_{ij} are the influence coefficients, q_j are the flow rates, and P^* are the maximum allowable pressure drops.

The deliverability function can also be formulated in terms of flow rate versus time data or for rate cumulative data. In the flow rate cumulative approach a decline curve for the well or reservoir is obtained (usually from material balance calculations or simulation studies). A typical exponential decline curve is shown in Fig. 11.30, where q_0 is the initial production rate and q_a some limiting allowable rate based on operational or regulatory considera-

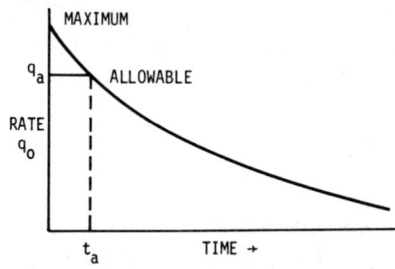

Figure 11.30: Exponential decline curve.

Figure 11.31: Rate cumulative curve.

tions over a time t_a. This curve can be expressed by the following equations:

$$q_j = q_a, \qquad 0 \le t_a \tag{11.34}$$

$$q_j = \frac{q_0 - S \sum\limits_{r=1}^{j-1} q_r \, \Delta t_r}{1 + (S/2) \, \Delta t_r}, \qquad t > t_a \tag{11.35}$$

where the variable S is defined from the rate cumulative recovery curve of Fig. 11.31.

Equations (11.34) and (11.35) must be incorporated as inequality constraints, since they only apply as upper limits; it is possible for the flow rate to be less than this number. The flow rates are therefore written:

$$q_j \le q_a \tag{11.36}$$

$$q_j \le \frac{q_0 - S \sum\limits_{r=1}^{j-1} q_r \, \Delta t_r}{1 + (S/2) \, \Delta t_r} \tag{11.37}$$

In addition to the individual well or individual reservoir constraint, there must be a limit on the total produced at a given time. This total can reflect the total for all the wells or the total for all the reservoirs. Thus:

$$\sum_{i=1}^{N} q_j^i \le Q_{T_j} \tag{11.38}$$

where i is summed over all the producing sources and Q_{T_j} is the maximum possible at time j. In the situation where additional drilling is undertaken in a given period of the planning horizon, the deliverability of these new wells will not be that of an initial well, since the average reservoir pressure will have dropped to some level below the original (except in very active water drive systems). If d_j represents the number of wells drilled in period j, then the total production in that period should be limited to that due to wells already producing and to those just drilled. The rate for new wells is usually assumed to be the mid-period rate—i.e., they are computed at $(j + \frac{1}{2})$ index or the midpoint of the period. The limiting constraint then becomes:

$$q_j \le \left(\sum_{k=1}^{j-1} Q_{kj} d_k \right) + Q_{jj} d_j \qquad (11.39)$$

where Q_{kj} represent the average production rate for wells drilled in period k and producing in period j. Note that by necessity, Eq. (11.39) refers to a reservoir rate, since it involves a summation of wells. This constraint is based on a declining production; in the event of a secondary recovery project, the constraint may be relaxed to enable the q_j-term to increase.

Pipeline Function

The function describing the pipeline behavior consists of a nodal balance on each segment and a limit on the pipeline capacity. If we consider a single segment from each reservoir, then the quantity of oil leaving the reservoir must equal that flowing in the pipeline segment. See Fig. 11.32. Thus:

$$q_j^m = q_{p,j}^m \qquad (11.40)$$

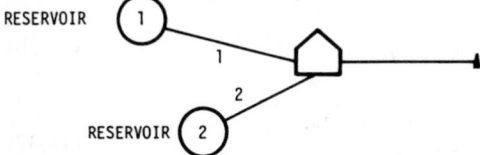

RESERVOIR ①

1

2

RESERVOIR ②

Figure 11.32: Pipeline segments.

where q_j^m represents production in year j from field m and $q_{p,j}^m$ represents line rate in segment m in year j. The capacity of each segment is limited by some maximum based on pipe dimensions. Thus:

$$q_{p,j}^m \le C_j \qquad (11.41)$$

It is possible to formulate the pipeline constraint to allow for expansion in capacity in a period. Bohanon[13] has used the following formulation to allow capacity expansion in a given period:

$$q_{p,j}^m \le C_{p,j} + \sum_{k=1}^{j-1} c_{p,k} \qquad (11.42)$$

where $c_{p,k}$ is the added pipeline capacity in year k.

Cost Functions

The cost functions will embody all the capital expenditures incurred in a given project. These will usually include drilling, expansion of facilities, platforms, process plants, and similar items. These cost functions can be expressed as a total cost figure versus the level or size of the item entering into the decision process—e.g., drilling cost versus drilling depth. The cost functions can be either linear or nonlinear.

Linear Cost Functions: Figure 11.33 illustrates a linear relation between total cost and level of variable. For the type figure, the total capital expenditure is simply

$$C_T^m = SQ \tag{11.43}$$

where S is the incremental cost and Q the parameter level. This variable type is easy to incorporate into the model.

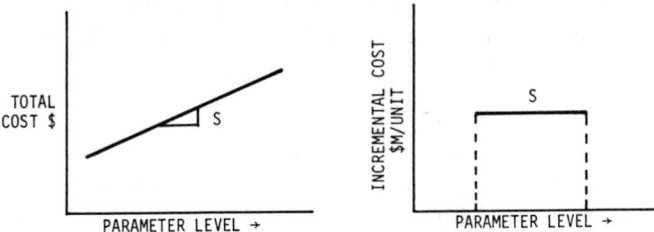

Figure 11.33: Linear cost function.

Nonlinear Cost Functions: Figure 11.34 illustrates a nonlinear cost function; to use this function a piecewise linear approximation[14] must be made as shown. The total cost for a given level Q_c of a variable as shown in Fig. 11.34 is:

$$C_T^m = \sum_{l=1}^{k} S_l Q_l \tag{11.44}$$

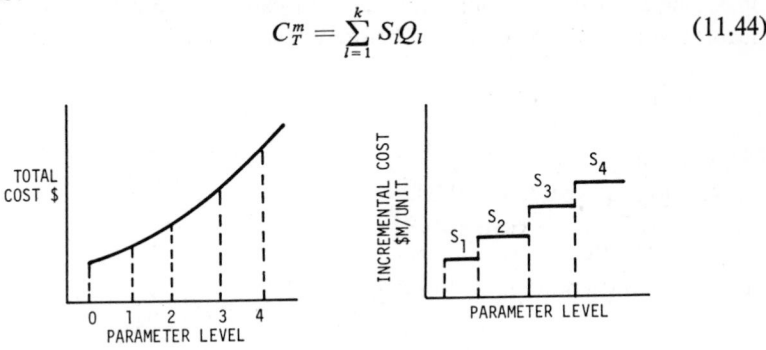

Figure 11.34: Nonlinear cost function.

A cost function as shown in Fig. 11.34 is a piecewise linear convex function, which implies that $S_1 < S_2 < S_3 < \ldots < S_k$. This type of function guarantees that a smaller level of a parameter will always be used before a larger level.[14] These types of functions occur in drilling cost–versus–depth correlations. In several capital expenditures the circumstances are such that the item is either present or absent. For example, a gas plant is built or it is not! You cannot have a portion of a gas plant operative. These cost items are therefore usually multiplied by a zero-one integer variable. If the value of this integer is 1, the item is included; if it is 0, the item is not. These variables are then constrained by the maximum number of the particular items allowed in a given period. A good example[13] is the fixed charge–incremental cost for a facility:

$$\text{COF}_j^m = C_F * \delta_j^m + S * Q_j^m \tag{11.45}$$

where COF_j is the cost of a facility in period j, which is equal to a fixed setup charge C_F plus an incremental charge $S * Q_j$, Q_j being the size of the facility; δ_j is a 0–1 variable. Another equation will normally be used to constrain the size of Q_j:

$$Q_j^m - Q_{max}^m * \delta_j^m \leq 0 \tag{11.46}$$

If the total number of facilities possible in any time j is K, then the 0–1 variables are constrained as follows:

$$\sum_{m=1}^{M} \delta_j^m \leq K \tag{11.47}$$

The objective function usually maximizes some profit quantity. This function comprises the revenue, investments, expenditures, and operating costs in a single entity. The objective function can be formulated as before-tax income or after-tax income. The before-tax cash flow (BTC) is made up of

1. Drilling cost (DC)
2. Operating cost (OC)
3. Facilities cost (FC)
4. Revenue generated (RG)
5. Miscellaneous costs (MC)

Then the total undiscounted cash flow is:

$$\text{BTC} = \sum_{j=1}^{T} (\text{RG}_j - \text{DC}_j - \text{OC}_j - \text{FC}_j - \text{MC}_j) \tag{11.48}$$

Each component of Eq. (11.48) can be derived from its appropriate Eq. (11.33) through (11.47). Friar[15] indicates one approach to formulating the net taxable income and after-tax cash flow.

The net taxable income must be obtained by partitioning the capitalized and expensed portion of drilling and facilities costs. If g is that fraction of drilling costs capitalized and h that portion of facilities cost capitalized, then $(1 - g)$ and $(1 - h)$ are the respective portions expensed. Thus, the net taxable income (NI) in a given year can be computed using the sum-of-the-years depreciation method:

$$NI_j = RG_j - (1 - g)DC_j - (1 - h)FC_j - OC_j - MC_j$$
$$- \sum_{l=1}^{j} \left[\frac{T - j + 1}{\frac{T - l + 1}{2}(T - l + 2)} \right] (gDC_j + hFC_j) \qquad (11.49)$$

where T is the total number of time periods in the planning horizon. To obtain the net income after taxes and depletion, we apply the company tax rate and depletion allowance to the net income:

$$\text{Taxable income} = NI_j - D_j \qquad (11.50)$$

where D_j is the allowable depletion allowance. The depletion allowance is obtained from the following:

$$D_j = \max \begin{cases} \text{Cost depletion} = \dfrac{U}{R}P \\ \text{Percentage depletion} = \min \begin{cases} 22\% \text{ Gross income} \\ 50\% \text{ Net income} \\ \max \begin{cases} 0 \end{cases} \end{cases} \end{cases} \qquad (11.51)$$

If the tax rate is α, then:

$$\text{Tax} = \alpha(NI_j - D_j) \qquad (11.52)$$

Thus, after-tax cash flow (ATC) is:

$$ATC_j = BTC_j - \alpha(NI_j - D_j) \qquad (11.53)$$

The components of Eqs. (11.53) and (11.48) are put together in Eq. (11.54):

$$ATC = \sum_{j=1}^{T} [RG_j - DC_j - OC_j - FC_j - MC_j - \alpha(NI_j - D_j)] \qquad (11.54)$$

The discounted after-tax cash flow is obtained by multiplying Eq. (11.54) by the appropriate discount factor λ_j for each period:

$$\text{ATC} = \sum_{j=1}^{T} [\text{RG}_j - \text{DC}_j - \text{OC}_j - \text{FC}_j - \text{MC}_j - \alpha(\text{NI}_j - D_j)]\lambda_j \quad (11.55)$$

Equation (11.55) constitutes the objective function which must be maximized with the constraints developed earlier in this section.

Mechanics of Solution

The optimization model obtained in even a simple multireservoir study is very large. Bohanon[13] quoted an example of 500 equations and 1000 variables for a four-reservoir problem. As the problems get larger, the optimization model grows larger at an alarming rate. The size of the problem sometimes justifies the development of a special linear programming code designed to effectively explore the structure of that particular problem. This "custom-building" then produces a very efficient model which can be reused many times and offers economies over the general-purpose linear programming codes available today from most software sources.

REFERENCES

1. H. M. WAGNER, *Principles of Operations Research* (Englewood Cliffs, N.J.: Prentice-Hall, 1969).

2. F. S. HILLIER and G. J. LIEBERMAN, *Introduction to Operations Research* (San Francisco: Holden-Day, 1969).

3. G. B. DANTZIG, *Linear Programming and Extensions* (Princeton: Princeton University Press, 1963).

4. G. HADLEY, *Linear Programming* (Reading, Mass.: Addison-Wesley, 1962).

5. H. A. TAHA, *Operations Research: An Introduction* (New York: Macmillan, 1971).

6. IBM CORP., *Linear Programming—Mathematical Optimization Subroutine System* (New York, 1971).

7. IBM CORP., *IBM Application Program 360A-CO-14X. Mathematical Programming System/360.* Version 2. Linear and Separable Programming, User's Manual (New York, 1971).

8. R. A. WATTENBARGER, "Maximization of Seasonal Withdrawals from Gas Storage Reservoirs," SPE Denver, 1969, Paper. SPE 2406. Presented at 44th Annual Meeting of SPE of AIME. Denver, Colo.

9. K. H. COATS, "An Approach to Locating New Wells in Heterogeneous Gas Producing Fields," *J. Pet. Tech.* (1969), **21**, 549–58.

10. A. S. LEE and J. S. ARONOFSKY, "A Linear Programming Model for Scheduling Crude Oil Production," *J. Pet. Tech.* (July 1958), 51–54.

11. A. F. VAN EVERDINGEN and W. HURST, "The Application of the La Place Transformation to Flow Problems in Reservoirs," *Trans. AIME* (1949), **186,** 305.

12. J. S. ARONOFSKY and A. C. WILLIAMS, "The Use of Linear Programming and Mathematical Models of Underground Oil Production," *Management Science* (July 1962), 394–402.

13. J. M. BOHANON, "A Linear Programming Model for Optimizing Development of Multi-reservoir Pipeline Systems," *J. Pet. Tech.* (1970), **22,** 1429–36.

14. G. B. DANTZIG, A. HAX, R. POMEROY, R. SANDERSON, R. VAN SLYKE, "Natural Gas Transmission System Optimization," AGA Pipeline Research Committee Report, AGA Catalog No. L20040.

15. L. C. FRAIR, "Economic Optimization of Offshore Oil Field Development," Ph.D. dissertation, University of Oklahoma, 1973.

16. J. H. HENDERSON, J. R. DEMPSEY, and A. D. NELSON, "Practical Application of Two-dimensional Numerical Model for Gas Reservoir Studies," *J. Pet. Tech.* (Sept. 1967), 1127.

17. J. H. HENDERSON, J. R. DEMPSEY, and J. C. TYLER, "Use of Numerical Models to Develop and Operate Gas Storage Reservoirs," *J. Pet. Tech.* (Nov. 1968), 1239–46.

18. G. B. DANTZIG, "Large Scale Linear Programming," Operations Research House, Technical Report No. 67–8 (Nov. 1967), Stanford University, Stanford, Calif.

19. G. ROWAN, and J. E. WARREN, "A Systems Approach to Reservoir Engineering Optimum Development Planning," *Jour. of Can. Pet. Tech.* (July–Sept. 1967), **6,** No. 3, 84.

BIBLIOGRAPHY

BENTSEN, R. G. and D. A. T. DONOHUE, "A Dynamic Programming Model of the Cyclic Steam Injection Process," *J. Pet. Tech.* (Dec. 1969), 1582–96; *Trans. AIME,* **246.**

COATS, K. H., L. A. RAPOPORT, J. R. McCORD, and W. P. DREWS, "Determination of Aquifer Influence Functions From Field Data," *J. Pet. Tech.* (Dec. 1964), 1417–23.

COOKSEY, ROBERT A., JAMES H. HENDERSON, and JOHN R. DEMPSEY, "Total Computer Simulation of a Gas Producing Complex," *J. Pet. Tech.* (Aug. 1969), 942–48.

DEMPSEY, J. R., J. K. PATTERSON, K. H. COATS, and J. P. BRILL, "An Efficient Model for Evaluating Gas Field Gathering System Design," *J. Pet. Tech.* (Sept. 1971), 1067–73.

FLANIGAN, O., "Constrained Derivatives in Natural Gas Pipeline System Optimization," *J. Pet. Tech.* (May 1972), 549–56.

FRANCIS, M. A., "Advanced Equations for Natural Gas Flow Prediction," SPE 4692, 48th Annual Meeting, Las Vegas, Nev., Sept. 30–Oct. 3, 1973.

GHEZ, FABIEN, "A New General Optimization Method-Application to Gas Gathering System," SPE 4693, 48th Annual Meeting, Las Vegas, Nev., Sept. 30–Oct. 3, 1973.

KNAPP, R. M., J. H. HENDERSON, J. R. DEMPSEY, and K. H. COATS, "Calculation of Gas Recovery Upon Ultimate Depletion of Aquifer Storage," *J. Pet. Tech.* (Oct. 1968), 1129–32.

STONER, M. A., "Sensitivity Analysis Applied to a Steady-State Model of Natural Gas Transportation Systems," *Soc. Pet. Eng. J.* (April 1972), 115–25.

TAYLOR, T. D., N. E. WOOD, and J. E. POWERS, "Computer Simulation of Gas Flow in Long Pipelines," *Soc. Pet. Eng. J.* (Dec. 1962), 297–302.

Index